D1547599

Injection Molds and Molding

A Practical Manual

Injection Molds and Molding

A Practical Manual

Second Edition

Joseph B. Dym

VNR VAN NOSTRAND REINHOLD COMPANY
New York

Library of Congress Catalog Card Number 86-19118

ISBN 0-442-21785-4

Printed in the United States of America

Van Nostrand Reinhold Company Inc.
115 Fifth Avenue
New York, New York 10003

Van Nostrand Reinhold Company Limited
Molly Millars Lane
Wokingham, Berkshire RG11 2PY, England

Van Nostrand Reinhold
480 La Trobe Street
Melbourne, Victoria 3000, Australia

Macmillan of Canada
Division of Canada Publishing Corporation
164 Commander Boulevard
Agincourt, Ontario M1S 3C7, Canada

16 15 14 13 12 11 10 9 8 7 6 5 4 3 2 1

Library of Congress Cataloging in Publication Data

Dym, Joseph B.
Injection molds and molding.

Includes index.
1. Injection molding of plastics. I. Title
TP1150.D93 1987 668.4'12 86-19118
ISBN 0-442-21785-4

In loving memory of my sister
Rebe Dym Gilbert

Preface to Second Edition

This second edition covers the progress made in the field of injection molding as well as some innovative features here published for the first time.

The computer-aided program (CAD/CAM) is presented for consideration of product design, mold design, and manufacturing of mold components. The advantages of this process are compared to those of conventional processes. The deflections of deep wall-molding cavities have been determined not only to prevent them from cracking but also to eliminate mold problems associated with weak or undersized cavities. In connection with this feature, emphasis is placed on how to specify wall thickness of products to prevent ill effects on mold cavities and their quality in general.

The mold cooling system has been rearranged to comply with information needed by computer-calculated cooling methods.

In runnerless molds, a nozzle is featured that not only eliminates clogging of the opening, but also practically eliminates cold slug well and thus improves the quality of the moldings. A similar feature is incorporated in runnerless probes. Problems of insulated runners are stressed, and means of overcoming them are pointed out.

The front of the plasticating screw has been modified so that it will more thoroughly mix polymers and thus improve the appearance and performance of products as well as save coloring material.

A plastic screw design is also described that will remove gases and moisture from polymers during machine operation. When machine capacity is close to job requirements and occasionally causes flashing as a result, a way to solve the problem with only a slight increase in cycle is furnished.

"Starve Feeding" with process control is presented not only as a means of saving plastic material, but also of providing automatic operation and reduction of rejects. Clamping a mold in a press is an important way of keeping each of its halves from damaging the other. The formula given to keep a mold firmly in place is based on its weight.

The extension of the useful life of a mold is the goal of the method given for effectively caring for mold surfaces.

The ill effects of small gate, long land, gate blemishes, and cleaner break of submarine gates are all analyzed with the aim of finding suitable means for eliminating them.

In general, it can be stated that the second edition of *Injection Molds and Molding* will considerably enlarge the knowledge of its subject.

JOSEPH B. DYM

Preface to First Edition

The product designer is always concerned about consistency of quality in plastic parts at economical rates, especially those that are injection molded. With this concern in mind, observations were made in numerous molding plants to determine the degree to which people at all levels of responsibility display the specific interest in the operation that would assure consistency in part quality. These observations led to the conclusion that a source of information is needed that would not only explain every element connected with the injection molding operation, but would also emphasize the potential causes of variation in part quality.

Injection Molds and Molding: A Practical Manual, is written with the view of stressing features pertaining to mold designing, moldmaking, machine performance, parameter setting, material handling, and overall processing steps that will lead to consistent molded parts produced at an efficient rate. It should prove valuable to mold designers, draftsmen, moldmakers, mold maintenance personnel, molding start-up personnel, molding supervisors, superintendents, quality control personnel, process engineers, industrial engineers, and those in related occupations.

To those experienced in the field, *Injection Molds and Molding* will serve as a means of rounding out their knowledge. For those entering the field, it can become a book for the classroom or for self-instruction. The book contains the needed information for the successful operation of injection molding; the application of the same by the reader should result in satisfying benefits.

JOSEPH B. DYM

Sincere appreciation is expressed to the following companies for providing information and illustrations.

Cincinnati-Milacron, Plastics Machinery Division
D-M-E Co.
Formative Products Co.
General Electric Co.
HPM Corp.
Master Unit Die Products Inc.
Mobay Chemical Corp.
Mold Masters Ltd.
Rockwell International Corp.

Contents

Injection Molds and Molding

A Practical Manual

1 Introduction

Injection Molds and Molding: A Practical Manual treats the subjects of mold design and molding operations as intimately interrelated. In the opinion of the author, a sound molding plant can be established only when those executing the function of the mold design and those carrying out the prescriptions for molding fully understand these two phases of the operation.

Many elements of molds and molding have requirements that are frequently overlooked or intentionally omitted, because of a lack of appreciation of their value in part quality. For example, a large percentage of cavities is fabricated from prehardened steel. During fabrication, processing stresses are induced in the steel, which, if not relieved by suitable means, will add to the stresses generated during the molding operation. The combination of stresses can lead to failure of a costly mold component. Although stress relieving from fabrication is advocated by steel suppliers, it is rarely carried out, mainly because the tool design drawing does not specifically call for it. Another example is that of the "cold slug" well that is usually combined with the sprue-pulling arrangement. The purpose of the well is to trap the cool material coming from the front of the nozzle before it has a chance to enter into the cavity. The size of the well depends on the inside dimensions of the nozzle and the manner of its heating. If this well is absent or is of incorrect volume, some of the cool material will enter the cavity, bringing about an uneven rate of cooling in the part, which will reflect itself in stresses, warpage, and dimensional variations.

These required features are typical of a great many that are listed in this text that are essential for a successful product performance. They are indicated as important requirements, their part in product quality is explained, and, when absent or incorrectly incorporated, their ill effects are clarified. The intimate knowledge of required features can be effectively applied to the operation of molding in general and especially to the debugging of a new mold prior to release for production. This initial proving-out of a mold is the opportune time to see that all features that determine the cost, performance, and overall quality of the product do their part in carrying out the functions needed for a successful part.

Certain aspects of the molding operation and mold design are emphasized and reemphasized herein in the hope that they will form an indelible impression on the mind of the reader and thus insure their implementation. Some areas of the overall molding field are considered from the viewpoint of the mold designer, product designer, and molder in order to make them aware of how much they can contribute to the production of a correctly molded part. Potential causes of problems, whether they are in the mold or molding parameter control, are analyzed, and appropriate solutions are suggested.

Calculations have been introduced to mold strength and press protection considerations, where judgment was the only deciding factor. Judgment based on successful experience is important, but as a rule does not provide optimum solutions. When sound judgment is checked against answers derived from calculations of appropriate formulas, an atmosphere of confidence is created in which very costly molds and equipment are used with a high degree of safety.

For mold setup purposes, some molding parameters are proposed for figuring their values instead of obtaining them by means of trial. All the calculations and figurings are introduced with the expectation that they will lead to accurate and predictable results as well as conserve time and material.

The advent of the inexpensive pocket calculator and its simple as well as rapid manipulation provides an inducement to use the calculation route. Material sheets are provided as a guide to the needs of each material for moldability and parameter setting.

The sequence of chapters is planned in the same manner as the events take place in actual practice. The starting point is the analysis of the part drawing for determining that the dimensions and configuration are designed for optimum performance. The next step is to visualize the mold and its features, which will make possible the production of parts to required specifications. The last step provides the outline of molding parameters that will utilize the mold in a fashion that will produce acceptable parts.

This sequence is especially pertinent to the injection molding of thermoplastics. As a matter of fact, 18 of the 21 chapters herein are devoted to thermoplastic molding, although many of them are also applicable to some degree to other injection processes. We have dealt with thermoplastic molding in this way because it is by far the largest activity in the molding field; its operating conditions are most demanding because of very high forces (20,000-psi injection pressure), high material temperatures (up to 800°F), and high frequency of cycling (3 to 4 cycle/min.). In addition, most of the basic principles are applicable to injection processes. Once the thermoplastic injection-molding phases are mastered, it becomes relatively simple to make the transition to the other injection processes.

The deviations of other injection processes are mostly along the lines of feeding the material to cavity, mold-temperature control, pressures involved, and part removal from the mold. The method of compensating in the mold for each deviation is described in the pertinent chapters.

During the last 15 years, the author has been involved in dealing with 30 or more molding shops and about 50 moldmaking enterprises extending over practically the whole of the United States. In addition, he has instructed key personnel for plastic operations in understanding and solving molding problems.

The author's discussions with people at all levels of operations led him to conclude that it would be most useful to have a source that would provide data and explanations for every step in the moldmaking and molding process. The usefulness of the data and explanations would be enhanced by elaborating on those requirements that have a vital bearing on the quality and consistency of the molded product. The treatment of molds and molding as an integrated unit is another aid in the goal of attaining a high-quality product. It is generally recognized that most mold designers have only superficial knowledge of the molding operations. This is not conducive to optimizing features that will aid the quality of the product. By presenting the information in such a manner that the mold features are the ingredients that will generate a part with good properties, low stress levels, and a high degree of consistency, the missing link in the mold designer's knowledge will be eliminated, and the result will be greater usefulness of plastics.

In addition, the molder's and moldmaker's understanding and appreciation of the details that lead to a successful product would result in clear-cut specifications for acquisition of a mold and in the long run save costly delays and expense in mold modification during mold proving.

What is meant by *details* can be seen from the following examples: when we speak of gate location, we consider it as stressed area and a possible source of weakness, as an appearance problem, as a source of dimensional variations because of thick and thin sections, and as a potential cause for surface blemishes. Or when a cavity is made of sections, we point out how to calculate the cross section of the matrix to prevent the "blowing" of cavity components. Or when stripping of molded parts is internal to the mold, we explain how to calculate the dimensions of bolts and rods that actuate such stripping. Or when moisture must be removed from raw material, we provide the needed temperature and time for each material to be dried and ready for molding.

These randomly selected examples indicate that this book is a source of information needed by those on the firing line in order to carry out a successful operation. It can be used as an instructional text by those entering the field of plastics molding, as well as by those in the moldmaking and molding field for executing all the phases of the operations to produce "good and consistent products for the end user."

All the information is based on theoretical knowledge and its practical application by the author for almost 50 years. Nevertheless, the author does not intend to suggest that someone in the molding field may not find some variation from the views expressed in the book. If this does happen, it may be the result of the many variables encountered in the molding operation, which by themselves or in combination with each other may bring a distorted result.

Some of the variables may be found in the materials. They are manufactured to a range of tolerances whose extremes may require changes beyond the scope of normal operations and therefore produce an apparent discrepancy.

When we look at molds, we normally examine the design and overlook the fact that its performance is only as good as the workmanship that went into it. If not properly maintained, the machine and its functioning during the molding operation can be responsible for many distortions of results. Finally, accuracy of parameter and consistency of repetition have a very large influence on the end result.

When one considers the potential variables, it is not too difficult to visualize deviations from normal conditions, which on the surface may appear as contradictions to the text statements but in reality are a product of variables that have not been uncovered or corrected.

No assertion in the text, nor proposed procedures, nor any other information herein should be construed as a suggestion for modifying prevailing shop rules, instructions, or practices.

Briefly stated, the ultimate aim of *Injection Molds and Molding: A Practical Manual* is to remove the tendency to treat moldmaking and molding as an art instead of dealing with them as subjects based on known principles and established sound practices.

2 Plastics Mold Design

NATURE OF PLASTICS

The term *plastics* is as broad as the term *metals.* Plastics cover a wide range of distinct materials with varying thermal, mechanical, electrical, and chemical properties. They are synthetic materials.

The first plastic was introduced at the beginning of the twentieth century. Today, there are 40 families of basic materials that are popularly used and produced in billions of pounds a year. To enjoy such rapid growth, plastics must possess characteristics that make them a good candidate for advantageous replacement of other materials and must have inherent properties for use on products for which only plastics will fit the application (for example, a heart valve).

Some of the outstanding characteristics of plastic are its light weight, resistance to corrosion, electrical insulation, heat insulation, colorability, and resistance to chemicals. Plastics are readily produced in high volume at a low cost of production; in most cases, the products can be used without additional operations. They also possess many other characteristics, some of which are inherent to specific materials such as lubricity, transparency, toughness, etc. It is a very impressive list, and if we add to it those properties obtained from combining the plastic material with fibers, minerals, and other additives, we can visualize an extension of applications that no other group of materials enjoys.

There are two basic groups of plastic materials, thermosets and thermoplastics. *Thermosets,* when subjected to heat and pressure, will change their chemical nature and retain their new composition even when exposed to high heat. *Thermoplastics,* on the other hand, will soften and become fluid under heat and will solidify when cooled. This process is reversible without change of the chemical composition.

The sources of raw materials for most plastics are coal, petroleum, limestone, salt, air, water, and cellulose from cotton and wood. These materials undergo a large number of decompositions and recombinations so that a commercial plastic is produced. Putting it another way, these raw materials are simply a source of the elements and basic compounds that are used in the compounding of the finished plastic.

Those in the industry refer to plastic materials as *polymers,* or as many *mers* or *monomers.* A *mer* is a repeating chemical unit in a high polymer. It can be com-

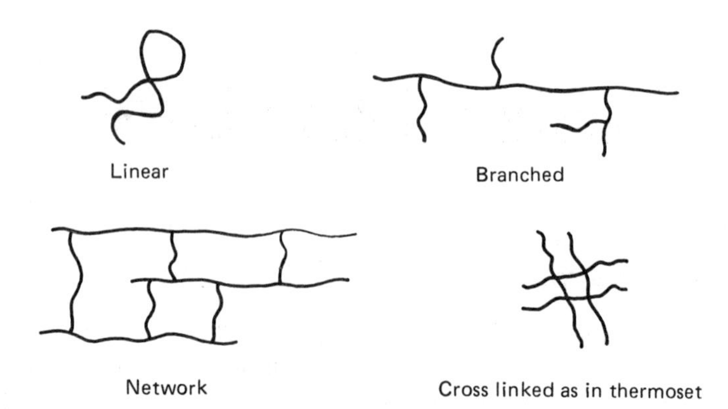

Fig. 2–1. Molecular chains.

pared for visualizing purposes to a link in an ordinary chain, where the chain is the practical constituent of a plastic material. The mer or *monomer* (single mer) is the building block of plastic materials and is a simple chemical compound whose usefulness becomes extended when it reacts with the same or other monomers and converts into a polymer. These small building blocks are also called *molecules,* and, when reacted with each other, they form a molecular chain. The process of forming a chainlike molecular structure from its own monomer or other monomers is known as *polymerization.* Polymerization is a chemical process, carried out under heat and pressure and frequently in the presence of other chemicals that play a part in the formation of the long-chained polymer. The molecular chains can be linear, have branches, or have a network in a plane or in space. The molecular chains are intertwined and enmeshed with each other to form a finished polymer.

When a polymer contains a single repeating unit such as polyethylene, it is known as a *homopolymer,* and when a polymer contains two or more other structural units such as ABS, it is a *copolymer.* All these terms are frequently used in connection with plastic materials, and, for this reason, the preceding brief explanation is given. See Fig. 2–1.

The chemistry of a plastic material and the processing steps leading to and including polymerization are complex and a sizable subject in their own right.

MOLD DESIGN

This brief explanation of the nature of plastics, leads to the next step of converting it into a useful product. To do this, we must understand how a plastic part is born.

A product design drawing is made showing the configuration and sizes of the part, the material from which it is made, and the specifications for use requirements.

The drawing becomes the basis for constructing the mold (die) in which the part can be produced by placing it in a suitable machine. The machine provides the necessary actions to fill the cavities in the mold and to let the material solidify into the desired shape, after which it will be removed from the opened mold as the finished product.

It is frequently expressed in the industry that plastic parts are only as good as the molds from which they are produced. If a design has no deficiencies and if its implementation is carried out with good workmanship, an acceptable part is produced. *All* parts would be acceptable if mold designers as well as moldmakers gave full consideration to all elements that enter into a mold, to make it capable of generating a product that meets required specifications.

A poorly working mold not only causes high production costs and requires an excessive amount of attention from key personnel, but it may also jeopardize the quality of the product and its intended service. Molds are expensive tools, and a small percentage of increase in their cost may mean the difference between a good one and one requiring constant nursing. On the other hand, an attempt to save on the tool cost can increase the manufacturing cost by as much as 50% and may even endanger the life of the product because of quality variation.

If a mold is worth making, it should be designed well and should incorporate the best workmanship to produce an economical part with properties to meet the needs of the buyer.

Like any other tools of conversion, molds have to be looked upon as a means to an end. They are a necessary way of attaining an end product, which is the molded part. This is the frame of mind with which we should approach the formulating stages of moldmaking.

Designing a mold, similar to designing a product, requires a certain amount of planning for and general visualization of what is to be accomplished. In order not to overlook any phase of the problem at hand, it is best to outline the factors that lead to a systematic consideration of all the important elements.

We must constantly keep before our eyes the one and only ultimate objective—i.e., a mold shall produce, without undue maintenance, a product with the anticipated properties, characteristics, and qualities at a price agreed upon by the user. A mold could be designed and built along the most ingenious lines, incorporate inventive features, and present the best workmanship, but, if it does not meet the stipulation of supplying a quality mold and a quality product at the quoted price, it will cause friction between the concerned parties.

A thorough understanding of all the conditions that prevail during the molding operation should insure that necessary precautionary measures are incorporated in the mold to safeguard the expected features in the product.

The function of a mold is to receive molten plastic material ranging in temperature from 350° to 700°F at pressures between 5000 and 20,000 psi. In the injection process, the plastic comes from a heated nozzle, and passes through a sprue bushing into feed lines (runners) and thence via a gate into a cavity. The cavities are maintained at temperatures ranging from 30° to 350°F, at which solidification takes place. They are provided with a means for controlling the temperature. The range of temperatures and pressure depends on the type of plastic material. The plastic is held in the cavity for a prescribed time until full solidification takes place. At this point, the mold opens, exposing the part to the ejection or removal action. The molding press shown in Fig. 2-2 is of the screw-plasticating type. Another type is the plunger machine.

The general principles of molding are very similar regardless of the type of press employed. All presses must meet the basic elements of molding: *time, temperature,* and *pressure.*

Time refers to the duration required to prepare the material for flow and to maintain it in the mold until solidification occurs. *Temperature* refers to the degree of heat needed for the material to flow within the mold and to the heat level in the mold needed to permit rapid and controlled solidification. *Pressure* refers to the pounds per square inch exerted (1) on the material to form the configuration of the part and (2) on the mold during part formation and solidification.

A description of the path of the material through the machine follows. Numbers in parentheses refer to Fig. 2-2. Plastic material in granular form is loaded into the hopper *(1)*. From there, it drops onto the feeding portion of the rotating extruder screw *(2)*. The screw moves the material through a heated extruder barrel *(3)* where it is plasticized or made fluid so that it can be fed into the injection chamber. The starting position of the screw is at the front of the extruder head *(4)*. As the material is advanced by the screw, it generates a certain amount of pressure, which displaces the screw, thus providing space for the discharging material. This space is called the *injection chamber (5)*. The plastic is now ready to be injected into the mold or die *(6)*. The two halves of the mold which are firmly attached to the die head *(7)* and movable platen *(8)* are closed by the main clamping ram *(9)* and held under a pressure of 2 to 7 tons psi of projected molding area (projected area of cavities and runners). At this point, the fluid plastic from the injection chamber is forced into the mold under pressure of 5000 to 20,000 psi and forms the configuration of the cavities. The mold is maintained at a temperature low enough to permit solidification or "curing" of the plastic in a short time. When rigidity is attained, the press opens the mold and actuates the ejection mechanism *(21)* so that the parts *(22)* are removed from the mold. All the movements within the press as well as the required time exposures are controlled by suitable timers and limit switches that actuate hydraulic valves to bring about the desired results. A more detailed description of the process and explanation of numbers is found in Chapter 14.

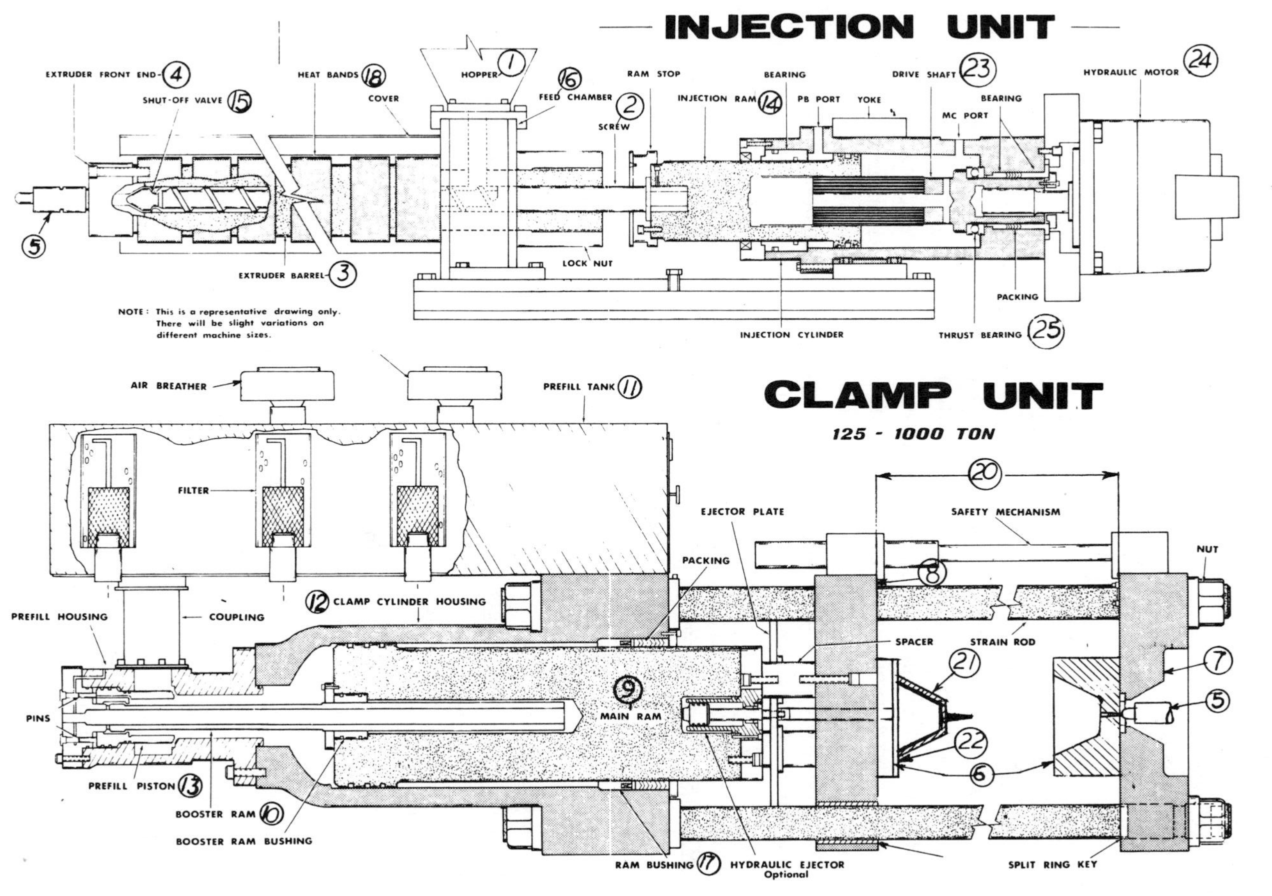

Fig. 2–2. Injection machine. *(Courtesy of HPM Corp.)*

Press sizes are rated in tons of clamping capacity and ounces of shot size, using polystyrene as the standard for shot capacity. The movable platen of the press is the source of clamping pressure; therefore, it represents the tonnage of the press. The extruder barrel is designed in heat capacity to plasticize a certain amount of styrene per hour. The maximum distance of screw travel establishes the space for the injection chamber and therefore creates the cubic inches for the maximum shot size.

From the general concept of the molding operation, we can recognize that it is important to design a mold that will safely absorb the forces of clamping, ejection, and injection. Furthermore, the flow conditions of the plastic path have to be adequately proportioned in order to obtain, in cycle after cycle, uniformity of product quality. Finally, effective heat absorption from the plastic by the mold has to be incorporated for a controlled rate of solidification prior to removal from the molds. All these and related subjects will be discussed in subsequent chapters. It should also be stated that some prospective users of plastic products look at them with a certain degree of suspicion because of some unfavorable experience. Such experiences can be avoided if (1) the product designer fully considers the peculiarities of the specific plastic material and dimensions the part accordingly and (2) if the tool designer incorporates all details that are conducive to good moldability. Last but not least, the molder must recognize the significant molding parameters that influence product quality and maintain them at prescribed levels.

All in all, the plastics operation is a collection of many small details that must be carefully executed to insure a successful product.

A conversion tool that must successfully carry out all the prescribed functions has in the past been designed by toolmakers whose judgments were the main guide points in selecting the proportions of mold elements. It is the purpose of this book to provide the necessary data for calculating strength and other limitations of mold components so that current designs of molds can be adequately engineered without resorting to empirical practices.

COMPUTER-AIDED DESIGN AND MANUFACTURING (CAD/CAM)

A further step in putting the mold design on a systematic basis has been the introduction of *Computer-Aided Design and Manufacturing* (CAD/CAM). When this step is considered, cost and other factors such as delivery of completed item, quality of mold, and so forth may justify additional expenditures, so that the direction to be followed will be self-evident. An abbreviated description of CAD/CAM will provide some insight to this system of designing and manufacturing a mold so that one can select a procedure that is most advantageous for prevailing conditions.

Producing a plastic part from the initial concept to the finished product involves many complicated mathematical functions and relationships that would be uneconomical or impractical to evaluate by hand. For this reason, CAD/CAM performs a vital part in the overall product development.

The CAD/CAM system is carried out by a designated operator whose duties are to become familiar with the flow of information through the sophisticated machines and at the same time to learn what each machine is capable of delivering. The minimum training of a CAD/CAM operator is four weeks. After an exposure of one year to this operation, one can expect the full benefits of computer-aided mold design.

The CAD/CAM installation presents a sophisticated and expensive tool that requires top-notch management and operator commitment to make it work.

The normal procedure for designing a molded product consists of the following steps. The product engineer designs and details all the required performance characteristics of the parts. Then the proposed design is submitted to the estimating department which determines the cost of parts to be manufactured as well as tooling expenditure connected with the project. If the projected costs are favorable for commercial production, the next move is to order mold design and mold building.

Considering the application of computers to these functions performed by humans, one usually looks to these machines to do the same amount of work in a much shorter time and in a very accurate manner. With the use of CAD/CAM, most of the essential operations are carried out in a similar manner as outlined in the normal procedure with the additional features described below.

1. Graphic display. The engineering design is transferred to the *graphic input screen.* Then, using sophisticated interactive software in which the operator supplies answers to the program's questions, the computer refines the design with the aim of reducing material content and/or conserving cycle time. The refinement may involve in such areas as reducing wall thickness where it is not needed, eliminating ribs, and in general modifying features that are not essential for performance of the product. See Figure 2-3. A reduction in wall thickness not only saves in material cost but also adds to the shortening of the molding cycle since that time decreases in direct proportion to the square of material thicknesses.

With the changes incorporated in the modified design, the dimensioning of the part is corrected on the screen, and this information is used as a basis for all subsequent operations. The improved design as shown on the screen can now be used at the push of a suitable button, to view the product from all angles, from both outside and inside, although at this stage we are still dealing with the conceptual image of the product. This is also the stage at which the computer can construct a solid view of the part and its appearance, so that the personnel concerned with that phase can finalize their judgment of the project. From a

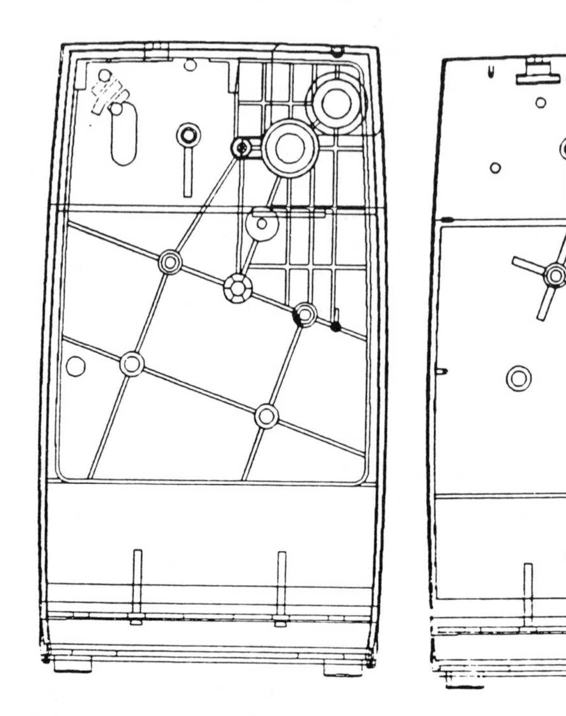

Fig. 2–3. Example of CAD/CAM. *a,* Housing design as presented in engineering drawings. *b,* Same housing analyzed by design software. *(Courtesy of G.E. Co.)*

molder's point of view, an accurate material weight can be obtained from this preliminary presentation. With the information so far recorded, all that is necessary now is to allow for shrinkage of the plastic material and prepare a mold design for the actual product.

2. Mold-base design. The second step is to apply the information from graphic presentation to design a mold base. Economic analysis programs are available for the number of cavities in the base and also the suitable press in which the job is to run. The projected number of pieces per year is the prime consideration for this purpose, although other factors such as process control to mold the part to tolerance, the additional cost of making cavities, and cost of a larger mold base are all taken into account in the overall picture.

In case of mold base selection, there is also another interactive program for the computer that permits the mold designer to select the proper mold base and related accessories from a company such as D.M.E. Co. After this selection is done, the cavity and core are dimensioned and fitted into the mold base. To view any selection of the cavities, cores, and mold base for analyzing purposes, one need only push a suitable button, proceed with the study to be undertaken and correct possible discrepancies. During an analysis of that type, the overall dimensions of the mold base are displayed on the screen to insure that the corrections the operator makes are within the limiting bounds of the base. The computer automatically provides a complete "bill of material" for all the components used in the mold, so that items not in stock can be ordered. After all the software answers about details of construction of the mold are provided and incorporated into the mold design, the only steps left to consider are mold cooling and its associated polymer flow.

3. Mold cooling. The removal of heat from the material injected into a mold is a vital ingredient in mold performance. The cooling efficiency of the mold is determined by type of coolant, flow rate of material, pressure, and temperature. The computer, which has been programmed for the laws of physics governing these parameters, can remember and calculate their values in a rapid manner. The same job done by hand would be prohibitive in consumed time. Here again we have a software to the computer that in question-and-answer format allows the designer to select the proper cooling design for the mold. The parameters include location of cooling lines, sizes and types of cooling lines, type of resin and its melt temperature, type of coolant, number of circuits, bubblers, baffles, heat pipes, distance of each from the molding surface, etc. After the cooling system has been adopted, the major design phase is completed.

4. Mold flow. Once the cooling system is determined, the only remaining task is to locate the gate and runner system before proceeding with the mold-flow analysis. The gate location should (1) provide preferably unidirectional flow, (2) be spaced for proper filling, (3) keep molding stresses to a minimum in certain locations of product performance, (4) have the weld lines located for least impact on product function, and (5) last but not least, have the approval of the product designer for appearance purposes.

The runner system is usually (1) required to have a favorable runner to part weight, (2) should not have a higher cooling requirement than the part, and (3) should provide a balance of pressure and temperature or combination of both to fall in the overall picture of the mold-flow system.

Certain molding problems were not easily understood before the advent of mold-flow analysis. Among these not fully comprehended phenomena were warpage of products, molded-in stresses, flashings, and excessive fill pressure. A flow-analysis software to the CAD/CAM will ask questions about the mold to be analyzed and about the description of flow patterns in which the various branches and their segment are identified. This is done by supplying the specific dimensions of each flow. It is followed by giving the resin, the injection time, and the mold temperature at various locations. With the supplied answers the computer will automatically calculate for each flow segment such parameters as pressure drop, shear rate, shear stress, melt temperature, cooling time to ejection, and other information pertinent to the flow analysis. In reality, *flow* in the mold may be described as the increment of pressure drop that is responsible for moving a segment of the flow forward and is resisted by the shear stresses acting on both faces. The computer solves simultaneous equations of heat transfer and plastic flow.

There are several programs available for simulating the melt flow during mold filling. Different mathematical models and assumptions are used in predicting what actually occurs in the mold. Once a program is chosen, it should provide the molder with such setup information as pressure to be used, mold temperature, melt temperature, rate of injection filling, and cooling time in the mold. It is possible to have all this information before any metal has been cut. When a mold is designed with computer-aided procedures and is put in production, one can expect the mold to produce a part with acceptable residual stress levels.

5. *Numerically controlled operation.* The all-important step is manufacturing the mold. The conventional cutting of cavity and core and fitting of these to the mold base consumes by far the greatest manpower in the overall picture of the process as a whole, from blueprint to production of parts. The CAD/CAM system has the capability to punch paper tape that can be used for operating *numerically controlled* (NC) machines. With the aid of paper tapes, NC machines can now follow the necessary tool paths for machining hole configuration, pockets, profiles, or complete three dimensional contours. Once this has been done, the NC programmer can verify the tool paths graphically to check for accuracy and possible interference. The information thus developed can then be put to use or stored for later application. Generally speaking the manpower reduction in making a mold by the CAM method is considerable, and the side benefits of close-tolerance machining is of appreciable value in fitting and assembling the complete mold.

Benefits and Costs of the System

As we can see from the text, the product designer, mold designer, and molder each play an important part in the CAD/CAM system. They have to choose from several alternatives and simulate the program that would provide the suitable answer. This process is decidedly preferable to the conventional one of building prototypes. Predicting the performance of a mold without cutting any metal is an advantage. The steps taken in CAD/CAM can be revamped to arrive at the optimum conditions of the product. Furthermore, any operations carried out by the computer are much faster and more accurate than when done by humans. By far the greatest number of man-hours involved in preparing for production of a molded part occurs in machining all the components in a mold. It has been estimated that in this phase alone about 25% in man-hours can be saved. For this saving alone it would pay to look into CAD/CAM.

There are other savings that deserve consideration. Refining the product design can save cycle time by an average of 25%. Material savings on the average will be about 5%. Mold delivery time can be cut to about 70% of present schedules. All these are current advantages that would justify looking into CAD/CAM procedures.

There are also some disadvantages. The equipment, including all the necessary software, is quite expensive. At present the cost of CAD/CAM runs anywhere from $200,000 to $500,000. Secondly, it requires training of the operators to manipulate the computers so that the best molded products result.

Regardless of these two drawbacks, CAD/CAM is a tool no progressive molder can afford to overlook for application to current mold designs.

3
Injection Molds for Thermoplastics

An injection mold normally consists of two basic sets of components. One is the cavities and cores; the second is the base in which the cavities and cores are mounted. This combination of components is a result of standardization in the moldmaking industry.

MOLD BASE

The exploded view in Fig. 3-1 shows parts of a mold base. The *locating ring* surrounds the *sprue bushing* and is used for locating the mold in the press-platen concentrically with the machine nozzle. The opening into which the ring fits is made to a tolerance of -0.000 and +0.002. The ring itself is made 0.010 smaller than the opening providing a clearance of 0.005 per side. A clearance above this amount may cause misalignment with the nozzle, which in turn would entrap part of the sprue, causing sprue sticking on the wrong side. The sprue bushing on the locating ring end has a spherical radius of 1/2 or 3/4 in. to fit the machine nozzle radius. The opening of the sprue, known as the "0" dimension, is specified to suit each job. The hole through the length of the sprue has a 1/2 in./ft taper or 1°, 11-1/2 min. per side. This hole must have a good reamed and polished finish to prevent sprue sticking. The sprue bushing goes through the front clamping plate or *backup* plate and *front cavity* plate, thus extending to the parting line. The parting line is formed by cavity plates A and B. Cavity plate A, as the name implies, retains the cavity inserts and supports the *leader pins,* which maintain the alignment of cavity halves during operation. One of the four leader pins is offset by an amount of about 3/16 in. to eliminate the chance of improper assembly of the two halves.

Leader pins are made 0.001 below nominal and to a tolerance of +0.0000 and -0.0005. The bushings are 0.0005 above nominal and to a tolerance of +0.0000 and -0.0005. This provides a maximum clearance between leader pin and its bushing of 0.0025 and a minimum of 0.0015. In actual practice, this amount of play may never exist because of the four pins and bushings and tolerance of their location. If by chance they do materialize and the job requires extreme accuracy of alignment, a set of tapered plugs and sockets can be incorporated into the base. See Fig. 3-2.

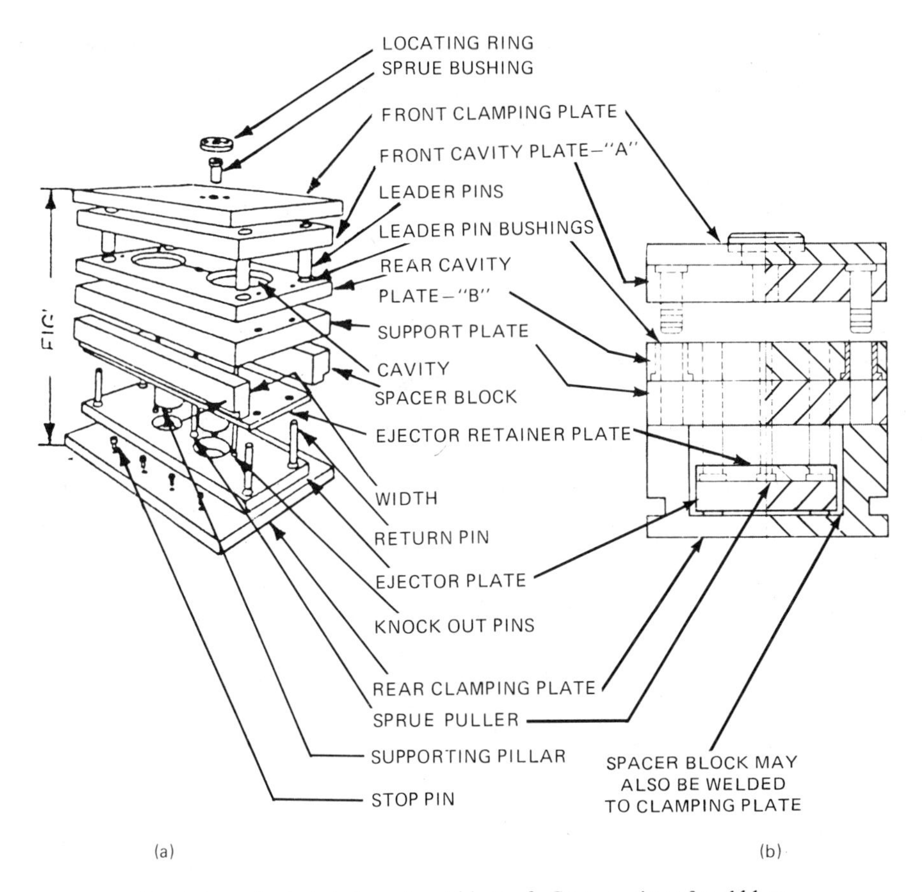

Fig. 3–1. *a,* Exploded view of mold base. *b,* Cross section of mold base.

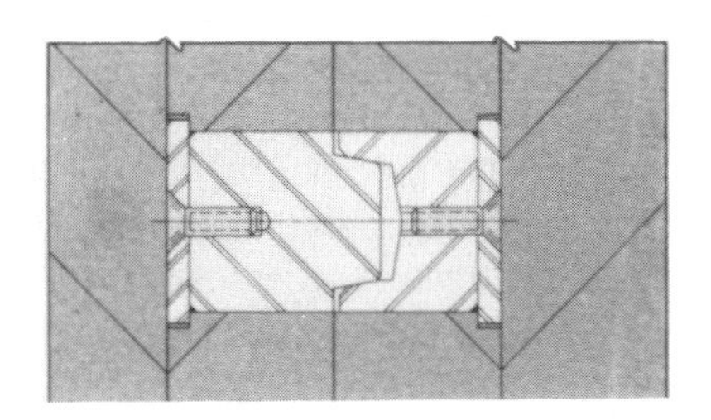

Fig. 3–2. Precision aligning plugs.

Occasionally, there are instances where the core protrudes a considerable amount, thus requiring precise guidance of mold halves before cavity and core touch each other. In such an event, the leader pins must be long enough to perform the job of guiding prior to contact of the mating halves. When needed, the protrusion of leader pins from the parting line can be as great as the height from parting line to bottom of rear clamping plate.

Mating with A plate is B plate, which holds the opposite half of the cavity or the core and contains the *leader pin bushings* for guiding the leader pins. The core establishes the inside configuration of a part.

B plate has its own backup or *support plate.* The backup B plate is frequently supported by pillars against the U-shaped structure known as the *ejector housing.* The U-shaped structure consisting of the *rear clamping plate* and *spacer blocks* are bolted to the B plate either as separate parts or as a welded unit. This U-shaped structure provides the space for the moving *ejector plate* or ejection stroke also known as *stripper stroke.* The ejector plate, ejector retainer, and pins are supported by the *return pins.* When in an unactivated position, the ejection plate rests on stop pins. When the ejection system becomes heavy because of required high injection forces, additional supporting means are provided by mounting added leader pins in the rear clamping plate and bushings in the ejector plate.

The stripper plates provide clearance holes around the supporting pillars. The rear clamping plate of the U-shaped structure usually has openings for *stripper rods,* which actuate the ejection plate. The ejection plates retain the heads of *knockout pins* as well as *return pins.* All these pins move with the stripper plate.

All the mold plates (excluding the ejector parts) and spacer blocks are ground to a thickness tolerance of ±0.001. A combination of tolerances could conceivably build up to cause an unevenness at the four corners. If large enough, such a condition could damage a platen when under full ram pressure. It is advisable to check the uniformity of all four corners prior to preparing the base to receive cavities.

Another item that requires attention is the contact area of the spacer blocks. The stress on these areas should be such as to prevent the embedding of these blocks into the plates, thus causing a change of the space for the ejection system. The safe tonnage that a mold base will take as far as spacer blocks are concerned can be calculated in this way. Let us take a 9-7/8 X 11-7/8 standard mold base made of low carbon steel. The weakest section of the spacer bar is at the clamping slot area. For this size mold base, the width of the block is 1-7/16, and the width of the clamping slot is 5/8. The area will be 1-7/16 minus 5/8 or 13/16 X 11-7/8 X 2 since there are two blocks.

$$\text{Area} \times \text{Allowable stress} = \text{Compressive force}$$

The allowable stress for low carbon steel is 25,000 psi.

$$13/16 \times 11\text{-}7/8 \times 2 \times 25{,}000 = 462{,}000 \text{ lb or } 231 \text{ tons}$$

Higher strength steel throughout the base can double or even triple the ability to absorb the compressive force. The addition of supporting pillars will also increase the compressive force in proportion to the area that they provide. Thus, two 2-in.-diam pillars would add:

$$\text{Area of a 2-in.-diam} = 3.14 \text{ in.}^2 \times 2 = 6.28 \text{ in.}^2$$

$$6.28 \times 25{,}000 = 157{,}000 \text{ lb. or } 78.5 \text{ tons}$$

The embedding problem does arise especially when changes are made in supporting bar dimensions or supporting blocks.

CAVITY AND CORE

At this stage, it becomes necessary to outline the cavity and core so that they can be presented to the product designer in order to obtain his or her agreement to the parting line and gate location. Figure 3-3 shows a cross section and top view of the part drawn to actual size. A freehand outline of cavity core is made using as a guide the fact that the outside of the part will form the cavity and the inside of the part, the core. This establishes the parting line and gate location. An analysis is made of whether, with this arrangement, the part will stay on the core during mold opening. The depth of the ribs and the binding between them as a result of shrinkage should overcome any tendency of sticking in the cavity. The finish in the slots for the ribs would be smooth as machined or as electric-discharge-machined (EDM) but would not have a directional polish. The arrangement in Fig. 3-3 is deemed satisfactory for presentation to the designer.

Let us now assume that the part is designed without the ribs and that the center ring has tapers in the direction of withdrawal as illustrated in Fig. 3-4.

With this design, the part would definitely stick in the wrong side. Since the specifications indicate that there should be no ejection marks on the outside surface, we will move the parting to a new position and depend on the outer diameter (O.D.) of the part to retain it on the core side. This will call for circumferential polishing at the O.D. to provide minute undercuts. Ejection pins are placed as indicated in Fig. 3-4. Even a slight ring scratch of 0.001 depth in the middle of the flange could aid in the retention of the part.

If it were permissible to have the ejection-pin impressions on the outside, then the arrangement in Fig. 3-3 could prevail, except that the half that was stationary would become the ejection half and would incorporate the ejecting

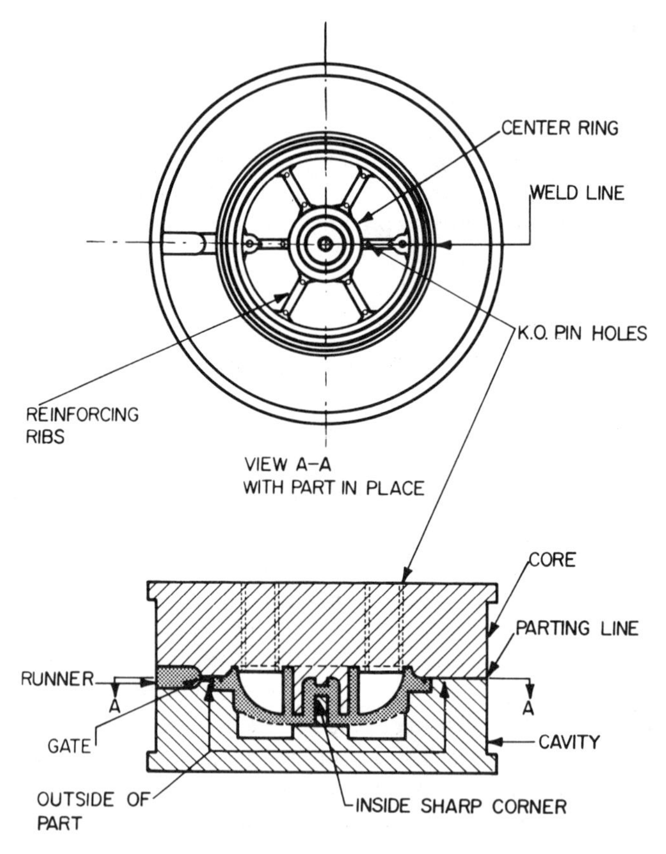

Fig. 3–3. Parting line. Cavity and core outline. *(Courtesy of Rockwell International Corp.)*

pins (Fig. 3–5). The photograph in Fig. 5–5 shows the actual mold and parts with runner system.

Not all parts lend themselves to this kind of treatment. As an extreme example, let us look at a 1/4-in.-diam rod, 2 in. long. The part is symmetrical, and there are no means of making it stick to the ejection side. This part (illustrated

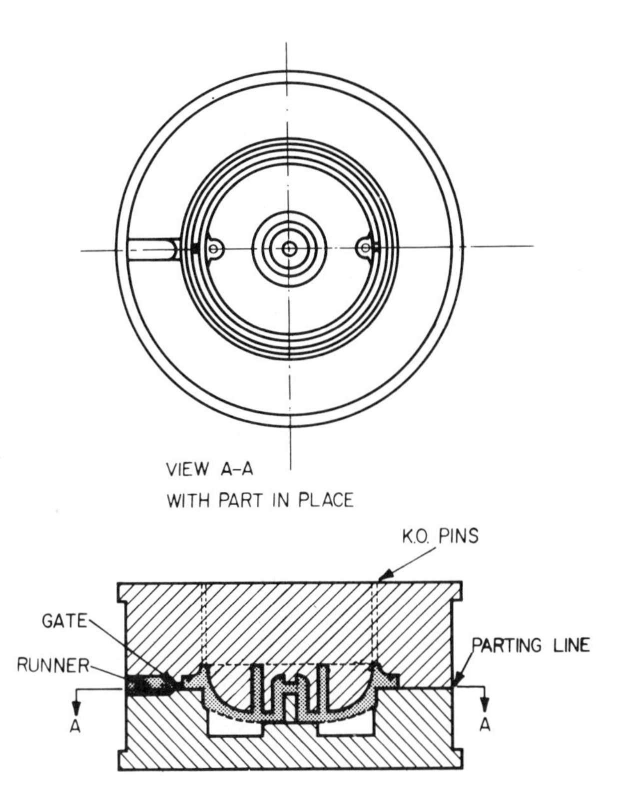

Fig. 3-4. Parting line. Cavity and core outline. *(Courtesy of Rockwell International Corp.)*

in Fig. 3-6) will stick to the side where the gate is. Should the end opposite to the gate tend to adhere to the stationary half, it could be provided with a dummy gate to retain it on the correct side.

Needless to say, there is a great variety of cases, each requiring individual treatment. With proper attention and thought, simple means can be devised to locate the parting line and provide a method of retention on the ejection side.

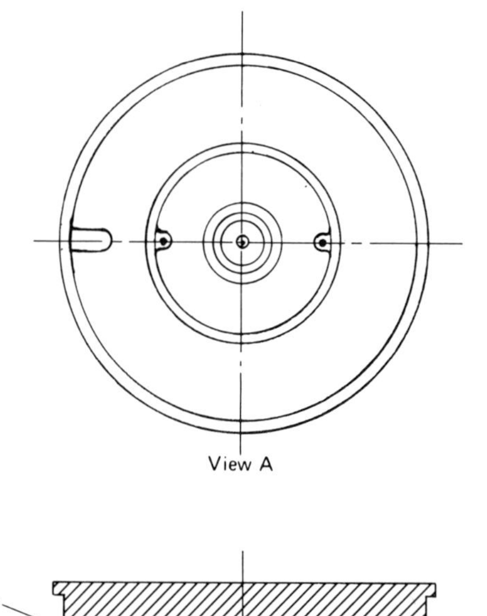

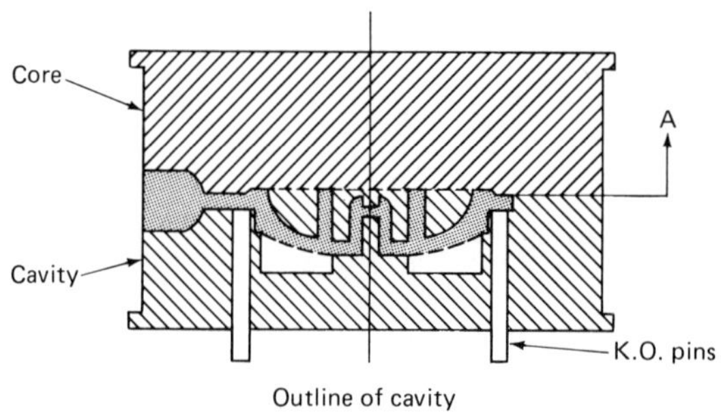

Fig. 3–5. Parting line. Cavity and core outline. *(Courtesy of Rockwell International Corp.)*

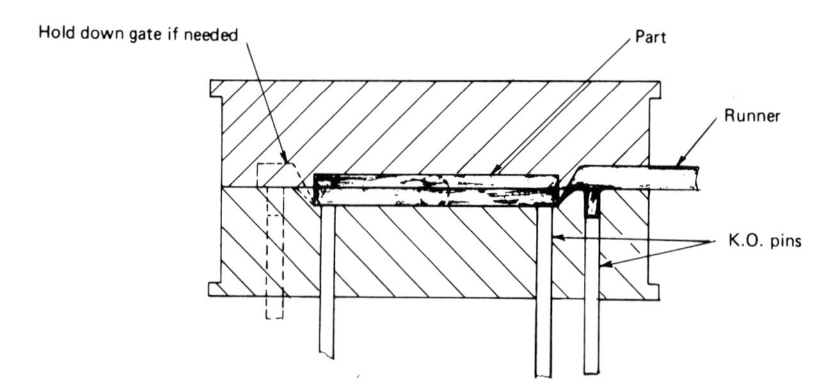

Fig. 3–6. Mold for round rod.

4
Consideration of Product

With a general concept of the molding operation in mind, we can proceed to the details of each specific phase. The first step is an analysis of the product drawing to see how compatible it is with sound molding practice. We will examine it to determine whether the part is dimensioned and shaped so that, when molded, it will fulfill its anticipated performance requirements. There are many details in a part drawing that can be readily modified (without affecting the end use) to make the design more suitable for molding purposes. Such modifications are generally overlooked due to a lack of recognition of the part they may play in overall quality.

DIMENSIONS FOR PRODUCT PERFORMANCE

A determination is made of how a part under consideration will be used and what performance specification the product is to meet. This is done by reviewing either an assembly drawing or a prototype of the complete product. The essential functions that the part is to carry out are established. The dimensions that have a vital contribution to the performance of the product are noted.

Figure 4–1 illustrates a part that was analyzed along these lines by viewing the prototype assembly. We note that some close tolerance dimensions are marked with P or M, denoting Performance or Mating requirements. This P indicates to the moldmaker (toolmaker) that performance dimensions require extraordinary care in the making of the cavity, whereas M (mating) dimensions are important to the extent that another part that works with it will readily assemble and fit in the overall picture. This or a similar type of designation will assure that the needed accuracy will be built into the tool where it is most essential and consequently should be reflected in dimensions of the part.

MATERIAL CONSIDERATIONS

At the time mold design is being executed, it is desirable to determine not only the specific grade of the material, but also its manufacturer. Some suppliers have varying recommendations as to processing for the same generic type of material. This information is included in Chapter 18. Of particular interest is the mold-temperature and melt-temperature range, which would indicate how much heat is to be dissipated as well as runner and gate data, which determine

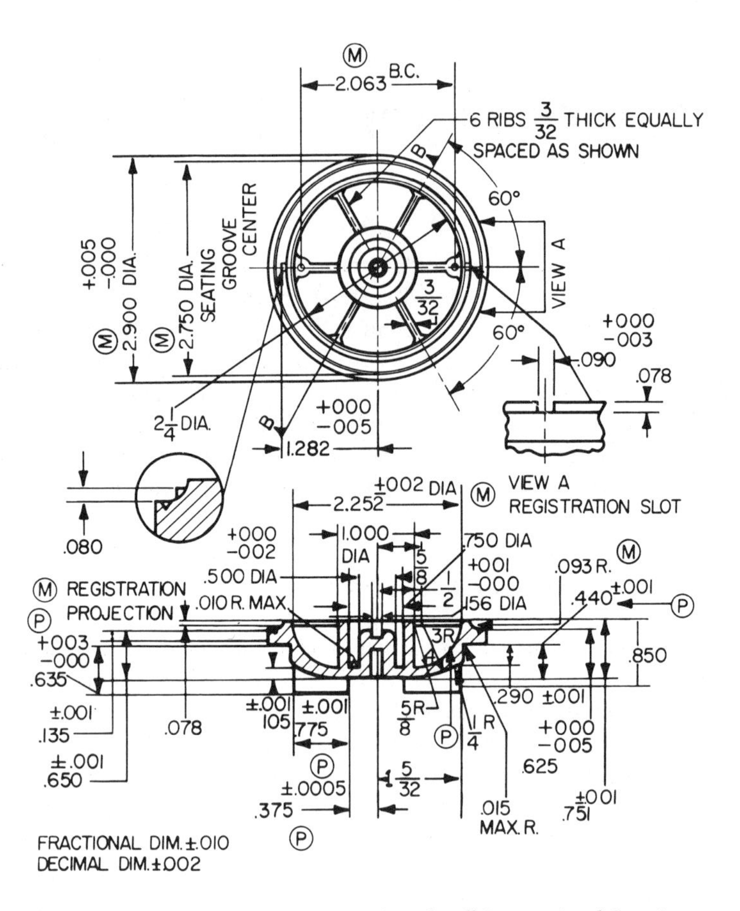

Fig. 4–1. Cover plate. *(Courtesy of Rockwell International Corp.)*

melt-flow characteristics. The following paragraphs will discuss "hot" and "cool" molds, "high" and "low" pressures, melt temperatures, etc. These statements should be taken in the context of "material processing data" wherein the range of temperatures and pressures are indicated for each material.

SHRINKAGE

By *shrinkage* is meant the dimensions to which a cavity and core should be fabricated in order to end up with a part of desired shape and size. The usual way to decide on the amount of shrinkage is by looking up the data supplied by the manufacturer of material. The supplier's information is obtained from a test bar molded according to an ASTM procedure. This means that the test bar is molded at a specific pressure, mold temperature, melt temperature, and cure time. The thickness of the test bar is normally 1/8 in. However, molded parts are very rarely produced under the same or even similar conditions as those used for test bars.

For precision parts with close tolerance dimensions, the shrinkage information from test bars as furnished by material suppliers is inadequate. We must, therefore, become familiar with the factors that influence the shrinkage so that we may arrive at a more exact dimension for a specific part.

Shrinkage is influenced by cavity pressure to a very large degree. Depending on the pressure in the cavity alone, the shrinkage may vary as much as 100%.

Part thickness will cause a change in shrinkage. A thicker piece (1/8 in. or more) will shrink on the high side of the data, whereas a thin one (1/20 in. or less) will shrink to the lower value.

The mold and melt temperature also influence shrinkage. A cooler mold will bring about a lower shrinkage, and a hotter melt will cause a higher shrinkage.

A longer time in the cavity will allow the part to come closer to mold dimensions, which means a lesser shrinkage.

Openings in the part will cause variations in shrinkage from section to section, because the cores making these openings act as temporary cooling blocks, which prevent change in dimension while part is solidifying. Gate size on the large size will permit higher cavity-pressure buildup, which brings about a lower shrinkage.

The shrinkage problem can be separated into these categories:

1. Materials with a shrinkage rate of 0.008 in./in. or less are readily predictable and are not difficult to adjust with molding parameters, such as cavity pressure and mold or melt temperature, or, as a last resort, with cycle.

2. Parts made of materials with high shrinkage (above 0.010 in./in.), but which are symmetrical and suitable for center gating, will also be readily predictable and adjustable with molding parameters.

3. Parts made of materials with high shrinkage rate, which are symmetrical but cannot be center-gated, may approximate a center-gate condition if multiple gating close to the center (3, 4, or 6 gates) is possible. In this case, the prediction of shrinkage is somewhat more difficult but still presents a chance of success.

4. The major problem exists with materials that have a high shrinkage rate, i.e., about 0.015 to 0.035 in./in. In most of these cases, the material suppliers either show nomographs in which all contributing factors are drawn and coordinated to supply reasonably close shrinkage information, or they point to examples with actual shrinkage information and molding parameters so they can be used for comparative interpolation. With most high-shrinkage materials (e.g., nylon, polyethylene, acetal), when the material is edge- or side-gated, a larger shrinkage occurs in the direction of flow and a smaller one perpendicular to it.

If, upon review of the shrinkage information, there is still a doubt about whether the precision dimensions will be attained, then there is one way left for establishing accurate shrinkage data, i.e., prototyping. In this method, a single cavity is built, and the critical dimensions are so calculated that they will allow for correction after testing, by providing for metal removal (machining). The test sample should be run for at least half an hour and under the same conditions as a production run. Only the last half-dozen pieces from the run should be used for dimensioning. It is best to make the measurements after a 24-hr stabilization period at room temperature. For example, in Fig. 4-1, the lower area shows in the center a projecting boss of 0.500 diameter, and on its top a recess. This recess has a requirement of +0.001 and -0.000 on its diameter, and of 0.156 and 0.135± 0.001 in its depth. In order to meet these close tolerances, the mold pin that reproduces the sizes in the plastic would be made larger in diameter (to 0.162) and longer in depth (to 0.140), test shots would be made, and the sizes checked. On the basis of results from the test shots, the pin sizes would be machined down in diameter as well as in depth to suit the needs.

Close tolerance locations of holes can also be prototyped in a simple manner by calculating the dimension and using straight pins for the cored holes. The corrected location is made by providing a shoulder pin that encompasses the existing hole as well as the pin with new and correct hole location (Fig. 4-2).

Precision gears, when molded from a material with a high shrinkage rate, should be fabricated in a prototype system. The following method has proven very successful on about 200 different gears. The mold is made complete except for the teeth. The outside diameter of the gear blank is made about 0.010 in. smaller than the root diameter of the teeth. This is the diameter with a calculated shrinkage included. The blank is molded, and the actual shrinkage is established from it.

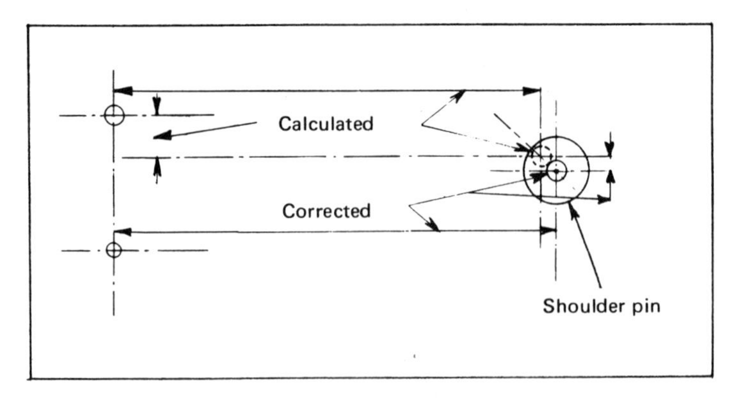

Fig. 4-2. Hole location.

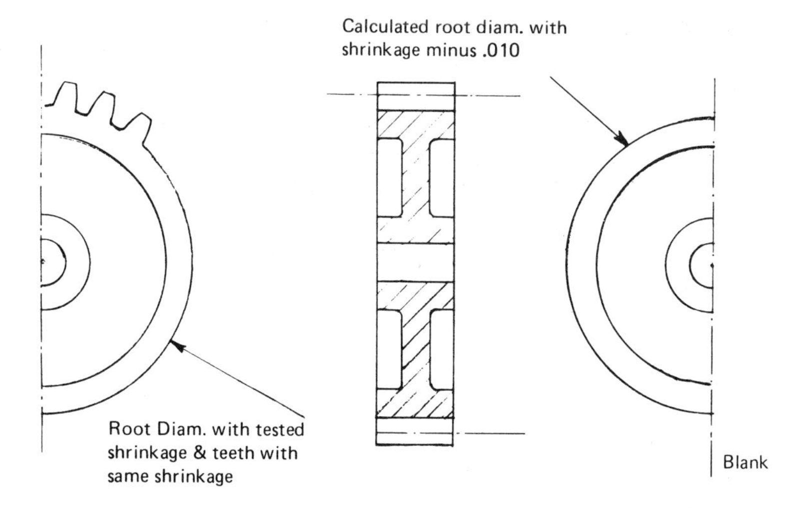

Fig. 4-3. Gear.

The next step is to use the shrinkage information to shape the teeth. The newly obtained tooth form is machined into the blank of the cavity (Fig. 4-3).

When a part involves very close tolerances on some dimensions, it becomes the responsibility of the product designer as well as the tool designer to review all shrinkage factors and to jointly decide on the method by which shrinkage is determined. In many instances, the absolute dimension is not as significant as repetition of the dimension from part to part.

POSTMOLDING SHRINKAGE

Previously, it was suggested that measurements of prototype parts should be made 24 hr after molding in order to check on shrinkage results. With crystalline thermoplastics such as acetal, nylon, thermoplastic polyester, polyethylene, and polypropylene, the ultimate shrinkage may continue for days, weeks, months, or even a year. The shrinkage noted 1 hr after molding may be only 75% to 95% of the total. According to data compiled by several material suppliers, shrinkage is a function of mold temperature, part thickness, injection pressure, and melt temperature.

The reason for postmolding shrinkage is that there is a molecular rearrangement and stress relaxation going on until equilibrium is attained, at which point shrinkage stops. Both the molecular rearrangement and stresses are brought about by molding conditions. The conditions that are most favorable for reaching the ultimate shrinkage in the shortest time are a relatively high mold temperature and a lower rate of freezing of the material. Each material has its own rate of postmolding shrinkage as a function of time. Curves showing the rate of shrinkage as a function of time for changing mold temperature and varying part thickness are available from material suppliers.

When the upper range of mold temperature (shown in material processing data sheets, Chapter 18) is applied to the molding operation, we have a condition that is most conducive for shrinkage stopping in the shortest time after part removal from mold. A slower rate of heat removal from the part is also desirable. Thin parts, which by their nature have a faster heat-removal rate, and consequently have a longer shrinkage stabilization time. It is to be repeated that the problem of postmolding shrinkage exists with crystalline or semicrystalline thermoplastics.

The configuration of a product, the end use, and assembly will determine to what degree the postmolding shrinkage will be a factor and what steps have to be taken to overcome a potential problem.

On some critical parts, an annealing-stress-relieving operation may be necessary to offset possible dimensional changes. It is to be remembered that each crystalline material has a different postmolding shrinkage stabilization time.

FLOW LENGTH

The length of the path that the material flows from machine nozzle to the extreme "out" position of a cavity will govern to a large degree the mold temperature. Thus, a short flow (less than 3 in.) will call for a cool mold, whereas a flow of over 6 in. in length will demand a temperature at the upper end of the range.

MOLD TEMPERATURE

Mold temperature is not only related to flow length but also has a bearing on weld line strength, level of molding stresses, and appearance (sinks). A higher temperature will give stronger weld lines, lower stress levels, glossier appearance, and deeper sinks.

MOLDING STRESSES

Molding stresses are concentrated around gates, openings (holes, etc.) inside sharp corners, sharp transitions in cross section, and weld lines. The stressed areas detract from the strength by a considerable amount (at least 10% to 20%); therefore, a load, heat, pressure, and ultraviolet and chemical exposure should not be permitted to act on the stressed surfaces.

PARTING LINE

Parting-line considerations depend on the function that a part is to perform. Thus, for example, if a part is to act as a shaft in a bearing, it would not tolerate a conventional parting-line projection, however small it might be. In such a case, flats (see Fig. 4–4) can be employed to overcome the parting-line defect and have a free running shaft.

Parting lines are not necessarily straight lines but depend on the contour of the molding. Figure 4–5 shows an irregular parting line. When a parting line involves two mating halves and the dimensional tolerance is close, an allowance has to be made for possible matching of the halves. This allowance should be approximately 0.005 to 0.010 in. less than the finished dimension.

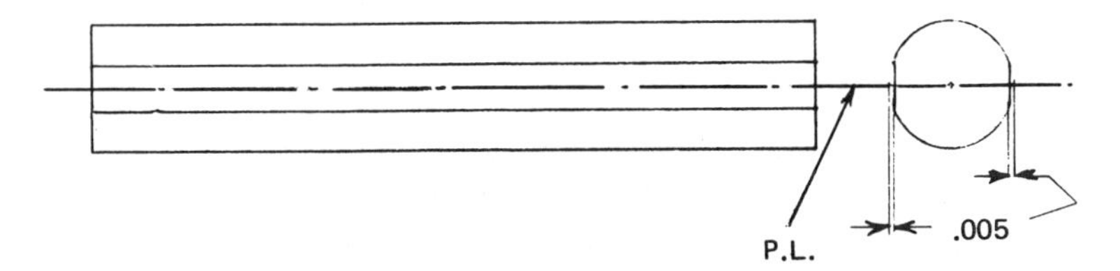

Fig. 4–4. Free running shaft.

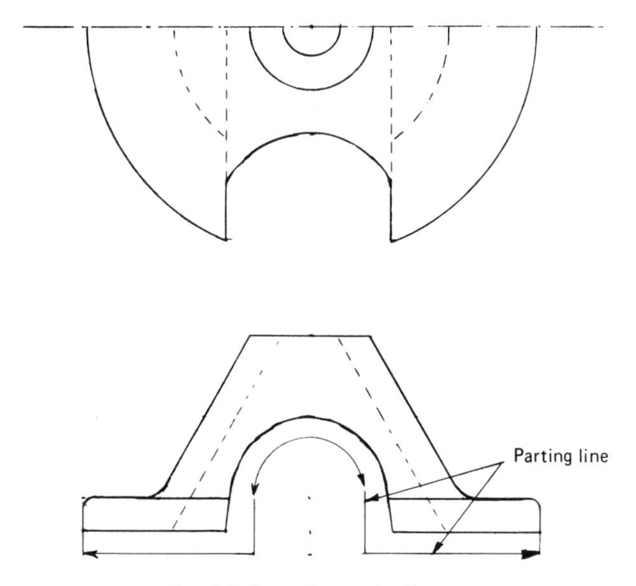

Fig. 4–5. Irregular parting line.

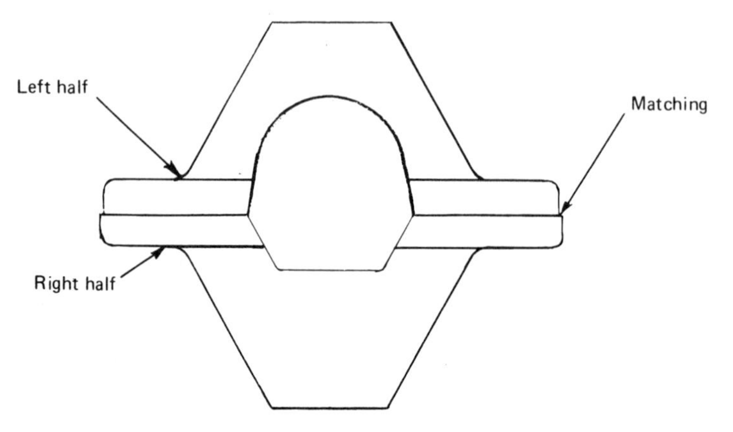

Fig. 4–6. Parting-line matching.

The matching of parting lines may also be required in applications where a slight mismatch at the parting line can cause discomfort (e.g., with hand grips) or be a cause of poor appearance (Fig. 4–6). In any event, problems connected with parting-line location should be solved to suit end-use needs and approved by the product designer.

GATE LOCATION

Gate location is not only a stressed area but it also presents an appearance problem; therefore, its location should be established in accordance with those considerations. It is preferable to have the gate located at the heaviest cross section.

WELD LINE

Weld-line strength and appearance should be kept in mind during mold design so as to minimize reduced strength and the appearance of "crack lines." A weld line is formed whenever an interference with the material flow exists—such as a core—making an opening in the part. The material, upon encountering the core, circumvents it by splitting the stream, and then envelops it and rejoins at the opposite side. This rejoining or self-welding is known as the *weld line.* Changing the location of the gate will have a different effect on the strength of a part. In Fig. 4–7, area A will be strongest due to gate location, material flow, and where the weld line is formed. Area B will be strongest for the same reason.

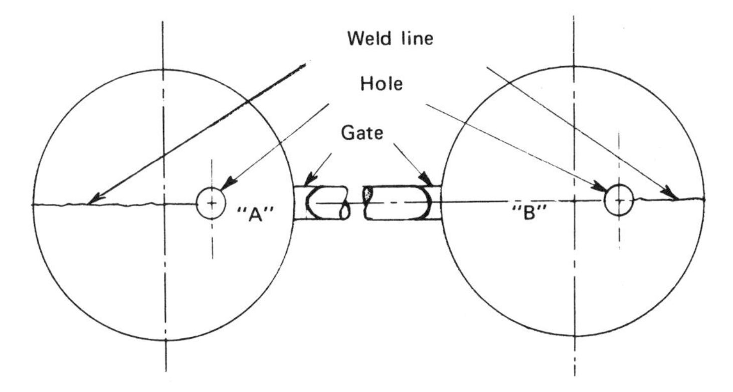

Fig. 4–7. Gate and weld-line location.

TAPER ADDITION, ALSO CALLED DRAFT ANGLES

When not specified, taper addition should be called to the attention of the part designer so that when it is incorporated, it will cause no interference. In any case, some taper should be provided on vertical surfaces to facilitate withdrawal. The amount of taper for each area involved should be indicated on the part drawing. It is common practice to dimension draft in degrees because standard end-milling cutters are available with various tapers per side, thus making it economical to produce such tapers in the mold. It is to be remembered that difficulty in ejecting the piece from the mold can cause warpage, strains, or generally impair the quality of a product. Tapers for ease of ejection should be looked upon as a very important phase of plastics dimensioning.

There are cases where straight holes or configurations are essential. In such events, there is normally a tolerance provided, which may be utilized for a taper. For example, in Fig. 4-8 the 1/4-in.-diam times 3/4-in.-deep hole has a limit of 0.245 minimum and 0.250 maximum. By making the core so that the bottom will be 0.250 and the top 0.246, some taper will be provided that will avoid binding on the core. Whenever this type of arrangement is permissible, it should be marked in this manner.

The question may well be asked, "How much taper is necessary for easy removal?" Let us refer to Fig. 4-9 and analyze the conditions during molding. Upon completion of the shot, the part will be shrinking due to solidification. The outside wall of the cup will be smaller than the mold because of a decrease in wall thickness, which can and does take place since there is no interference with shrinkage. On the other hand, the inside of the cup cannot decrease in size because it is stopped by the core. It will bind on the core, and the amount of binding will depend on the shrinkage of the material.

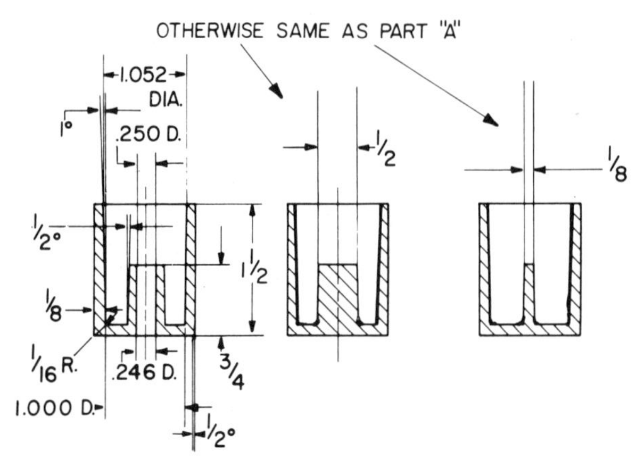

Fig. 4–8. Tapers and projections. *(Courtesy of Rockwell International Corp.)*

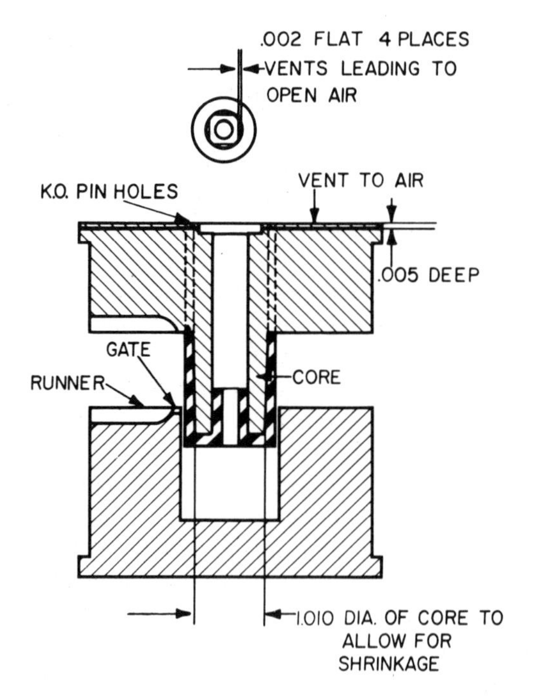

Fig. 4–9. Part A being removed from cavity. Venting core pin. *(Courtesy of Rockwell International Corp.)*

Let us now view the cup made from a material whose mold shrinkage is 0.010 in./in. The inside of a cup being 1.00 in., the core will be 1.010 in. to allow for shrinkage. When the cup leaves its core, the dimension of the top of the cup should be slightly larger than 1.010 in. so that it will be free to move. If the depth of the cup is 1.5 in., and we allow a clearance of 0.025 in. per side when the part is leaving the core, it would then call for 1° taper on the inside. The 1° taper is equivalent to the 0.025-in. clearance on top of the cup. The outside, being smaller than the cavity due to shrinkage of the wall, will work satisfactorily with 1/2° side taper.

Some of the factors that will allow for a smaller taper on a core are full rigidity of part during ejection, low coefficient of friction between the plastic and the mold steel, smooth finish of the core with directional polish to coincide with the withdrawal of the part, and minimal gripping force resulting from low shrinkage. The cavity tapers as a rule can be smaller than those of the core due to the shrinkage of part thickness, which provides some space between cavity and part and needs little additional space for free movement.

In most cases, 1/2° taper on a core and the same or less on the cavity will suffice, but careful analysis of all the factors is necessary before deciding on any set taper. Generally speaking, the higher tapers are preferred, provided that they will not unduly contribute to waste of material nor create heavy sections. If the tapers are a cause of a considerable waste of material, then the cost of material should be balanced against the cost of plate and sleeve stripping. The latter will function satisfactorily with a minimal taper.

PROJECTIONS, BOSSES, AND INTERNAL VENTING

Projections and bosses, which are deep in relation to cross section, require some consideration because deep projections will tend to be porous and will have poor filling qualities unless steps are taken to evacuate the air that is being displaced. This can be accomplished by means of vented knockout pins, moving cores, etc.

If the depth in relation to projection thickness is large—e.g., more than four times the thickness—it will be difficult to remove the air from the space while the plastic is flowing in. Providing flats on the knockout pins will correct this difficulty (Fig. 4-10). These flats are small enough to keep material from flowing in, yet large enough that air and gases can escape. This becomes necessary where the escape means at the parting line are not within the reach of internal configurations, making it imperative to provide internal exit for air and gas.

IDENTIFICATION OF SUPPLIER AND CAVITY

The location of supplier and cavity identification should be questioned so that no interference is experienced during application of the part to end use.

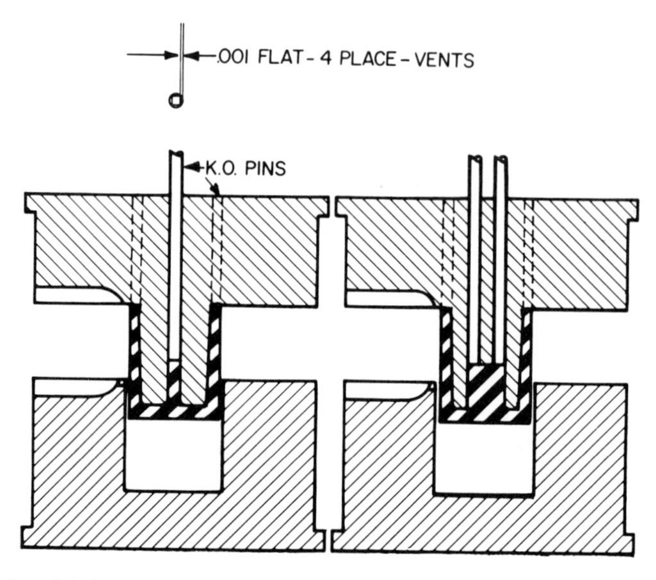

Fig. 4–10. Part B being removed from cavity. Venting "projections." (*Courtesy of Rockwell International Corp.*)

EJECTOR PINS

Ejector pins have a tendency to produce a very slight flashline, which in some areas of a part may be objectionable. Therefore, their location and the amount of recess formed by them in the part should be agreed upon with the product designer.

VISIBILITY OR CLARITY REQUIREMENT

Parts made for visibility demand special attention in terms of gating and flow over a core. The danger exists that in some cases, the flowing material will entrap air or gases, causing a marred surface (Fig. 4–11).

Figure 4–11 shows a simple cylinder with the end closed. Let us assume that the end is used for viewing purposes; therefore, the material cannot be fed through this area. The gate or feeding point, if placed at the bottom of the cylinder, may cause the material to flow around the circumference first and complete the fill at the top or end of the cylinder. This would create a weldline and gas pocket at the top, where it would be objectionable. To overcome this, the dimensioning should be so arranged that the least resistance to the flow would be offered at the top, thus filling this portion first and completing the

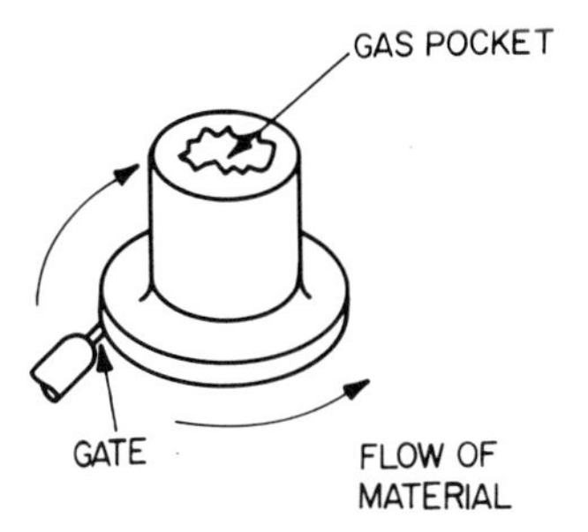

Fig. 4–11. Clear cover. (*Courtesy of Rockwell International Corp.*)

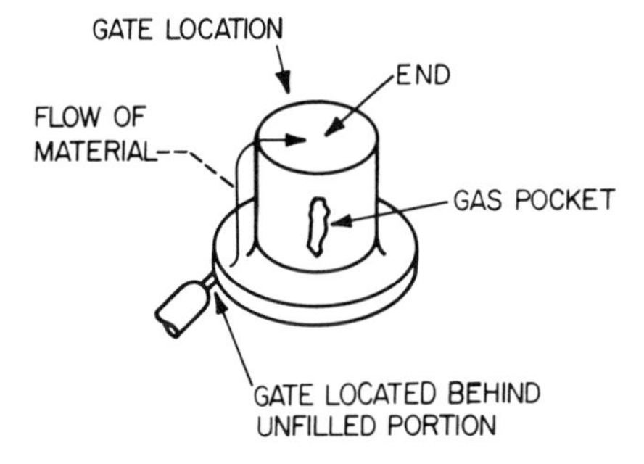

Fig. 4–12. Clear cover. (*Courtesy of Rockwell International Corp.*)

shot around the circumference. The details of such an arrangement would depend upon the configuration requirements of the part. In any case, the designer should be conscious of this type of problem. (See Fig. 4–12.)

REQUIREMENTS FOR ALIGNMENT

Some parts are used with each other in a manner requiring that their surfaces match, as is the case with right- and left-hand halves of a housing. This requirement should be taken into account by dimensioning the parts from points where matching takes place, rather than from a theoretical centerline as is usually done. A note to the moldmaker should call attention to the matching needs and point out that measurements in steel follow the method outlined in the design.

PRODUCT DESIGN POINTS

There are a great number of features on a product drawing that do not specifically spell out how they are to be carried out in the mold design. Some of these

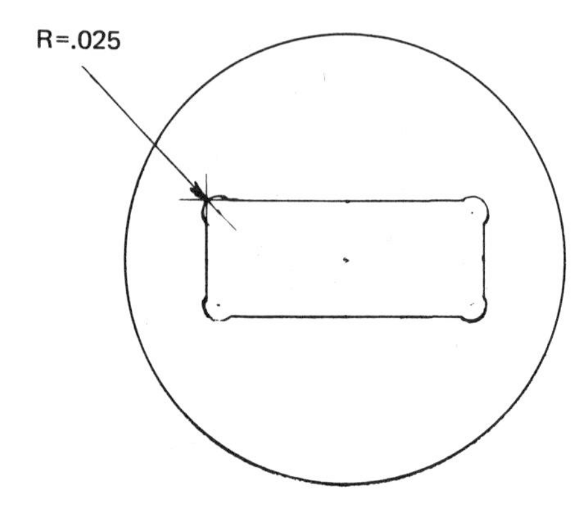

Fig. 4–13. Inside corners for strength. For clearance an .025 R eliminates sharpness effect.

adversely affect the strength and quality of the molded product. In most cases, these problem features can be modified by the designer in order to minimize the adverse effect on the properties of the part.

Let us examine some of these points and indicate how the problems can be reduced to tolerable limits:

1. Inside sharp corners are normally shown as two intersecting straight lines without specific indications as to the functional requirement of the degree of sharpness (Fig. 4–13). Inside square corners are stress concentration areas. They are closely related to a notch in a test bar. The "Izod impact strength" of notched and unnotched test bars shows the relative impact strength of each material at the two conditions. Thus, for example, polycarbonate has an impact strength of the notched 1/8-in. test bar of 12 to 16 ft lb/in. of notch, whereas the same bar unnotched did not fail the test. Polypropylene has an impact strength 30 times greater than that of the unnotched versus the notched bar. Nylon shows a drastic increase in impact strength as the radius increases from sharpness to 3/64 R. For nylon, a radius of 0.004 shows an impact strength of 1 ft lb/in. of notch; a radius of 0.010 shows 2 ft lb/in. of notch; a radius of 0.020 shows 4 ft lb/in. of notch, and a radius of 0.045 shows 12 ft lb/in. of notch.

A similar trend exists on most materials. These examples point out that brittleness increases with decrease of radius in a corner. Visually, a radius of 0.020 on a plastic part may be considered sharp, and its influence on strength is much more favorable than a radius of 0.004. To the moldmaker, a sharp corner is usually easier to produce; to the plastic part, however, it is a source of brittleness and, in most cases, highly undesirable. The inside sharp corners on plastic part drawings are a frequent occurrence. It is the mold designer's responsibility to call attention to such strength degradation and invite appropriate corrective measures.

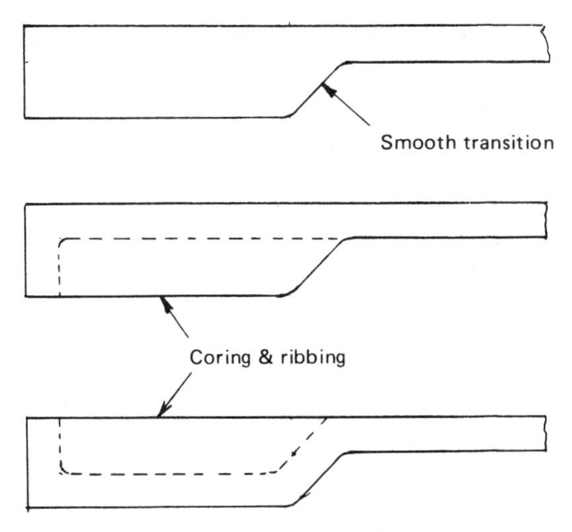

Fig. 4-14. Wall transition.

2. Thick and thin sections in a part lead to problems in molding. A uniform wall throughout a part brings about good strength and appearance. Thick and thin sections will have molded-in stresses, different rates of shrinkage (causing warpage), and possibly void formation in the thick portion. Since the parts in a mold solidify from the outer surfaces toward the center, sinks will tend to form on the surface of the thick portion. When thick (3/16 in. and over) and thin (1/8 in. or less) portions are unavoidable, the transition should be gradual, and, whenever possible, coring should be utilized (Fig. 4-14).

3. Sinks are not only caused by the reasons enumerated in (2); they also occur whenever supporting or reinforcing ribs, flanges, or similar features are used in an attempt to provide functional service without changing the basic wall thickness of a product. If an appearance of a sink on the surface is objectionable, then the ribs and transition radius should be proportioned so that their contribution to the sink would be minimal. Figure 4-15 is a guide to the dimensioning of ribs.

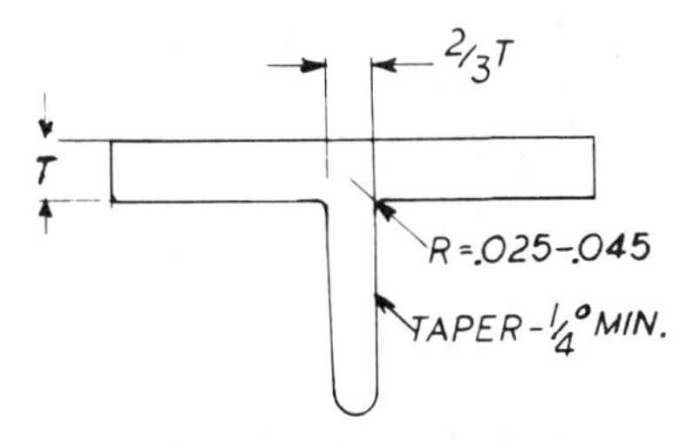

Fig. 4-15. Rib and wall.

4. Molded-in metal parts should be avoided whenever alternate methods will accomplish desired objectives. If it is essential to incorporate such inserts, they should be shaped so that they will present no sharp inside corners to the plastic. The effect of the sharp edges of a metal insert would be the same as explained in (1), namely, brittleness and stress concentration. The cross section that surrounds a metal insert should be heavy enough that it will not crack upon cooling. A method of minimizing cracking around the insert is to heat the metal insert prior to mold insertion to a temperature of 250° to 300°F so that it will tend to thermoform the plastic into its finished shape. The thickness of the plastic enclosure varies from material to material (Fig. 4-16). A reasonable guide is to have the thickness 1.75 to 2 times the size of the insert diameter.

5. Application temperature. Most plastic parts are used in conjunction with other materials. If the use temperature is other than room temperature, certain compensatory steps have to be taken to avoid problems arising from the difference in the thermal coefficient of expansion of the different materials. Most of the plastic materials expand about 10 times as much as steel. Thus, careful analysis of the conditions under which the metallic materials are co-employed for functional uses is called for. In the automotive industry, many long plastic parts are used in conjunction with metal frames. If proper compensation is not made for the difference in thermal coefficients of the materials, buckling or looseness may take place, causing noise or poor appearance.

6. Threads. Plastic threads have a very limited strength and may be further degraded if the thread form is not properly shaped. The V-shaped thread at the outside of a female thread will present a sharp inside corner, which will act as a stress concentrator and thereby weaken the threaded cross section. The rounded form that can be readily incorporated in the molding insert will appreciably improve the strength over a V-shaped form (Fig. 4-17).

When self-tapping, thread-cutting, or thread-forming screws are used, their holding power can be increased if, at joining time, either the screws or plastic

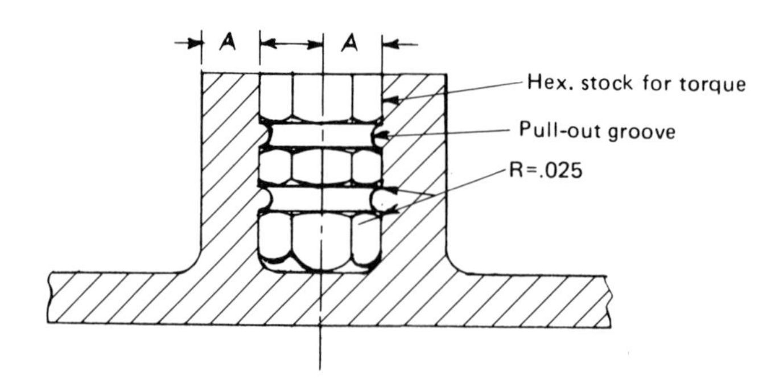

Fig. 4-16 Molded-in inserts.

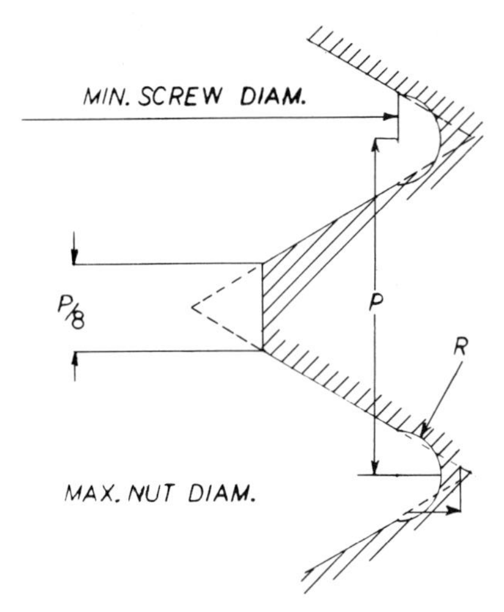

Fig. 4-17. Internal thread.

parts are heated to a temperature of 180° to 200°F. This will provide thermoforming action to some degree and keep the stress level caused by the joining action at a low point.

These possible sources of problems in the molded part should be marked on the part drawing and explained to the product designer for his or her corrective action or awareness of possible product defects due to design limitations. This is simply a step necessary in the chain of events, where the aim is to produce a tool that will provide parts for a good working product. In the final analysis, it should be recognized that although the mold design, mold workmanship, and molding operation are carried out to the highest degree of quality, they cannot overcome any built-in weakness of product design.

5
Mold Size and Strength

Once the part design drawing has been reviewed and corrected for the elements that would affect part strength and appearance, the next step can be taken, i.e., outlining the concept of the mold.

NUMBER OF CAVITIES

As a starting point, the number of cavities has to be established. This is usually determined by the customer, who balances the investment in the tooling against part cost. From the molder's point of view, the number of suggested cavities is based on an empirical rule that the yearly activity of parts is divided into an ordering frequency of about every 50 days; this 50-day requirement should be produced in 200 hr of three-shift operation. For example, if the yearly activity is 100,000 pieces, the 50-day requirement will be 20,000. The estimated cycle is 60 shot/hr or 60 pieces/cavity/hr. In this case, the number of cavities is

$$\text{Cavities} = \frac{\text{Pieces needed for 50 days}}{200 \text{ hr} \times \text{Pieces/cavity/hr}}$$

$$= \frac{20{,}000}{200 \times 60} = 1\text{-}2/3 \text{ or } 2 \text{ cavities}$$

This rule has been checked over a large number of products and has proven to be accurate. Regardless of how the number of cavities is established, the important criteria for selecting a suitable cavity material are the severity of usage and the duration of product activity. A severe usage is called when a mold is running for about half of the available time, whereas running for 10% of the available time is viewed as a very mild one.

Cavity Material

Selection of a metallic material for a particular configuration of a part is determined by its ability to maintain dimensions, shape, surface, and parting-line integrity throughout the expected life of the tool. Each case has to be decided on its own merits and requires individual analysis. As a guideline for a cavity material selection, a few examples are cited and the reasoning behind them is explained.

Chapter 10 deals with the properties of various materials and their suitability for certain applications. Let us review some plastic products, and see what cavity materials were selected and why.

Figure 5-1 shows a part in which optical characteristics are required. For this application, the cavity and core material will be a heat-treatable stainless steel of the 440-420 series. It will be heat-treated to 50 RC. This steel will maintain the original polished surface without danger of oxidation, rust, or other blemishes that can develop in other materials.

Figure 5-1 shows an undercut in the form of an *O* ring groove. This groove is generated by three "tulip" segments that move in the direction of the shoulder bolts when the stripper plate is actuated. The segments spread out, release the part, and return to the original position during mold closing. Another feature

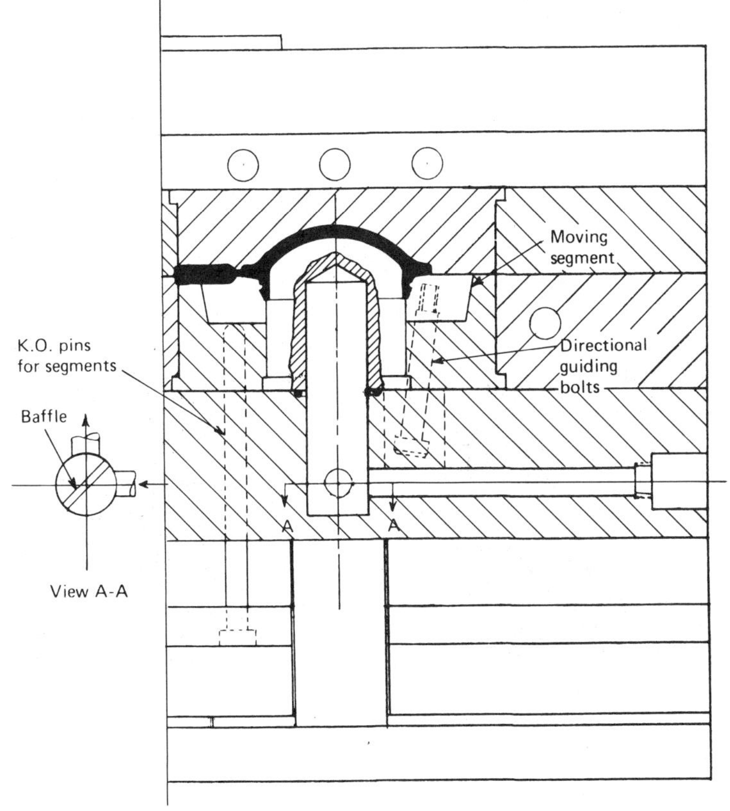

Fig. 5-1. Optical cover.

worth noting is the supply of the cooling medium to the core at 90° for separate cooling of each core. The thickness of the baffle is governed by the need to have the opening segment with an area that will assure a turbulent flow for the most effective heat transfer. If the baffle is inserted with a light press fit and it is made of a good heat-conducting material (brass, copper), the heat transfer throughout the inside of the core will be uniform.

Figure 5–2 shows a mold for a part of high activity and several years' duration. The plastic material in the figure has a filler and exerts an abrasive action on the cavity. The part is expected to maintain a good surface finish at all times. For the cavities of this mold, a P-21 steel, prehardened to 300 Bhn (Brinell hardness number), is selected. The steel of the cavities is retained in the original state for the first production run. During this run, all the dimensional as well as smooth operating requirements are worked out. Following this step, the cavities are nitrided to a depth of 0.005 to 0.010. During the nitriding, a core hardness of about 45 RC and a surface hardness of 70 RC are obtained, making the cavities most durable for wear against an abrasive material and capable of retaining the shape because of its high core hardness.

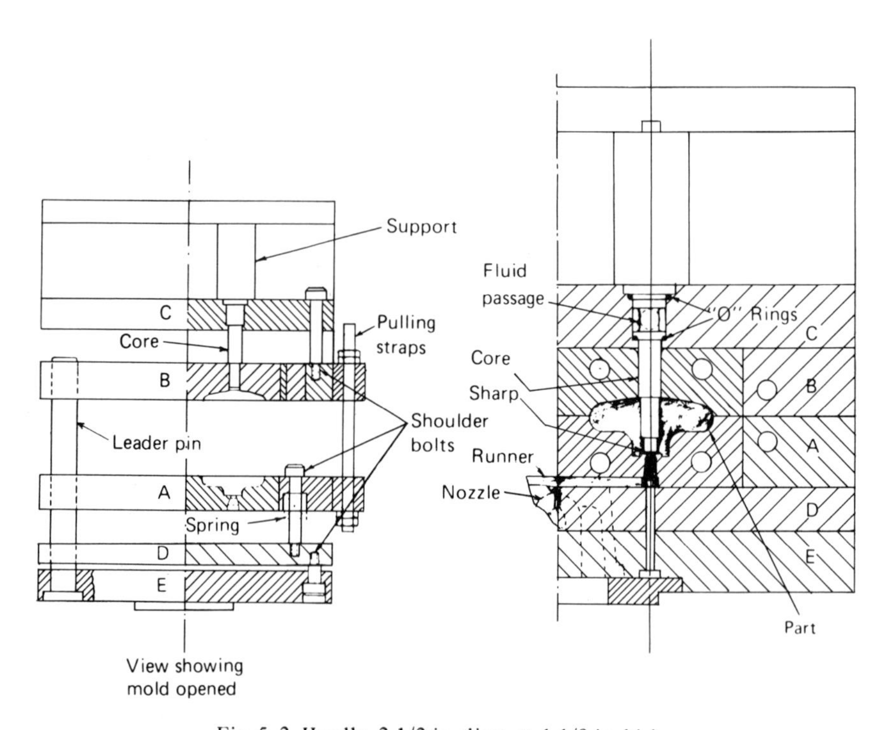

Fig. 5–2. Handle, 2-1/2 in. diam. × 1-1/8 in. high.

Figure 5–2 shows the principle of a three-plate mold in combination with a stripper plate. When the clamp starts opening, the pressure of the springs separates the A plate from the D plate and during this action breaks the small diaphragm gate from the part. The runner and gate are held in place by a sucker pin. The diaphragm gate forms sharp corners at the outer diameter entering into the part. These corners are generated by a sharp edge of the core pin on the inside. This sharp edge is a stress concentrator and causes brittleness at the corner, creating a condition for easy break-off. The diaphragm gate was applied to avoid weld lines for appearance and strength purposes. As the press opening continues, the pulling straps separate the D plate from the E plate, causing the removal of the runner system from the sucker pins and its dropping out of the mold. The shoulder bolt in the A plate limits the movement of the A plate. The opened space between plates D and A is wide enough to permit dropping of the runner by gravity out of the mold. With the progressing ram movement, the B plate is stopped by the pulling straps, while the C plate keeps moving, pulling the core pin out of the part until stopped by the shoulder bolt in the C plate at the end of the ram stroke.

The core pin is made of beryllium copper for good heat conductivity. In its head, there is a groove around which coolant flows for heat dissipation. If the inlet to the groove of the core pin is sized for turbulent flow, then the area of the groove (i.e., height × depth) should be one-half of the inlet area. Here again, it is essential to have the dimensions for turbulent flow, which means most effective utilization of the circulating fluid.

The opening between plates A and B is wide enough to allow easy dropout of parts from the mold. The space between plates B and C is to assure the complete removal of the core pin from the part. Pillar support under the core pins insures good retention in the plate.

Figure 5–3 illustrates a mold for a box-shaped cover open on two ends. The steel selected for this application was the H-13 grade, heat-treated to 54 to 56 RC. This steel has minimal distortion in heat treatment, a characteristic that is of greater significance on larger parts. When subjected to concentrated pressure such as that found at parting lines of plastic molds, caused by a retained piece of flash, it will not crack or cause serious damage. This steel has excellent abrasion resistance and good polishability, and, in general, has been found to be the best steel for tough plastic-molding jobs.

Figure 5–3 is an excellent example of a cavity built of sections, where deflection of the side walls is of major concern and requires careful analysis. The shape of this cavity is such that one end of the *U* is solid, and the opposite end is free to deflect. It is like a beam supported at one end only. The shape of the box is such that the block separating the cavities has to be cut away to a large degree, making the condition less favorable when considered from a deflection point of view. Because of strength considerations, fluid lines were only incorporated

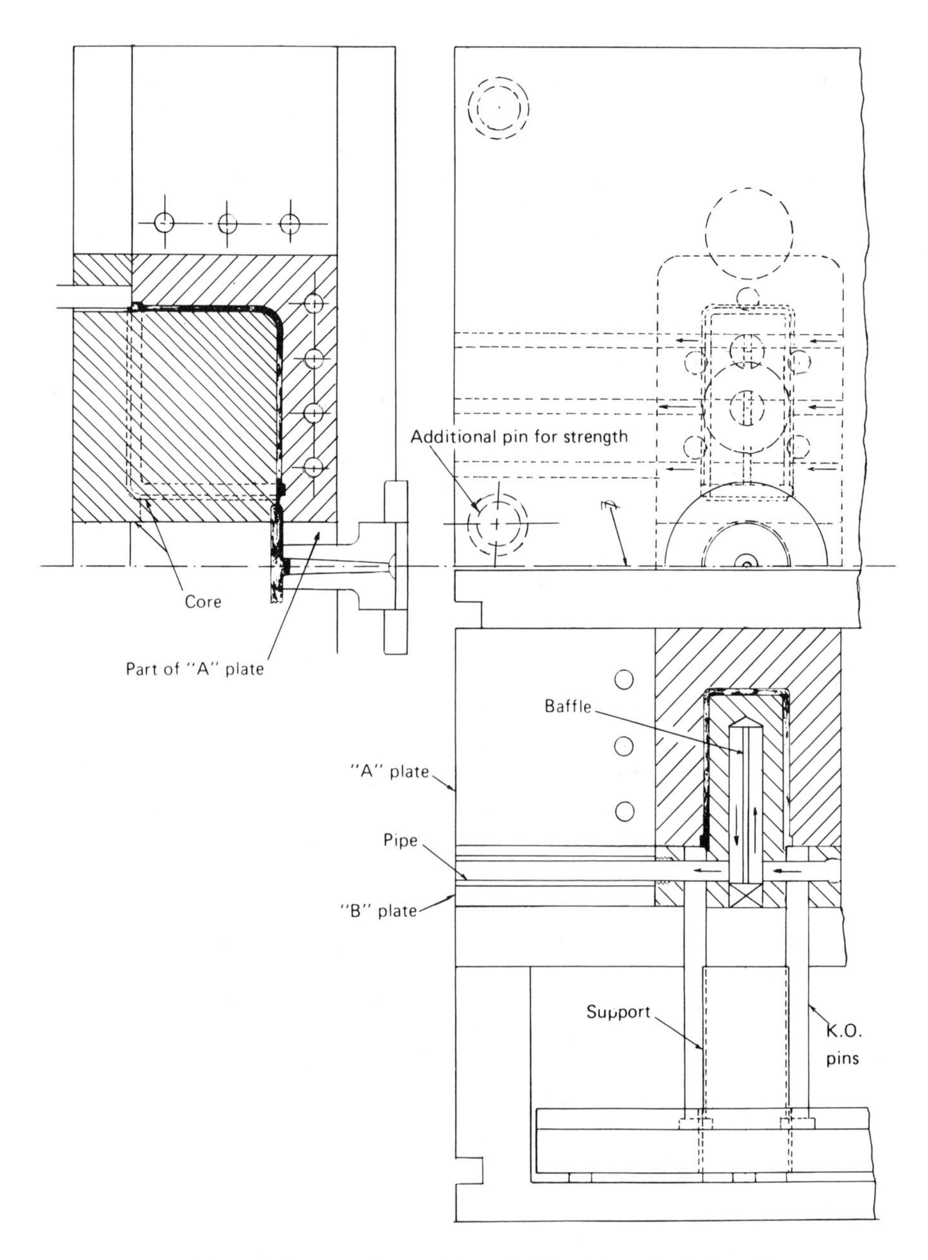

Fig. 5-3. Cover, 4-1/2 in. high × 3-3/4 in. deep × 2-3/16 in. wide.

into the bottom of the cavity, and the sides are expected to dissipate the heat with the aid of circulating lines in the A plate.

To provide more resistance to deflection, additional guide bushings and leader pins are provided at the end of cavities but incorporated in the A and B plates in line with standard pin location.

The cores are cooled by three baffled openings arranged for separate inlet and outlet flow. K.O. pins are of large diameter, and their segment engages the edges of the part.

Figure 5–4 illustrates a small part with close tolerances on the diameters and concentricity in the amount of ±0.001 in. The material has a dual shrinkage, namely, one in the direction of flow and another perpendicular to it. Therefore, allowances had to be made for corrections after a trial run of the mold, and cavity material was selected so that machining could be carried out to bring parts within tolerance. For this application, the H-13 grade of steel prehardened to 44 RC was used. In this state, the steel is machinable and can be used in a machined condition and still provide adequate abrasion resistance, toughness, and hardness. The main benefit is that after adjustments for dimensions are made, we have zero distortion. The plastic material in this case has lubricating properties, making the application of the prehardened H-13 steel most suitable.

Figure 5–5 is a photograph of the coverplate moldings and the mold. The material of the part is glass-filled, and a review of part drawing Fig. 4–1 points out the close tolerance dimensional requirements. With these facts in mind, the cavity steel for this application is H-13 heat-treated to 54 to 56 RC and, after

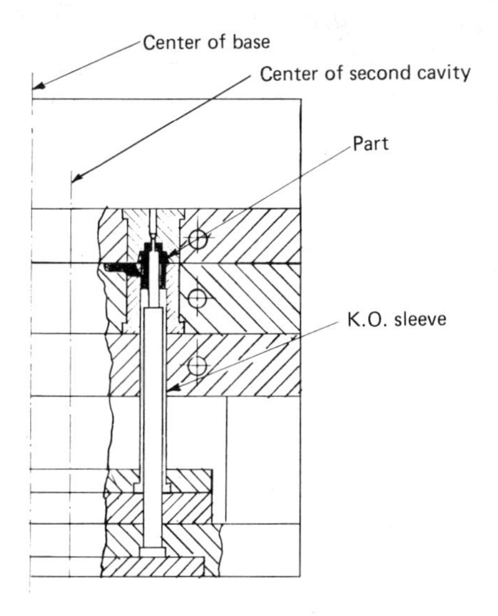

Fig. 5–4. Bushing, 7/16 in. diam. × 5/8 in. high.

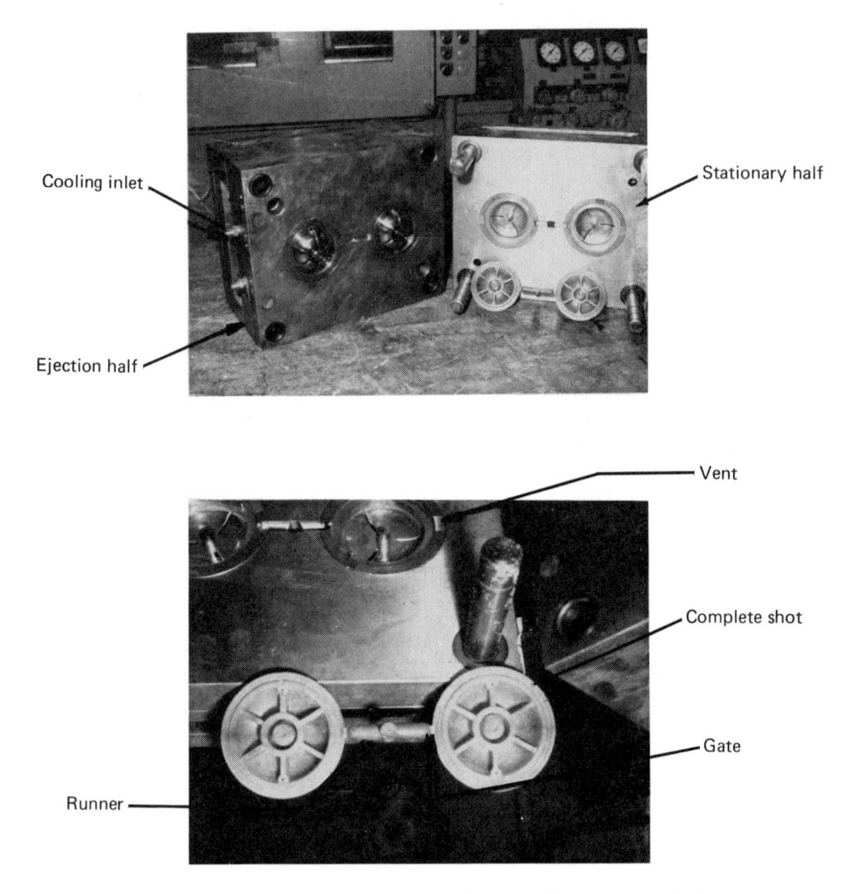

Fig. 5-5. Cover plate. *(Courtesy of Rockwell International Corp.)*

trial runs, plated with 0.001 thickness of chrome for added abrasion resistance. In this case, the cavities are relatively small, and the chances of steel change in heat-treating are negligible. Furthermore, the glass-filled material has a much lower rate of shrinkage than its unfilled counterpart (the shrinkage is 0.001 in./in.). Therefore, the mold dimension can be readily predicted.

Figure 5-6 is a layout of a four-cavity base for the part shown in Fig. 4-1. The steps that were taken in conjunction with this layout follow. First, the core and cavity outline were sketched around the part. Next, the width of the face at the parting line was calculated, to which the shoulder was added, and thus the outside dimensions of the core and cavity blocks were obtained. The next step was to locate the K.O. pins on the core and to decide on the manner of arranging cooling passages. Now it is possible to establish the thickness of A and B plates.

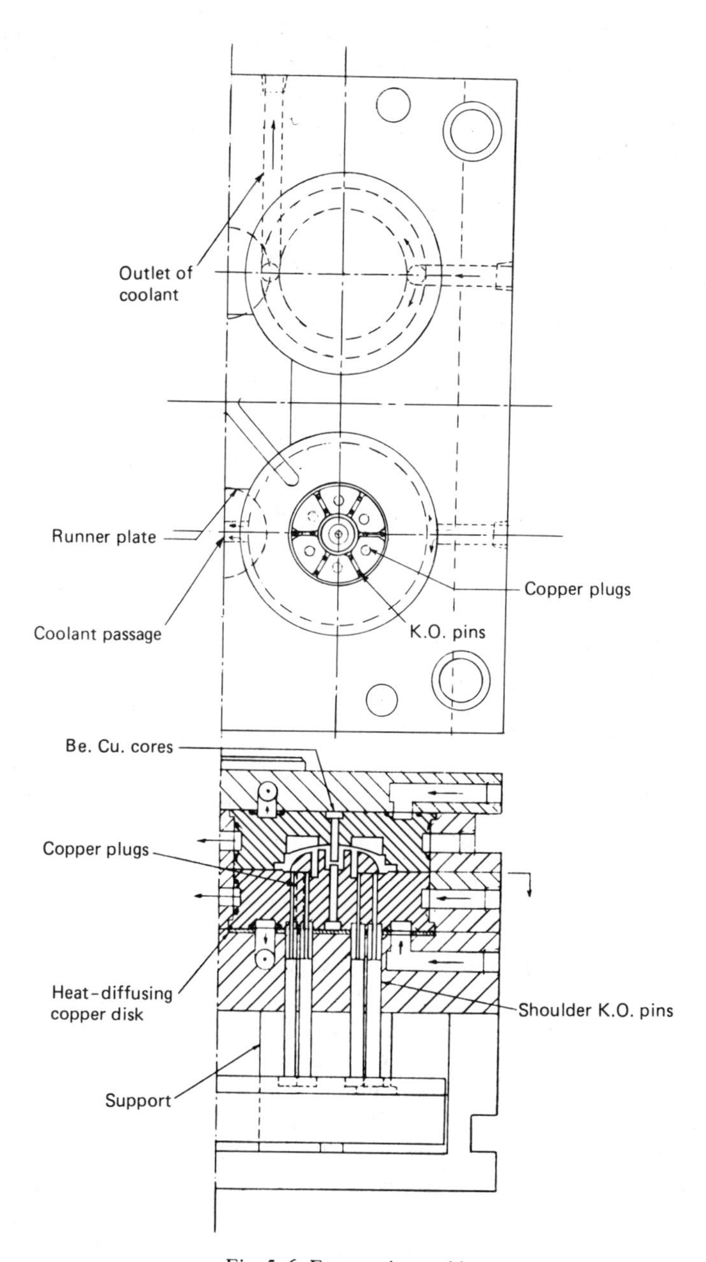

Fig. 5-6. Four-cavity mold.

In order to establish the width of the mold, it is usually best to space the core and cavity so that the outside edge is about in line with the width of the stripper plate. Since these cavities require sealing against leakage of the body, the spacing between cavities will call for a width between them of 1 to 1-1/2 in. so that retaining steel will not distort nor deflect during molding. The above moves and reasoning led to a mold base of 13-7/8 in. X 15 in.

The cooling of the core is accomplished by a groove on the circumference of the body and another groove on the bottom. The bottom side of the core block is supported by a brass or hard copper disk which is in contact with six 1/4-in.-diam copper rods inserted into the protruding cores for dissipating the heat.

The K.O. pins being 1/16 in. diam are strengthened against buckling by keeping their length to a minimum and by having the remainder of the overall length the considerably larger diameter of 5/16 in. The core pins in the center are made of beryllium copper and aid in their cooling. All grooves are dimensioned to assure turbulent flow conditions.

The mold designs in Figs. 5–1 through 5–6 are presented as a guide for determining the selection of cavity material. Since these figures also incorporate features not usually found in a mold design, it is desirable to describe and evaluate them. Whenever there is an opportunity to apply any of those features, the end result should be a better working mold. Obviously, these figures are not complete designs, nor do they include all the details pertinent to the making of a mold. Moreover, it was necessary to reduce the scale of drawings to fit the size of the book. Nevertheless, the points of interest are presented in such a manner that they can be readily visualized. Anyone who desires a complete set of drawings of a mold for guide purposes can obtain them from the suppliers of mold bases.

STRENGTH OF CAVITY

The dimensions of the outside of a cavity should be so proportioned that they will safely withstand the compressive forces of the clamping ram and the hydraulic pressures exerted by the molten plastic.

The parting line of a cavity is subjected to the repeated impact action of the closing ram. This type of action can cause peening of the inside edges, which in turn will bring about changes in dimensions, retention of the molded part in the wrong side, and possibly flashing at the parting line. If the compressive psi on the parting-line surfaces is low, the blow will be light and the peening action negligible. To bring about a low stress at the parting line, we proceed as follows: we determine the tonnage required for the job, select a tolerable compressive stress level, and calculate the corresponding contact area at the parting line. As discussed in Chapter 6, we find that we determine the clamping tonnage for a mold by multiplying the total projected molding area by 2 to 7 ton/psi. For polycarbonate, this value is 5 ton/psi. Once the clamping force is established, we select the allowable stress for the cavity material and calculate the corresponding cavity face width.

For example, let us demonstrate how the width of the parting-line face is determined. As a basis for calculations, we will use the part shown in Fig. 4–1 made of polycarbonate in a four-cavity mold. The projected area of this mold (Fig. 5–6) will consist of

1. Four times the area of part = $4 \times 0.785\, d^2 = 4 \times 0.7854 \times 2.9^2 = 6.60 \times 4 = 26.40$ in.^{2} (where d is diameter of part)
2. Four runners, 0.25 in. diam $\times$ 2 in. long, or a projected area of $4 \times 0.25 \times 2$ in. = 2 in.^{2}
3. Area of the sprue at the parting line = 0.5 in. diam or $0.7854 \times 0.5^2 = 0.20^2$

Total projected area = 26.40 + 2 + 0.20 = 28.60

Tons of clamp = Projected area $\times$ 5 ton/in.^{2} = $28.60 \times 5 = 143$ ton

The nearest press size available for this requirement is a 200-ton clamp. We now have to elect a tolerable compressive stress on the parting line. Allowable stresses for steels of 300 Bhn (Brinell hardness number) in this particular application is 3.5 ton/in.^{2}; for steels with a hardness of 44 RC or higher, it is 5 ton/in.^{2} The cavities for this mold are made of H-13 heat-treated to RC 54, and the allowable stress for it is 5 ton/in.^{2} Since the available press for the job is a 200-ton clamp,

$$\text{Compressive force} = \text{Contact area at parting line times allowable stress}$$

$$200 = \text{Area} \times 5 \text{ ton/in.}^2$$

$$\text{Area} = \frac{200}{5} = 40 \text{ in.}^2 \text{ of total contact surface}$$

For each cavity, it will be 40/4 or 10 in.^{2} of parting-line area. To determine the width of the face, we state that the contact area of 10 in.^{2} is equal to the area of the outside diameter (D) of the insert minus the area of the part cavity or, for practical purposes, the part diameter d. If $d = 2.9$ its

$$\text{Area} = 0.7854\, d^2 = 0.7854 \times 2.9^2 = 6.6 \text{ in.}^2$$

The contact area of 10 in.^{2} = $0.7854\, D^2 - 6.6$ or

$$0.7854\, D^2 = 16.6^2$$

where $D^2 = 21.1$ and $D = 4.6$ in.

$$\text{Face of the cavity} = \frac{D - d}{2} \text{ or } \frac{4.6 - 2.9}{2} = 0.85 \text{ in.}$$

An allowance has to be made for the area of the vents that are needed at the contact face. When runners are at the parting line, the runner plate will provide the additional contact surface to offset the vent areas. In three-plate molds where the runners are away from the parting line, the contact surface should be widened by the amount that the vents have reduced such an area. This is especially significant for cavities where the configuration is such that a vent is required at about every inch of mold circumference.

The calculated value for surface width should be given full recognition, and steps should be taken to insure that those placing the mold in the press will be alerted in regard to the safe tonnage to be used on a particular mold. Such information as "200 Ton Max. Clamp" should be stamped into the base in the proximity of the eyebolt so that no one handling the mold will overlook this precaution.

Will this wall thickness meet the requirements of a pressure vessel?

According to Lame's equation for thick-walled cylinders with ends closed and made of a "brittle" material, the thickness is

$$t = \frac{d}{2}\left(\sqrt{\frac{S+P}{S-P}} - 1\right)$$

From *Machinery's Handbook* ("Cylinders, Strength to Resist Internal Pressure"):*

$$t = \frac{2.9}{2}\left(\sqrt{\frac{60{,}000 + 20{,}000}{60{,}000 - 20{,}000}} - 1\right) = \frac{2.9}{2}(1.41 - 1) = 0.593$$

where S = allowable stress level of cylinder material; in this case, using a safety factor of 5, it is

$$\frac{300{,}000}{5}\text{ psi} = 60{,}000\text{ psi}$$

P = pressure of injected material = 20,000 psi.

The strength requirements from the two angles are satisfactorily met. In all the calculations, it was taken for granted that the ram pressure was applied to the cavities only. This was accomplished by having the cavity insert protrude above the A or B plate about 0.005 in.

Let us now assume that for some valid reason a two-cavity mold is ordered and the press in which it would be run would still be the 200-ton size. In this

*Schubert, Paul V., ed. *Machinery's Handbook,* 20th rev. ed. New York: Industrial Press, 1975.

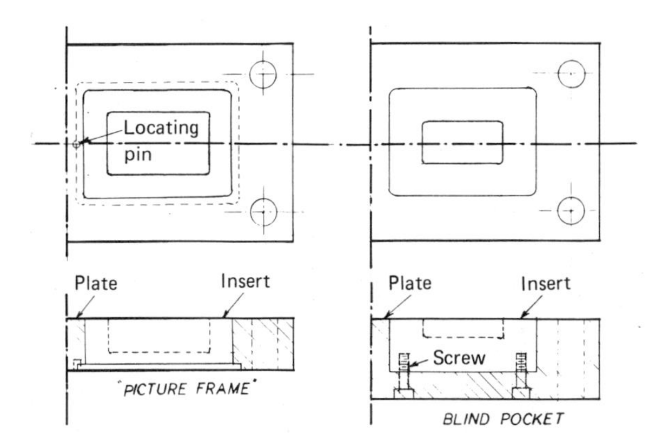

Fig. 5–7. Cavity mounting with inserts.

case, the width of the cavity face could still be the same as previously calculated except that cavity inserts would be mounted flush with the A and B plates so that the plates would absorb part of the force.

The problem of mounting a cavity, as shown in Fig. 5–7, will be favorably met in either a machined-through picture-frame pocket or the blind pocket, whichever is most suitable for mold temperature control as well as other considerations. From a purely strength consideration, the calculated dimensions will incorporate into the cavity itself the ability to safely absorb all the forces to which it may be subjected during molding.

Let us now examine another type of cavity, which will present other problems. We have in mind a cavity made up of sections, which might be done for reasons of economy in manufacture or for other valid causes. For the sake of simplicity, we will deal with a mold for a box that is 5 in. long, 6 in. deep, and 2.25 in. wide. See Fig. 5–8.

No. cavities = 2
Material: Polycarbonate
Projected area of part: $5 \times 2.25 = 11.25$ in.^{2}
Projected area of runners and sprue: 1.5 in.^{2}

The specification for material is H-13 steel, heat-treated to 54 RC. Calculations are made the same as used previously.

$$\text{Press tons} = \text{Projected area} \times \text{ton/in.}^2$$

$$= 2\,(11.25 + 1.5) \times 5 \text{ ton/in.}^2$$

$$= 127.5 \text{ ton}$$

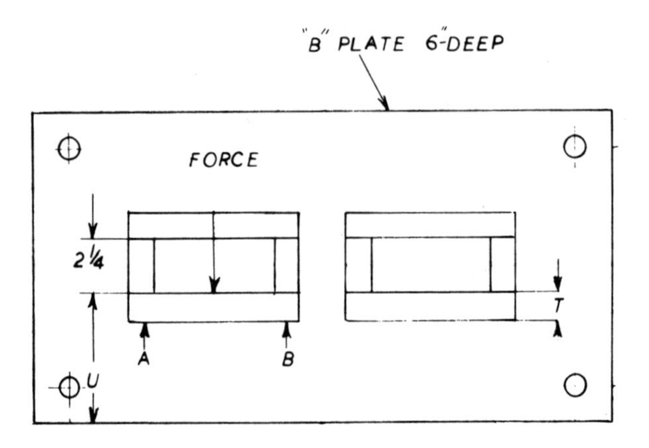

Fig. 5-8. "Picture frame" strength.

The press size available for the purpose is 200-ton.

$$\text{Contact area of cavities} = \frac{200}{5 \text{ (allowable stress)}}$$

$$= 40 \text{ in.}^2$$

$$\text{Per cavity} = 20 \text{ in.}^2$$

Referring to Fig. 5-8, the cavity contact area will be

$$(2 \times 5 \times T) + (2 \times 2.25\, T) + (4 \times T^2) = 14.5\, T + 4T^2$$

This should equal the needed contact area of 20 in.^{2} or

$$14.5\, T + 4\, T^2 = 20$$

In *Machinery's Handbook* ("Equations") or a similar reference book, we find the solution of quadratic equations with one unknown. The form of the equation is

$$ax^2 + bx + c = 0$$

$$x = \frac{-b \pm \sqrt{b^2 - 4\,ac}}{2a}$$

Our equation is

$$4\,T^2 + 14.5\,T - 20 = 0$$

$$T = \frac{-14.5 \pm \sqrt{14.5^2 - 4 \times 4\,(-20)}}{2 \times 4}$$

$$T = \frac{-14.5 + 23.03}{8} = \frac{8.53}{8} = 1.065 \text{ in.}$$

The negative sign in front of the square root is not applicable to this calculation.

Figure 5-7 shows two methods of mounting cavities in a plate. One is the picture frame in which the opening is cut through the plate; the other is the blind pocket in which a recess is machined to receive the cavity. For this size cavity; the picture-frame design is selected due to the high cost of accurate machining of a blind deep pocket.

The hydraulic pressure that would be exerted on the 5-in.-long, 6 in.-deep cavity blocks is:

$$\text{Pressure} = \text{Area exposed to plastic} \times \text{psi of injection}$$

$$= 5 \times 6 \times 20{,}000 = 600{,}000 \text{ lb}$$

This force will tend to deflect to a maximum in the middle and cause material to enter in between sections of the cavity. The amount of deflection that will not permit the plastic to force its way in between sections of cavity has to be very small. We therefore will call for a maximum deflection of 0.0001 and calculate the combined width of cavity and A plate that will make this possible. To keep plastic out of the deflected opening, 10 times the selected deflection would be adequate. We must keep in mind that the force of 600,000 lb, which is causing the deflection, will decay on the inside upon solidification of the plastic. There will be a springback force exerted by the steel wall of equal magnitude (600,000 lb), which will cause a pressure on the sides of the plastic box, making it difficult to open the mold. With the preselected deflection, the interference would be 0.0001, and, since the thickness of the plastic box wall may shrink by 0.0001, there would be zero interference, thus making the part free to move during mold opening.

We look at this load as acting on a beam supported along edges of *A* and *B* (Fig. 5-8) causing the maximum deflection to occur in the middle between the points of support. The actual load that is exerted on the cavity plate is that of a uniformly loaded beam. This load acts on the "A" plate as a concentrated load in the center of the beam. Since the mounting plate will be depended upon to supply the bulk of resistance to deflection, we will use the formula for the beam with a concentrated center load. Under the concentrated load condition, the deflection is 1.6 times greater than is the case with the uniformly loaded beam.

From *Machinery's Handbook* ("Beams)" we find—for a beam fixed at both ends and the load at the center—a formula for deflection:

$$\text{Deflection} = \frac{Wl^3}{192\,EI}$$

W = load in lb = 600,000

l = distance between supports = 5 in.

E = modulus of elasticity for steel = 30×10^6 lb/in.^{2}

I = moment of inertia of the cross section subjected to bending. The cross section is a rectangle = $U \times 6$ in.

Moment of inertia is a property of the cross section that resists bending. By substituting the values in the formula, we have

$$\frac{1}{10{,}000} = \frac{600{,}000 \times 5^3}{192 \times 30 \times 10^6 \times I}$$

or

$$I = \frac{10{,}000 \times 600{,}000 \times 5^3}{192 \times 30 \times 10^6} = 131$$

We find in *Machinery's Handbook* ("Moments of Inertia of Cross Sections") a formula for a rectangular shape where the "distance from axis to extreme fiber is comparable to the *U/2* shown in Fig. 5–8:

$$I = \frac{bd^3}{12}$$

In this case, b = 6 in. and $d = U$. Substituting numbers in the formula, we have

$$131 = \frac{6 \times U^3}{12}$$

$$U^3 = 262$$

$$U = 6.4$$

$$2U + 2.25 = \text{width of mold base}$$

$$12.8 + 2.25 = 15 \text{ in. approx.}$$

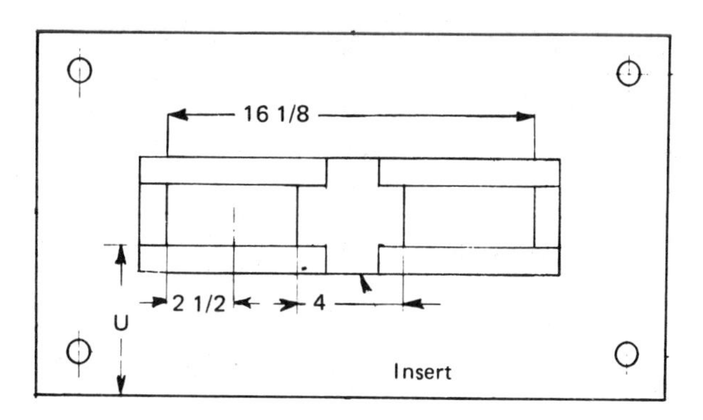

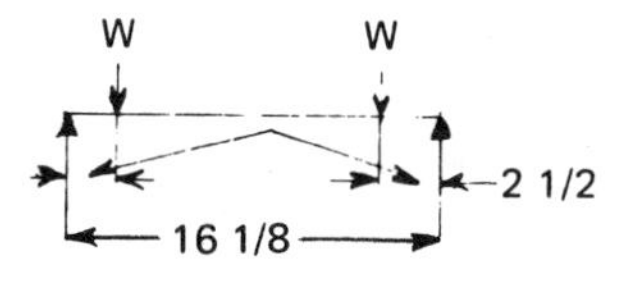

Fig. 5-9. Extended "picture frame" strength and beam diagram.

The standard mold base width that can be considered for the purpose is 14-7/8 in. or 15-7/8 in. In practice, whenever an excessive deflection is uncovered, the tendency has been to substitute a plate with a higher Rockwell hardness, only to find out that the end result is the same. This can be corrected by increasing the width of the cross section that is in the direction of the acting force. The resistance to deflection changes with the third power of the width, which means that a relatively small change will bring about a much higher resistance to deflection. Thus, for example, a 10% increase in width U will cut the deflection by 33%.

It is to be emphasized that in the deflection formula, there is no factor of tensile strength or any property related to hardness and the only variable that can be changed is I, the moment of inertia in which the width U appears as a factor.

Let us now assume that it is decided to build the sectional mold by cutting one large opening in the base for the two cavities and making an insert that will form the two ends in one piece. See Fig. 5-9.

What will this arrangement do to the deflection? Diagrammatically, this condition will appear as shown in the beam diagram in Fig. 5-9. For this condition, *Machinery's Handbook* ("Beams") shows a formula for deflection.

$$\text{Deflection at point of load} = \frac{Wa^2}{6EI}(3l - 4a)$$

$$\text{Deflection at center of beam} = \frac{Wa}{24\,EI}(3l^2 - 4a^2)$$

In this case, $a = 2.5$, $l = 16.125$,

$I = 131$, and $W = 600{,}000$ (using the preceding example).

$$\text{Deflection at load} = \frac{600{,}000 \times 2.5^2}{6 \times 30 \times 10^6 \times 131}(48.38 - 10)$$

$$= \frac{6.25 \times 38.38}{300 \times 131} = \frac{240}{39{,}300} = 0.0061$$

$$= 0.0061 \text{ in.}$$

The deflection of a beam with fixed ends is 25% of the preceding value or approximately 0.0015 in.

$$\text{Deflection at center} = \frac{600{,}000 \times 2.5}{24 \times 30 \times 10^6 \times 131}(765 - 25)$$

$$= \frac{2.5 \times 740}{1200 \times 131} = \frac{1850}{157{,}200}$$

$$= 0.0118 \text{ in. or 25\% of this is } 0.0029$$

The deflection at the point of the load is found to be about 15 times greater than in the preceding construction. The deflection of 0.0015 would permit some materials to flow in; the opening would become progressively worse with subsequent shots, and, with it, the wall thickness and associated problems would increase. Materials that would not flow in such narrow spaces would in all likelihood present difficulty in mold opening, because an interference of 0.0015 in. per side would be a sizable amount for the stripping force to overcome. There have been cases where the deflection of the plate was large enough to cause a sizable increase in wall thickness of the part. When cavity pressure decayed, the steel was springing back to the original position with a force equal to that which caused it to deflect. The pressure from deflected steel on the increased wall of the part was so great that mold opening was impossible, and only by complete disassembly of the mold into the basic components was it feasible to separate the halves. After clearing out the plastic, the mold would be reassembled and reestablished to operation. Such problems can be avoided by calculating the required proportions for strength, etc.

A third arrangement, as shown in Fig. 5-10, would reduce the side pressures resulting from deflection stresses by half of the arrangement shown in Fig. 5-8.

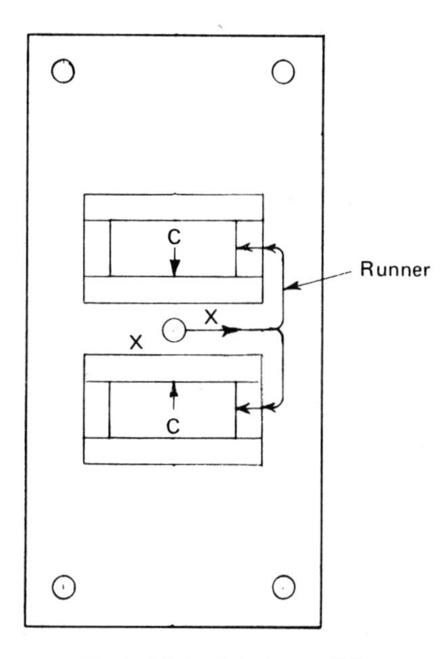

Fig. 5-10. Cavities in parallel.

The inside forces at X neutralize each other, thus leaving only the outside walls exerting the side pressures. This type of cavity placement would also save 20% mold base weight when the major consideration for mold base selection is the deflection factor. This placement of cavities will exert only one-half of the side pressure, thus making mold opening relatively easier. There is, however, a potential source of problems if the gate is placed in the center of the cavity at C. Such gate location may cause a deflection of the core, resulting from the hydraulic pressure exerted on the 5 X 6 side prior to the filling of the remainder of the cavity. At the first stage of filling, the injection pressure may only be 5% of the full psi of injection. Even at the low level of 1000 psi, the force exerted on the side of the core will be 1000 X 5 X 6 = 30,000 lb, a force large enough to cause movement of the core. The core movement would mean thin and thick walls and difficulty in mold opening. The obvious solution to this difficulty is to place the gate in an area where the possibility of deflection is minimized or eliminated, such as in the center of the bottom or in the middle of the 2.25-in. width.

Mounting the cavity in a blind pocket and calculating the deflection under these conditions will point out the relative merits of the several methods of cavity and plate arrangement (Fig. 5-11). The side that is subjected to the hydraulic pressure of injection may be considered as a thick plate fixed in position and supported at all edges. At the top or parting line, some support is received

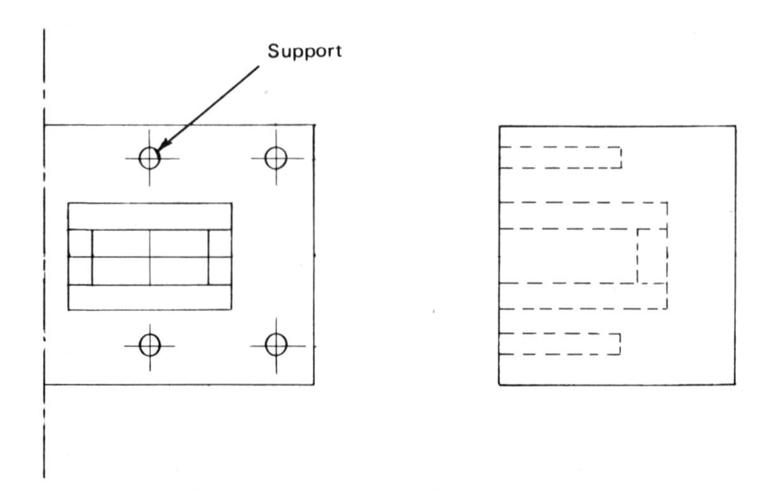

Fig. 5-11. Blind pocket strength.

from the core side of the mold as shown in the sketch or by guide pins located in the middle of the 5-in. side. For a thick plate of rectangular shape, with all edges fixed and a uniformly distributed load over the surface of the plate, we find in *Machinery's Handbook* ("Plates, Flat") a deflection formula

$$\text{Deflection} = \frac{0.0284\,W}{Et^3\left(\dfrac{L}{l^3} + \dfrac{1.056\,l^2}{L^4}\right)}$$

L = long side of rectangular plate = 6 in.

l = short side of rectangular plate = 5 in.

$t = U$

$$\text{Allowable deflection} = \frac{1}{10{,}000}$$

$$E = 30 \times 10^6$$

$$\frac{1}{10{,}000} = \frac{0.0284 \times 600{,}000}{30 \times 10^6 \times U^3\left(\dfrac{6}{125} + \dfrac{1.056 \times 25}{36 \times 36}\right)}$$

$$U^3 = \frac{5.68}{0.068} = 83.5$$

$U = 4.38$ including cavity thickness.

The width of the A plate will be 2 × 4.38 + 2.5 or 11.26 in.

The area of this plate is only 75% of that shown in Fig. 5–8 and emphasizes the value of blind-pocket mounting wherever there is a chance of any part of the cavity deflecting.

The four different cavity-mounting systems were analyzed, and needed wall thickness in the retaining plate was calculated in order to stress the importance of carefully reviewing the effect of the injection molding forces on the cavity construction and its mounting method in the mold base. Experience in the field indicates that there is not enough attention given to potential molding difficulties at the time of mold design. Thus, additional expenditures are incurred after the discovery of deficiencies during molding operations.

The calculations for deflection of different constructions definitely indicate the preferred arrangements in the cavity plates from a strength point of view. Cavities are normally mounted in a plate so that the female half will project about 0.005 in. above the plate, while the male half may be even with its own plate. This insures good contact between the cavity halves at the parting line, thereby producing a flash-free line in this area.

Another very important mold construction that requires accurate sidewall dimensioning of the cavity is the one in which the depth of the product is no greater than the average radius, measured from the bottom up, as indicated in Fig. 5–12. The procedure for establishing the data needed to calculate the wall thickness of a cavity is as follows:

1. *Pressure profile in cavity* (Fig. 5–12). First we determine the pressure required at the center-gated bottom to move the plastic from the sprue bushing to the top of the cavity. This is obtained from the curve of Fig. 5–13 where for a certain wall thickness a corresponding pressure per inch is given. The pressure per inch times the length of flow of material equals the total pressure required to fill the cavity. This value is now the vertical part of the pressure profile. It consists of one part that will densify the plastic at the top of the cavity (assumed to be 2000 or 3000 psi) and another part that decreases to zero at the top of densifying pressure and reaches its maximum at the total pressure required to fill the cavity. This information is now superimposed on the drawing of the product, making it possible to arrive at any cross section for a pressure that prevails in that area.

2. *Maximum deflection.* The products under review can be considered tubular in shape except that in the mold construction there is a taper to ease withdrawal of the product from the cavity. The cavity with an integral bottom can be looked upon as a cylinder of similar construction where the bottom forms a restricting influence on the deflection of the cylindrical part. This deflection starts out practically at zero at the bottom and progresses to the maximum at the depth equivalent to the cylinder radius r (see Fig. 5–12). At this depth the

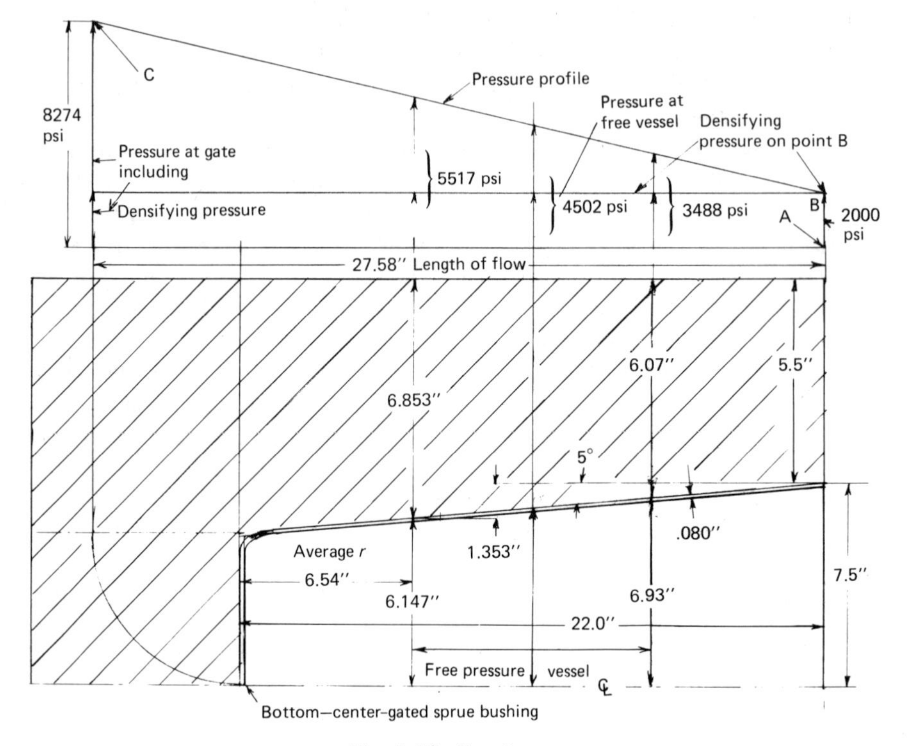

Fig. 5–12. Case 1.

deflection would be in proportion to the pressure in the area times the radius divided by the modulus of elasticity and the result multiplied by a constant.

3. *Constant* C_1. When a cylinder wall exceeds one-tenth of the cylinder radius, it has been found that the stresses during deflection are quite complex. An empirical formula was developed that is a multiple of the cylinder radius and gives answers for C_1 in relation to the maximum pressure of profile in cavities. The constant C_1 also applies to any pressure calculated along the height of the analyzed item. See Table 5–1.

The relationship of

$$C_1 = \frac{.654\, n^3 + .4}{n^3 - 1}$$

when drawn as a curve points out that a transition occurs from *thin* tubing to a *heavy-wall* tubing until C_1 equals 1. At that point, there is a gradual change of the multiplier n until it reaches $n = 3$. After this, for all practical purposes the difference between additional increase in n and the constant C_1 is almost negligible.

The cavity wall increases in thickness due to a pressure rise. Therefore, the maximum pressure of 16,000 psi on the inside of the gate is assigned to $n = 3$. The gauge pressure reading is normally 20,000 psi, but the pressure loss in the machine nozzle and sprue bushing amounts to about 4000 psi. In fact, it should be emphasized that none of the pressures mentioned in this discussion are gauge pressures but are derived from constructing a profile of pressures in the cavity.

For each 1/10 of decrease in n, a decrease of 1000 psi is recorded. The bottom of pressure may be considered to occur at $n = 1$. This corresponds to the 2000 psi required for a cavity with little resistance to flow that can densify the material in the mold. In using the formula for the constant of C_1, to compare thick and thin tubing of it should be noted that the stresses on inside and outside are more complex for thick tubing than for thin tubing. In thin tubing where the thickness does not exceed 1/10 of the radius of the tubes, inside and outside stresses are about equal. A formula for deflection of thin tubing is

$$d = p \times r/E \times r/t$$

in which all designations are the same as before except that t does not exceed 1/10 r. It is evident from this that deflection in thin tubing is much higher than that present in thick tubing.

When the wall of the cavity is less than the appropriate amount, the hydraulic pressure of the plastic tends to move the core in relation to the cavity due to a weak spot in the cavity or variation in wall-thickness. The result is a defective product and a shortened tool life. The most pronounced defect is a gas pocket formed on the side of the product wall. This not only will produce a porous structure at the outline of the gas pocket but will also reduce the part's current-insulating capacity. Two other byproducts are nonuniform wall of the part and generally a poor appearance of the molding. For all these reasons, this defect will make the product unacceptable from the user's point of view.

The type of product configuration under consideration is manufactured by feeding the material in at the bottom center with an expected flow in the direction of its length in an umbrella fashion. The mold construction for such products must have a sidewall thickness of sufficient strength that during the molding action it will not cause defective parts and contribute to a long-lasting tool life. This deflection of mold walls should be less than one-thousandth of an inch in order to avoid molding problems. It should be noted that when cavity expands during molding and the pressure of molding is decayed, the springback of the mold steel causes a stress in the plastic that may bring about difficulty in mold opening and associated defects in the product.

The following calculations to determine the wall thickness of a mold for products that are relatively deep and center-gated are based on the work done with pressure vessels and adapted to the conditions of molding operations. The general formula for application is

$$d = p \times r \times C_1/E$$

where d = deflection of the wall in inches
p = molding pressure in psi
r = radius of molded product in inches
E = modulus of elasticity in psi = 30×10^6
C_1 = constant that depends upon the pressure prevailing in the mold

Let us follow two examples through all the methods for obtaining the necessary data; then application of the formulas will be self-explanatory.

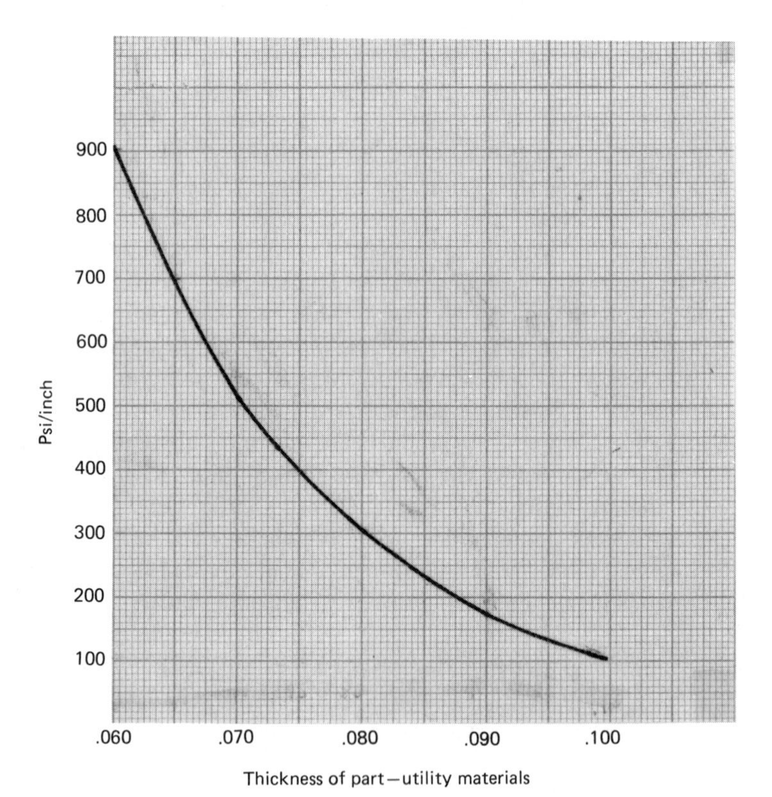

Fig. 5–13. Case 1. Pressure to move plastic from center-gated sprue bushing to top of cavity.

Fig. 5–12 shows the product in case 1. The product is made of .080 thickness of material. The curve in Fig. 5–13 gives us the pressure per inch required for this particular thickness, which is 300 psi/in. The length of flow that the plastic will be exposed to is 27.58 in. Therefore, the pressure from the inside of the gate to the end of the product is 27.58 X 300 = 8274 psi. The pressure is at its highest at point C, which is 8274 psi at the inside of the gate. It consists of a densifying component (2000 to 3000 psi) and one that connects point C with point B giving a full profile of cavity pressure. Looking at the deflection formula, we have established the pressure at the depth of $r = 6.54$ in. from the pressure profile which is 5517 psi and calculated the product radius at the same point as being 6.147 from the dimensions of the drawing. The constant C_1 can be calculated by substituting into the equation $C_1 = (.654\,n^3 + .4)/(n^3 - 1)$ in increments of 1/10 values of n from $n = 1.7$ up to 3.0. The variable n is a multiple of r; carrying out this multiplication gives the outside radius of the cavity. See Table 5–1 (p. 64). The deflection can be figured with all data available:

$$d = p \times r \times C_1/E$$

$$= 5517 \times 6.147 \times .815/30 \times 10^{-6} = .00092''$$

The values of $C_1 = .816$ and $n = 1.94$ were interpolated from the table using the 5517 pressure. The deflection is less than 1/1000 in. The product 1.94 X 6.147 would give a 11.93 radius 6.54 in. from the bottom. Adding 1/2 in. for water circulation, we have an O.D. radius of about 12-1/2 in.

The upper part of the product can be considered as also having a restraining bottom formed by the interlocking of press and shell of the mold (see Fig. 5–12), so meeting mold-strength requirements will be satisfactory, especially since the pressure in that region is considerably reduced. The deflection on the top considered at the distance $r = 6.54$ will be:

$$d = 3488 \times 6.93 \times .895/30 \times 10^{-6} = .00072 \text{ in.}$$

As before, the $C_1 = .895$ and $n = 1.75$ were interpolated from Table 5–1 using the pressure 3488 psi. This also gives a deflection of less than 1/1000 in. The product 1.75 X 6.93 would give a 12.13" radius 6.54" from the top. Adding 1/2 in. for water circulation, we have an O.D. radius of about 12-5/8 in. The deflection calculations show that in practice the expansion of the radius will be about equal along the height of the cavity by making the O.D. radius about 13 in. (rounded).

The free pressure vessel (Fig. 5–12) has its own constant and applies to cylinders much longer than those found in molding operations.

Table 5–1. Calculating Outside Radius of Cavity.

r multiplier	1.7	1.8	1.9	2.0	2.1	2.2	2.3	2.4	2.5	2.6	2.7	2.8	2.9	3
Constant C_1	.92	.87	.83	.80	.78	.76	.74	.73	.72	.713	.706	.668	.666	.665
Pressure profile in 1000^3 psi	3	4	5	6	7	8	9	10	11	12	13	14	15	16

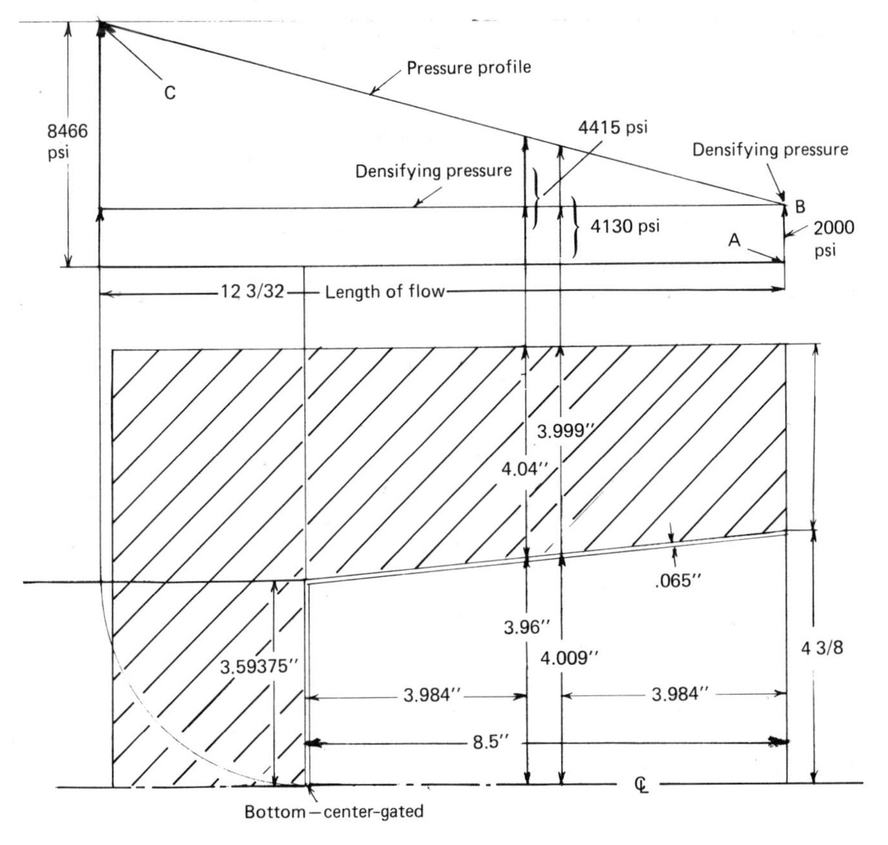

Fig. 5–14. Case 2.

The second example is of a bucket with the dimensions shown in Fig. 5–14. The formula in this case is also $d = p \times r \times C_1/E$. The pressure profile drawn above the cavity as before, again shows the pressure necessary to make the end of the flow dense, which is 2000 psi. The total pressure required to push the plastic through the cavity is calculated from the following. The product wall has a minimum thickness of .065 in. We will therefore use this thickness to establish the psi/in. of flow, since that would be the maximum pressure produced. The total distance the material must cover in the cavity will be 12-3/32 in. At this wall thickness, the pressure will be 700 psi. The total pressure is 8466 psi, and the average radius is 3.984.

The pressure is at its highest at point C, which is 8466 psi at the inside of the gate. It consists of a densifying component (2000 to 3000 psi) and one that connects point C with point B, giving a full profile of cavity pressure. Thus from the bottom of the 3.984 radius, there is a pressure of 4415 psi and a radius of 3.96 in. at the same location (3.984 in. is the average radius of the cavity).

The deflection is

$$d = p \times r \times C_1/E$$

$$= 4415 \times 3.96 \times .85/30 \times 10^{-6} = .0005 \text{ in.}$$

The values $C_1 = .85$ and $n = 1.85$ were interpolated from Table 5–1 using a pressure of 4415 psi. The deflection is less than 1/1000 in. The wall thickness will be $1.85 \times 3.96 = 7.326$ in. Adding 1/2 in. for water circulation, we have an O.D. radius of about 7-7/8 in.

The deflection on the top also considered at the distance of 3.984 in. will be

$$d = 4130 \times 4.009 \times .87/30 \times 10^{-6} = .00048 \text{ in.}$$

The values $C_1 = .87$ and $n = 1.8$ were interpolated from Table 5–1 using a pressure of 4130 psi. This also gives a deflection of less than 1/1000 in. The product 1.8×4.009 would give a 7.22 in. radius 3.984 in. from the top. Adding 1/2 in. for water circulation, we have an O.D. radius of about 7-3/4 in. A block of steel of 8-in. radius will satisfactorily meet this requirement.

Maximum pressure occurs at the bottom of the cavity. For this reason, construction of the cavity should be such that the restraining bottom is an integral part of the shell. There are cases, however, in which this is not practical. For these applications, the method outlined in Fig. 5–15 is suggested.

The examples used to calculate the deflection were round in shape and provided with a taper for molding purpose. The same procedure for figuring the deflection can also be applied to a variety of inside configurations (Fig. 5–16) as long as the pressure exposed on the sidewall is equal to the *projected r* times the unit pressure in psi. Where the inside configuration varies, the radius that appears in the formula has to be the dimension denoted in the diagram.

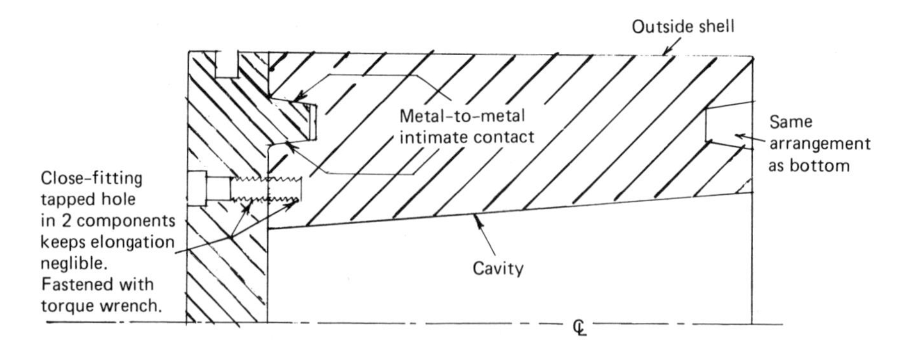

Fig. 5–15. Construction method when restraining bottom not part of shell.

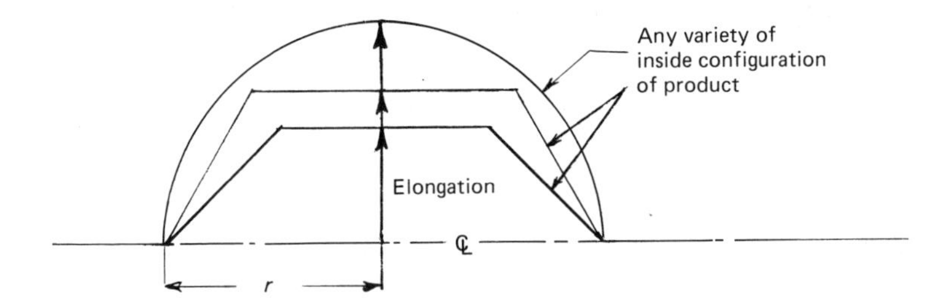

Fig. 5–16. Variation in inside configurations.

In the discussion of cavity dimensions, the constant of C_1 was used and no limits of the inside lengths, were specified in applying this constant. In order to arrive at these dimensions, the following method was used. A cavity during molding operation may be compared to a beam supported at both ends with a concentrated load at the center. In such a case, the deflection would be

$$d = \frac{Pl^3}{48\,EI}$$

where P = load
l = length of beam
E = modulus of elasticity
I = moment of inertia.

In this equation, keeping all the factors the same except for the length, we can establish the following proportion relating the constants to the lengths:

$$\frac{C_2}{C_1} = \frac{l_2{}^3}{l_1{}^3}$$

or

$$l_2 = l_1 \sqrt[3]{\frac{C_2}{C_1}}$$

where

$$C_2 = \frac{1.3n^2 + .4}{n^2 - 1}$$

where n = multiplier of the radius (as before).

The ratio C_2/C_1 in the area of mold operation is 2.3. Therefore,

$$\sqrt[3]{\frac{C_2}{C_1}} = 1.3$$

$$l_2 = l_1 \times 1.3$$

and since $l_1 = r$, it would be $l_2 = 1.3r$.

This means that the beginning of factor C_2 would start at $r + 1.3r$, provided that the design has one restraining end. With both ends of the restraining type, the length would be twice that of l_2 above, or 4.6 r. Whenever the lengths exceed this limitation, the constant C_2 would take effect in the formula of deflection.

STEEL AND SIZE OF MOLD BASE

The size and type of mold base is determined by placement of cavities, by method of feeding the cavities, by ejection employed, by type of pockets desired, by temperature control, by type of cam action, or by any unusual factor that becomes necessary for a specific part. Making a layout of these and the other elements thus far established will indicate the type and overall size of the mold base. Taking the four-cavity mold discussed earlier as an example, we obtain the outside width and length of the cavities. To the width of the cavities, we add 1.75 in. per side so that they are placed close to the support blocks of the ejector housing. On the end, we add an additional inch for the return pins or 2.75 in. per end (See Fig. 5-6.) These overall dimensions are checked against standard available mold bases, and the selection is made to satisfy the outline of the layout.

There are normally three grades of steel employed in mold bases. They are:

1. The lowest-priced steel grade in a mold base is a medium carbon type with tensile strength of 55 to 75 $\times$ 10^3 psi. This grade is suitable for application where the cavities are in themselves strong enough to withstand the conditions of application. The main function of a mold base in the preceding case is to keep the two halves aligned and the ejection side rigid enough to permit ease of ejection on a cycle of two or more times a minute. Where the cavities are mounted in a cut-through plate (see Fig. 5-7), care must be exerted that the surrounding frame is thick and wide enough to safeguard the guiding features of the halves.

Where blind pockets (see Fig. 5-7) are employed, this steel is suitable for a majority of applications.

2. The next higher grade of steel employed for bases is an AISI type 4130, heat-treated to a hardness of 300 Bhn with a tensile strength of 126 to 155

$\times\ 10^3$ psi. This grade is usually considered for cases where the cavities are constructed in sections, and it is the function of the plates to retain these sections without allowing them to separate under the forces of injection pressures. It is also applied for cases where cooling lines and other machining requirements weaken the cavity plate to a point that a material with higher physical properties is prescribed.

3. There are occasions when it is desirable to machine cavities into the cavity plates instead of fabricating cavities and inserting them into mold-base plates. This may be the case for a product with a yearly activity of less than 10,000 pieces and a configuration that is relatively easy to machine from a mold-base plate. For such application, the mold-base cavity plates may be specified to be an AISI 4135 steel heat-treated to 300 Bhn with a physical strength of 129 to 155 $\times\ 10^3$ psi. It is a suitable steel for polishing and higher hardness heat-treating if necessary.

PILLAR SUPPORTS

The general construction of a mold base usually incorporates the U-shaped ejection housing. If the span between the arms of the U is long enough, the forces of injection can cause a sizable deflection in the plates that are supported by the ejector housing. Such deflection would cause flashing of parts. To overcome this problem, the span between supports is reduced by placing pillar supports at certain spacings so that the deflection is negligible in size.

For the determination of pillars and their spacing, the beam formula can be applied. For this purpose, we consider a 1-7/8-in.-thick plate (Fig. 5–17) as a beam supported at 8.5-in. centers and with a uniform load. For this loading system, we find in *Machinery's Handbook* ("Stresses of Beams at the Center")

$$\text{Stress at center} = \frac{WL}{8Z} = S$$

where W is the load that the plate can support

L is the length between supports = 8.5 (Fig. 5–17)

Z is the section modulus or a property of the cross section that resists flexure.

In *Machinery's Handbook* ("Section Modulus"), we find the formula:

$$Z = \frac{bd^2}{6} = \frac{bB^2}{6} = \frac{15 \times 1.875^2}{6}$$

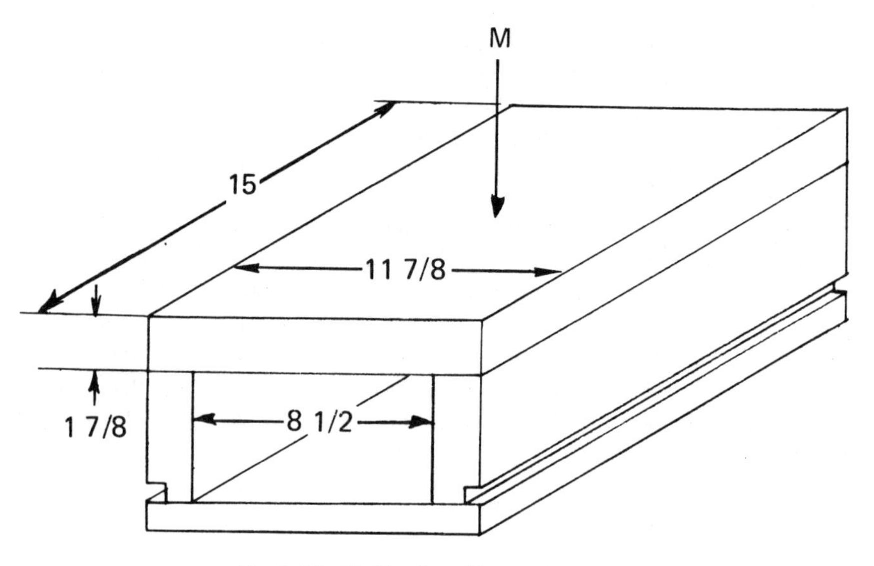

Fig. 5–17. Outline for pillar requirements.

in which $d = B = 1.875$

$$b = 15$$

$$Z = \frac{15 \times 3.52}{6} = 8.80$$

S is the allowable safe stress and the suggested value by the mold base manufacturer is 12,000 psi. Referring to Fig. 5–17, we have

$$12{,}000 = \frac{W \times 8.5}{8 \times 8.8}$$

$$W = \frac{12{,}000 \times 8 \times 8.8}{8.5} = 99{,}275 \text{ lb}$$

is the permissible load on the support plate.

When the mold is closed, the cavities will exert a concentrated pressure on the support plate. For this condition, a safe concentrated stress in compression of 7000 psi is allowed. The compression formula from the handbook is

$$S = \frac{P}{A}$$

where $S = 7000$ psi Allowable stress

$P = W = 99{,}275$, and

$$A = \frac{P}{S} = \frac{99{,}275}{7000} = 14.13 \text{ in.}^2$$

Thus, the total area of the back of the cavities can be only 14.13 in.^{2} By adding one row of support pillars, the L dimension is 4.25 in the load formula, thus doubling the load capacity of the plate and also doubling the permissible cavity area to 28.26 in.^{2} and at the same time maintaining the allowable stress of 7000 psi. The number of rows of pillar supports decreases the distance between the resting points of the beam, thereby increasing the area for the concentrated pressure of the cavities.

ALIGNMENT OF HALVES

The leader pins and bushings in the mold base can have a maximum misalignment of 0.0025 and a minimum of 0.0015. These values are theoretical and assume that there is practically no tolerance on the location of pins and bushings. In reality, those figures are smaller and can be measured on each base by placing one half in a fixed position and moving the opposite half against an indicator. Although the information on mold-base misalignment may be available and is within required tolerances, it is best to incorporate additional means for alignment where the part tolerances are tight. If the requirements are such that concentricity of a part made between the A and B sides is 0.004 or less, then additional alignment features should be provided. On small parts, an interlocking cone can be provided integrally with the core and cavity. On larger parts, conical plugs can be incorporated in the mold base (Fig. 3–2).

6
Mold and Press

The mold and press should supplement each other for a goal of optimizing their performance. This is in reference to product quality, consistency, and efficient cycle time. It is true that a mold will work satisfactorily in a number of press types and sizes, but only one type that has shot capacity, clamp size, speed of injection, and overall cycle operation for which the mold was intended will give most favorable results.

The press specifications that should be kept in mind during mold design are described in the following sections.

MACHINE SIZE

The most economical press size is one with the smallest capacity in terms of ounces and clamp tonnage for a specific job. The shot size is established by first determining the weight of the complete shot and comparing it with the nearest rated capacity of a machine in ounces. The machine ounces are based on styrene. For materials other than styrene, the capacity of machine is downgraded to between 90% and 65%, depending on how fast a cubic inch of material under consideration can be brought to melt temperature in comparison with styrene. The plastic raw-material suppliers normally indicate in their literature the downgrading or percentage of relative shot capacity reduction for each material.

The space that any material of a certain weight will occupy in the measuring chamber can be figured by dividing the specific gravity of the material under consideration into that of styrene. For example, polycarbonate with a specific gravity of 1.2 will occupy a space of

$$\frac{\text{Specific gravity of styrene}}{\text{Specific gravity of polycarbonate}} = \frac{1.06}{1.20} = 0.877 \text{ or } 87.7\%$$

of the space that styrene would require.

For polypropylene,

$$\frac{1.06}{0.905} = 1.175 \text{ of the space that styrene would require or } 117.5\%$$

The 17.5% greater volume means that a full machine capacity shot of polypropylene from a volume point of view alone would require a bigger machine than the one with the nominal capacity, by the amount of 17.5%.

Most of the time, the prime concern is the volumetric capacity of the measuring chamber for a full shot. We must also direct our attention to the minimum shot the machine can handle. For heat-sensitive materials, the lower limit is recommended to not be less than 20% or one-fifth of rated capacity. With a low volume of material per shot, there is too long a residence time and thereby a long time exposure to a high temperature, which can cause degradation in appearance and properties.

Most machines have a pointer attached to the injecting screw and a stationary scale to indicate the distance of the injection stroke. If it is desired to convert the weight of a shot in grams into inches of screw travel, it can be done in these steps:

First, we convert into cubic inches

$$\text{in.}^3 = \frac{\text{Gram}}{\text{Specific gravity} \times 16.39}$$

or when the weight is in ounces,

$$\text{in.}^3 = \frac{\text{Ounce}}{\text{Specific gravity} \times 0.58}$$

We proportion the cubic inches of the material to the volume of the injection-metering section and multiply it by the full screw travel:

$$\frac{\text{in.}^3 \text{ of material}}{\text{in.}^3 \text{ of metering section}} \times \text{Screw travel} = \text{Distance of travel for material at hand.}$$

For example, for converting the preceding cubic inches into screw travel distance, we use material = polycarbonate; weight of shoe = 300 grams or 10.65 oz, specific gravity = 1.2.

$$\text{in.}^3 = \frac{300 \text{ gram}}{1.2 \times 16.39} = 15.25$$

$$= \frac{10.65 \text{ oz}}{1.2 \times 0.58} = 15.2$$

A 200-ton, 14-oz press, as an example in which we have 25.1 in.^{3} of metering volume and corresponding to it 8 in. of screw travel. For the 300-gram shot, the screw will travel

$$\frac{15.2}{25.1} \times 8 = 4.84 \text{ in. approx.}$$

The reason for indicating the screw travel as approximate is that the material in the measuring chamber is expanded due to the heat, while the calculated volume is based on the solid material. The difference between the expanded volume and the solid one will call for a slight adjustment in distance of travel.

The next step in finding machine size is to establish the clamp capacity in tons. This is done by figuring the projected area of cavities and runners and multiplying it by 2 to 7 ton/in.^{2}. For a great many materials, the 2 ton/in.^{2} is adequate. For polycarbonate, however, the value is 5, and for nylon 7 ton/in.^{2}. These numbers have been obtained from observations in actual operations. The usual parameters were 20,000 psi injection pressure (the maximum rate of injection the machine is capable of delivering) and mold temperatures in the neighborhood of 100°F or higher.

There are times when the needed tonnage of press figured by the use of 2 to 7 ton/in.^{2} points to a 5% to 10% higher requirement than is available at a specific plant. Questions then arise: how much tolerance is permissible on these figures, and under what circumstances can a tolerance be applied? That is, how did these figures of 2 to 7 ton/in.^{2} come about?

Let us work with a 200-ton, 14-oz press as an example, and see what information we can derive. The material that we will use will call for 2 ton/in.^{2}

$$200 \text{ ton} = \text{Projected area in in.}^2 \times 2 \text{ ton/in.}^2 \text{ or}$$

$$\frac{200}{2} = \text{Projected area} = 100\text{-in.}^2 \text{ of molding surface and runner projected area}$$

If we take this mold with 100-in.^{2} area and inject into it a hydraulic fluid at 20,000 psi, we will have a force F exerted on the mold, which would require an equal force to clamp it.

$$F = 100 \times 20{,}000 = 2{,}000{,}000 \text{ lb or } 1000 \text{ ton}$$

are exerted on the mold; therefore, the same force is needed for clamp capacity. This large force has been found to exist when a mold is sitting open (for any reason) for about 15 min. and the plastic material is permitted to soak up heat. The overheated plastic will act like a hydraulic fluid. When the shot is made, the mold will flash all over, indicating that the clamp, which under normal operating conditions keeps the mold tight, could not prevent the mold from opening. This is because the injecting force exceeded the clamping force due to the excessive fluidity of the plastic.

We have determined that a hydraulic fluid injected into a mold with 100-in.^2 projected area developed a force of 1000 tons and needed a ram of at least equal capacity to keep the mold from opening.

Now let us see how much injection pressure with a hydraulic oil on the same mold can be contained by the 200-ton clamp of the press we are working with.

$$200 \text{ ton} = \text{Projected area} \times \text{Injection pressure in psi}$$

$$200 \text{ ton} = 400{,}000 \text{ lb} = 100 \times \text{psi}$$

$$\text{psi} = \frac{400{,}000}{100} = 4000 \text{ psi}$$

The injection pressure of 4000 psi exerted by a hydraulic fluid on the mold area of 100 in.^2 will be contained by the 200-ton press. This tells us that a 200-ton clamp will hold a mold with a projected area of 100 in.^2 and an injection pressure of 4000 psi in the cavity, no matter how fluid any plastic material may be.

Let us now see what happens when we are injecting a plastic material. Again, we will start with a 20,000-psi injection pressure, which is indicated on the gauge on the control panel. For all practical purposes, we can consider it as the pressure at the nozzle. However, there is some small pressure loss in the injection system. While the plastic is within the nozzle, it has high fluidity, but, as soon as it enters the mold, instant skin formation occurs all along the passages. This skin increases progressively in thickness with time and brings about a decreasing opening in the passages. While all this is going on, the pressure in the cavity needed to make the parts dense is building up. This cavity pressure is one of the components of the total injection pressure and is also the component that the clamping ram has to overcome. The cavity component is about 50% to 66% of gauge pressure, and the remainder is the pressure drop in the sprue bushing, runner, and gate.

In order to reduce the ability of cavity pressure to open the mold, we have to decrease the fluidity of the plastic as well as the temperature of the mold in increments of 15°F. The fluidity of the plastic is kept low by not allowing it to soak up heat and by maintaining the cylinder temperature on the lower end at a value that will allow production of satisfactory parts. The changes in the two directions will most likely cause a higher pressure drop outside the cavity, and the total needed injection pressure may be beyond the range of machine capacity. To overcome this, the pressure drop between the nozzle and cavity can be brought within tolerable amounts by a slight increase in runner diameter as well as in gate depth (see Chapter 7).

Another possible aid in decreasing the need for larger clamp size is to inject about 95% of the material at the highest speed and let the remaining 5% be delivered by the low-volume pressure-holding pump. This merely gives the material additional time to lose heat so that it will be less fluid when full pressure in the cavity is reached.

This arrangement of cavity filling would be predicated on a timer function that would not be too accurate for the purpose. A better way to accomplish a similar filling would be to have a limit switch, actuated by the moving screw and located at the end of screw travel, that is adjustable about ½" to 1" from the end to zero position. Some presses are equipped with that type of limit switch, which is known as the "injection high override." In this case the limit switch in reality bypasses the injection high timer and controls the filling of cavity by position of the screw with respect to "coming home."

At first the high volume, fast injection pump is ordered to dump its volume to the tank. At the same time, the hold pump with considerably lower volume is put in place to slow down the delivery of final volume to fill the cavity. While the slower filling takes place, the remainder of the shot has time to solidify around the edges and thereby reduce the danger of flashing at the parting line. The reason for this type of action is explained in the preceding discussion in which we indicated that a 200-ton press would require 4000 psi to keep the mold of 100 in.^{2} closed if the injected liquid was of oillike consistency. A mold being filled in less than one second with oillike consistency and at higher pressure than 4000 psi would cause the mold to open up and thereby causes flashing at the parting line. The injection high override allows the high rate of injection for the bulk of the cavity to be limited to a relatively low pressure and then completing the remainder at lower speed. The pressure of the low-volume holding pump can equal the setting of the high-pressure pump; sometimes it is lower or even higher. That would depend on freezing of the gate that would stop the material flowing.

The machines that do not come with this type of a limit switch can be equipped with relative ease by the user. This feature permits molding large areas in lower capacity presses. It is true that slowing down filling at the end of screw travel will increase the cycle time, but in most cases this will only amount to 10% to 15% of injection time. This type of action is usually preferable to converting to a press with twice the tonnage.

In summary, the plastic with a relatively low heat content and the mold at a loweredd temperature will produce a state at the parting line that is least conducive to opening of the mold and thereby require a lower clamp force.

On parts with wall thickness of 0.060 or less, the ability to change conditions as described is very limited, whereas on thicker walled parts there is considerable working room.

When a mold is in the early design stages and the calculated requirement of the clamp is close to the rated capacity of press to be used, it is advisable to proportion the known contributors to pressure drop such as runners, sprue bushing, and gates on the higher side so that lower injection pressures can be used, thereby reducing clamp size requirements. The lowered pressure drop can be attained by using a recessed sprue bushing, increasing runner diameter by 1/32 in. and gate depth by 20% over the normal values.

In the vast majority of cases, pressures of 2 to 7 ton psi will provide safe and practical working information.

CLAMP AND PLATEN DATA

Figure 6–1 shows a typical platen drawing. Figure 6–2 illustrates stroke of press and space for mold thickness. The operator's manual of each machine should contain platen information as follows:

1. pattern of clamping holes
2. platen size and inside dimensions between tie bars
3. pattern of knockout bar arrangement
4. maximum and minimum mold thickness that the press will accommodate
5. maximum and minimum press daylight
6. type and distance of stripping

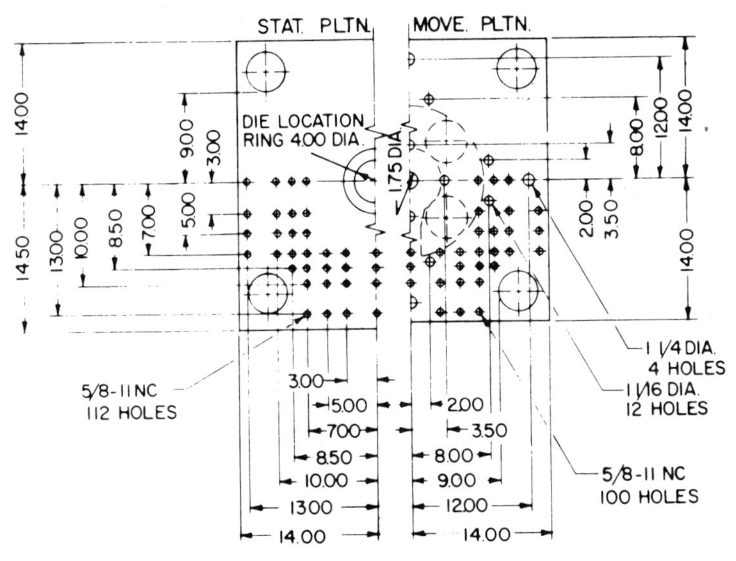

Fig. 6–1. Standard platen. *(Courtesy of HPM Corp.)*

MAXIMUM DAYLIGHT ARRANGEMENT

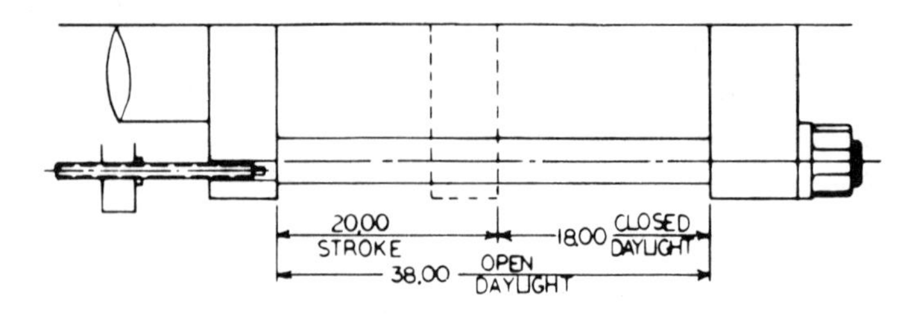

(1) Moving platen connected directly to ram. Mechanical knock-out screws operating thru moving platen (daylights as shown). (2) Hydraulic ejection (optional) can be operative in a reduced maximum daylight arrangement with ejector rod extending thru center knock-out hole. (Requires reducing daylight dimensions by 3.0").

DAYLIGHT AVAILABLE WITH EJECTOR BOX INSTALLED

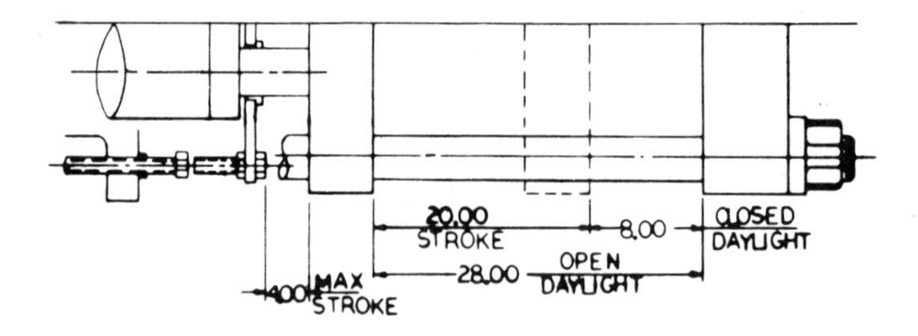

(1) Standard machine ejector plate is operated by mechanical knock-out screws. (2) Center hydraulic cylinder operation of ejector plate is optionally available. (Knock-out screws are not provided when hydraulic ejection is purchased).

Fig. 6-2. Daylight and stroke limits. (*Courtesy of HPM Corp.*)

7. stroke of press or distance the moving platen will travel
8. volume of metering section of cylinder for shot
9. stroke of injection cylinder
10. maximum injection psi on material
11. maximum force of stripping.

These press specifications have to be coordinated with mold details to insure proper functioning of the mold when placed in the intended press. In the following paragraphs, each specification listed will be reviewed to illustrate how to make appropriate allowances in a mold.

Mold mounting. Once the size of a press has been selected, the method of mounting should be checked against the platen hole pattern especially in cases where the mold size covers the platen area between the tie bars. The preferred mounting is to have the slots running parallel to the horizontal position of the machine. Where heavy molds (1500 lb and over) are involved and it is essential to run the clamping slots vertically (for fit reasons), provisions should be made on the mold for a bottom support to prevent slippage and, consequently, possible damage. It is especially important on the moving half of the mold.

Spacing of tie bars. The spacing between tie bars will have a bearing on whether conventional placement of a mold into press is feasible. Some shops use cranes and place molds from top of press, whereas others use lifts to slide them from the side. In any event, this phase requires checking. Placing of the mold into the press should not be a major project.

If any projecting elements from the base itself will interfere with standard placement of the mold into the press, such elements should be easily removable and provided with a means of foolproof relocation and positive alignment. In most plants, the placement of a mold into the press is looked upon as a standardized operation performed within well-defined time values. Any needed attachment should therefore be designed to fit into this concept of carrying out the operation of mold placement.

K.O. rods (stripper). The location of the knockout rod in the mold should match that of the press for either horizontal or vertical placement. The number of knockout rods should be such that the danger of bending the stripper plate is minimized. The load distribution on the stripper plate should be carefully analyzed during mold design while deciding on the number of rods. The possible interference of stripper rods with support pillars should be checked out.

Mold thickness. The mold thickness and required daylight to remove the parts efficiently from the press should be checked against the press daylight, stroke of the press, and stripping stroke to insure that the mold will perform as anticipated. A view of the mold in closed and open positions on the drawing board can uncover, in some cases, unexpected deficiencies.

Maximum and minimum press daylight. The maximum daylight determines how thick a mold the press will accommodate. The thickness, plus the necessary stroke for removal of parts, will determine whether the available daylight will permit proper functioning of mold. The minimum daylight determines how thin a mold can be used and still have full pressure applied to it.

Type and distance of stripping. When stripping is performed hydraulically, it is done by a cylinder that is mounted in the center of the ram. Its distance for stripping is governed by the stroke of the K.O. cylinder. It is normally a double-acting cylinder so that the stripper plate can be retracted in any desired position of press movement. It can be timed so that stripping will take place in a preferred position of ram travel.

The mechanical stripping does not have the flexibility of the hydraulic system. It is actuated by a stripper bar located behind the moving platen during press opening. When the mold is closed, the K.O. bar is in the extreme "out" position, and, as the ram moves backward, the K.O. bar hits two stop rods at a predetermined position causing the forward movement of the stripping plate. The stop rods have an adjustable setting as dictated by the distance of stripping and indirectly by the space provided for movement of the K.O. bar. The stroke or distance of stripping is shown in the operator's manual of each press.

Stroke of press. The depth of a part controls the needed stroke of the clamp. The stroke should be somewhat greater than twice the part depth, or greater than open daylight minus mold thickness. Open daylight for a specific job should be greater than twice the part depth plus mold thickness. On deep parts (3 in. and over), the press stroke should always be checked out to make sure that it is sufficient for the part.

Volume of cylinder metering section. Under press-size determination, it is shown how to compare the volume requirement of any material with that of styrene, since styrene is the standard for shot-size volume. The suppliers of several plastic materials suggest down-sizing of shot capacity by an appreciable percentage. While this may be true on some machines, it is not necessarily true for others. For example, if the cylinder is relatively long, the heating system efficient and protected against heat loss, and the heat throughout the system gradually and systematically raised, then we can create conditions in which the machine will approach rated capacity with most materials. An analysis of heat from hopper to nozzle will aid in establishing whether any downgrading is necessary. (See Chapter 14.)

Stroke of injection cylinder. With the knowledge that the volume changes with the specific gravity, we can readily establish the distance the screw will travel for any material. The cylinder area is constant for any volume; the only variable is the distance of screw travel. For example, if an 8-oz shot of styrene calls for 5 in. of screw movement, the comparable distance for polycarbonate will be

$$5 \text{ in.} \times \frac{\text{Specific gravity of styrene}}{\text{Specific gravity of polycarbonate}} = 5 \times \frac{1.06}{1.20} = 4.42 \text{ or}$$

about 4-7/16 in.

Maximum injection psi. Not all presses have an injection pressure of 20,000 psi maximum. Some are lower and some few are even higher. With this in mind, the machine specification should be checked, and, in case of lower injection psi, the mold passages may have to be increased so that the pressure drops will allow enough remaining pressure for compressing the material in the cavity.

Stripping tonnage. Many molds are designed so that one-half actuates the movement of plates in the opposite half. The force that causes such actuation is normally the tonnage that brings about the opening of press and mold. The actuating members are usually rods of suitable size and number that are capable of taking the strain produced by the stripping tonnage. The strength calculation of these rods are shown in Chapter 9. To protect the rods from being overloaded, a note of caution should be stamped in the vicinity of the lifting eyebolt and should read, "Max. Stripping 10 Tons."

EYEBOLT HOLES

Eyebolt holes are normally on the side of the clamping slots and should be provided on both halves opposite each other; they should be placed in areas where balanced lifting of mold base is possible. Holes should also be tapped on surfaces perpendicular to the slots.

The forged steel eyebolts have a safe load-carrying capacity as listed:

1/2 in.	2600 lb
3/4 in.	6000 lb
1 in.	11,000 lb

Only forged steel eyebolts should be specified for safety reasons, preferably those with a shoulder for better stability.

MOLD CONNECTION FOR FLUID

Mold temperature connections should be placed away from the operator side and recessed wherever feasible so that danger of their damage is eliminated. Whenever quick disconnect couplings are used, care should be taken to see that the openings in the fittings will not restrict the flow to the mold and to insure that the proper velocity for turbulent flow is maintained.

CLAMP TONNAGE AND MOLD SIZE

If we inspect the stationary platens of injection machines in an operating plant, we will find that a number of them have indentations and impressions. These are a result of some projections from the mold base and, in some cases, of the mold base being too small for the clamping force, thus causing a concentrated stress in the platen that brings about the flow of the platen metal. The platen impressions are dangerous because they reduce the contact area for the mold thereby increasing the potential for further indentation. These indentations, if permitted to increase in number, may ultimately cause cracking of a platen, which would not only take the press out of operation but also cause a large and expensive replacement. Practically all presses have provisions for reducing the clamp tonnage, but the problem is to recognize the danger and the limits within which it is safe to concentrate a load on the platen. Most platens are made of cast steel with a yield strength of about 25 ton/in.^2 Allowing a safety factor of 7, we have a permissible load of 3.5 ton/in.^2 With this information, we are able to calculate the minimum number of square inches a certain press size will safely accommodate or to determine to what tonnage to reduce the clamp in order to protect the platen against damage.

A mold of 10 X 12 is to be placed in a 500-ton press. First, we will establish the minimum square inches of mold base that will safely absorb the clamp force.

$$500 \text{ ton} = \text{in.}^2 \times 3.5 \text{ (permissible stress)}$$

$$\text{in.}^2 = \frac{500}{3.5} = 143 \text{ in.}^2$$

The mold in question is $10 \times 12 = 120 \text{ in.}^2$ This shows that the mold should be operated with a reduced tonnage.

$$\text{Tonnage} = (\text{in.}^2 - \text{Locating hole}) \times 3.5 \text{ (safe permissible stress)}$$

$$\text{Tonnage} = (120 - 12.5) \times 3.5$$
$$107.5 \times 3.5 = 376 \text{ ton}$$

To protect press platens against damage, it is advisable to check the contact area of the mold and the platen to see that the safe permissible load is not exceeded.

7
Moldability Features

Moldability can be defined as a group of features that are incorporated in a mold for the purpose of insuring repeatability of operation, properties, and other engineering requirements of products. The description of such features follows. It is to be noted that they have a major influence on the overall quality of plastic parts.

REMOVAL OF PARTS FROM MOLD

Adhering of parts on the ejection half of the mold requires the placing of cores and other retaining means on the moving half so that there is no chance of parts hanging up in the cavity. Even a slight tendency to stick in any portion of the cavity will cause warpage, stresses, and dimensional distortion of parts. Such tendency may indicate a need for additional taper, polish lines in the direction of withdrawal, or manipulation of mold temperature. The hanging-up in a cavity frequently calls for extracting the plastic by actual digging with the chance of damaging a mold. This leads to interruption of cycle and variation in quality. In dimensioning the cavity and core, close attention is to be given to the feature of insuring the unstressed retention of the part on the ejecting side. This is normally accomplished by the plastic shrinking tightly over the cores and adhering to them. In such cases, it is desirable to have a relatively rougher surface on the cores than is incorporated in the cavities. In some configurations, it becomes desirable to provide narrow undercuts of 0.002 to 0.005 deep in the area of ejection pins. Each shape requires individual analysis for the purpose of retention of the moldings on the moving side.

MATERIAL PATHS AND PRESSURE DROP

As the material flows through the different conduits, its condition at the destination (cavity) is determined to a major degree by the injection pressure that compresses it into the desired shape. The effective pressure that exerts the densifying force on the molded product in the component that can be recorded in the cavity by a transducer placed under the head of an injection pin. This cavity-pressure component is part of the total injection pressure indicated on the machine pressure gauge, minus all the pressure drops of the numerous passages.

The cavity pressure is universally recognized as the most significant factor of part quality. There are many process controls on the market that provide narrow variation of molding parameters and will produce consistent parts. All control suppliers agree that cavity pressure plays a very important part in establishing the repeatability of the molded product. The normal method of reading the injection pressure is on the pressure gauge that is connected to the injection side of the hydraulic cylinder, which actuates the movement of the material into the cavity. This gauge reading includes pressure losses experienced within the machine plus those encountered in the nozzle, sprue, runners, gate, and cavity itself. All the components exclusive of cavity pressure can be an appreciable percentage of the total available pressure if the passages are not of suitable size. The danger therefore exists that the remaining pressure for densifying the material in the cavity might be inadequate. By analyzing the melt rheology formula, we can readily see how to dimension the various passages that will lead to tolerable pressure drops. In practice, this means that if we keep the length of passage for material flow as short as possible and have the radius of passage opening on the large end of recommendations, we end up with a favorable pressure drop. For details see "Melt Rheology" (pp. 108–115). The supplier's recommendation incorporated in the material processing data sheet takes into account the sizing of passages for low pressure drop.

THE SPRUE

A complete runner system, sprue, gates, etc., is shown in Fig. 7-1. The sprue forms the transition from the hot molten plastic to the considerably cooler mold. The sprue is part of the flow length of the polymer and has to be of such dimension that the pressure drop is minimal and that its ability to deliver material to the extreme "out" position is not impaired. The starting point for sprue size determination is the main runner, and the outlet of the sprue should not be smaller than the runner diameter at the meeting section. Thus, a 1/4-in.-diam runner would call for a 7/32-in.-diam "O" opening, for an average sprue length of 2 to 3 in. It has been established experimentally that for shots of 6 in.^{3} up to 20 in.^{3}, the 7/32 "O" dimension will satisfy the low pressure-drop needs. For larger shots, a 9/32 "O" opening would be indicated.

Figure 7-2 shows a recessed sprue bushing used in conjunction with a 1-in. diam, extended machine nozzle that is heated by a band-type heater locked in place with a sheet metal wedge. This type of bushing may be applied for sprues that in a standard bushing would be longer than 3 in., and for the three-plate construction as an aid for easy and simple dropping of the runner system. This recessed sprue bushing will contribute zero pressure drop, reduce the travel of ram opening and closing, and minimize the amount of regrind per shot. All these advantages are worthy of consideration. A heated modified sprue bushing with

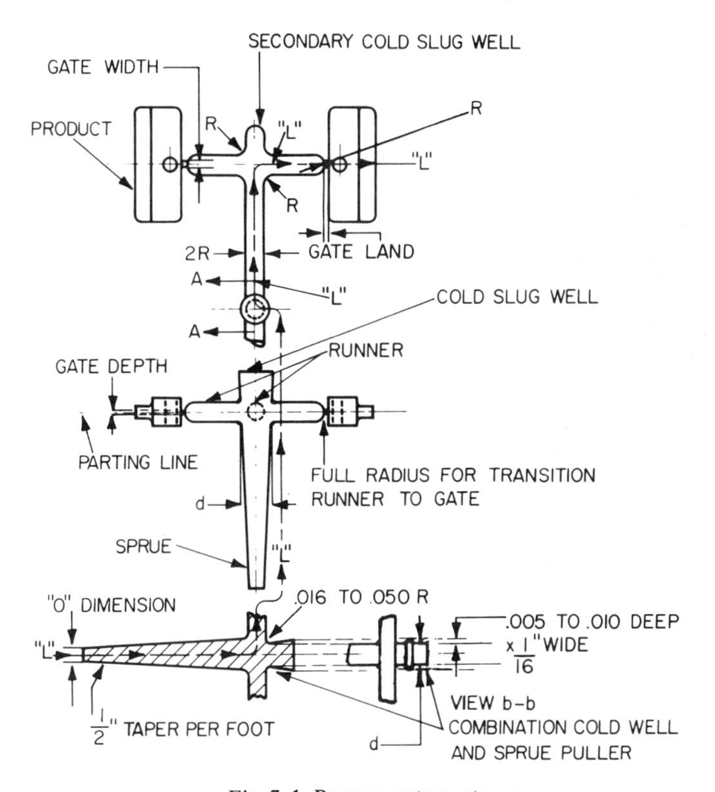

Fig. 7-1. Runner system, etc.

a small opening at the outlet would accomplish a similar result for long sprues. Care must be exerted to heat-insulate the bushing from the mold (Fig. 7-3).

RUNNER SYSTEM

The material processing data give a range of runner sizes for each material. The smaller sizes can be applied for cases when the length of runner does not exceed 2 in. and the volume of material is less than 15 in.^{3} For economical reasons, it is preferable to keep the runners on the smaller end, since it not only reduces the amount of regrind but also accelerates the freezing of the gate which means it affects cycle time. The pressure drop must be kept in mind. It becomes a matter of proportioning runners in relation to spacing of cavities, wall thickness of parts, length of cavities, and corresponding gate sizes. The following example would indicate the application of this statement to an actual case. A polycarbonate part is .090 thick, has a flow length in the cavity of 2-1/2 in., and its gate is located 2 in. from the sprue. From material processing data, we find a recom-

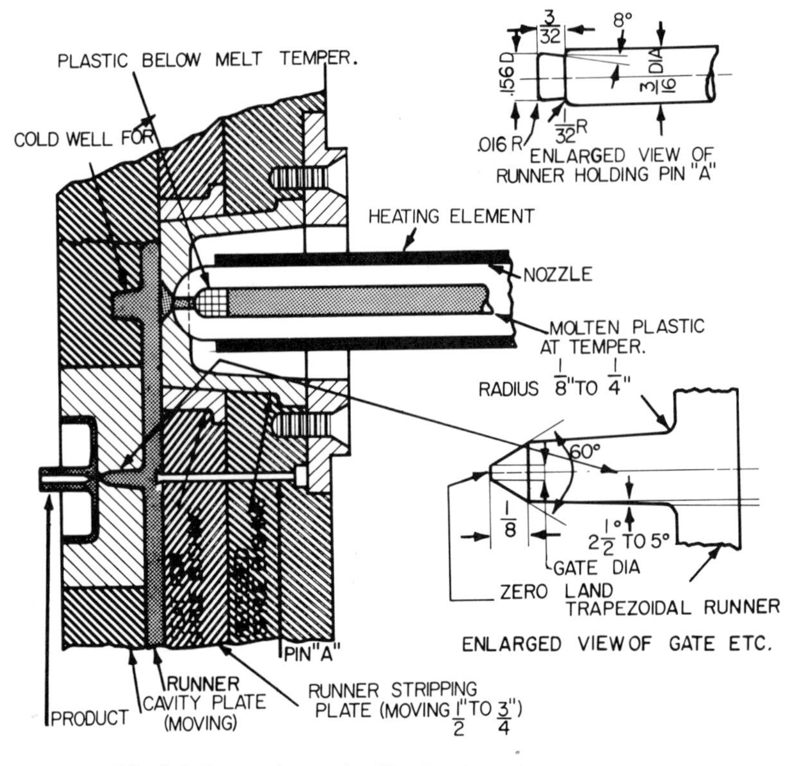

Fig. 7-2. Recessed sprue bushing for three-plate runner system.

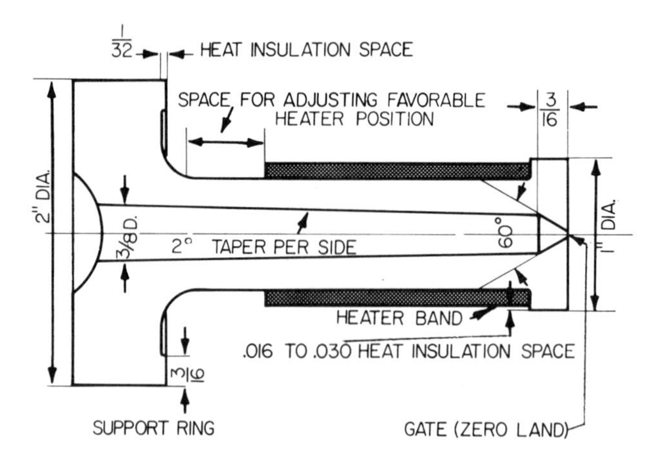

Fig. 7-3. Heated sprue bushing.

mended runner size of 1/4 to 3/8-in. diam, gate thickness 50% to 66-2/3% of part thickness, and width 2 times depth. Thus, the gate would be 0.06 X 0.120 and the runner 1/4 in. for this example. For a flow length in a cavity of, for example, 6 in., the runner size should be 9/32 or 5/16-in. diam. This would provide heated material at a higher rate as needed for the longer flow.

The surface finish of the runner system should be as good as that in the cavity. The profilometer reading should not be more than 50 rms. A good surface finish not only keeps the pressure drop low but also prevents a tendency of the runner sticking to either half of the mold. Such sticking would aggravate the highly stressed area of the gate portion to an even higher stress level.

The preferred cross section of a runner is circular. For a specific flow of material, the round runner will give the lowest heat loss and least pressure drop. The circular shape makes it necessary to split the runner into the two mold halves. There are, however, many mold designs that make it desirable to incorporate the runner in one plate only. In that case, a trapezoidal cross section is used of a size that will surround a corresponding round diameter. See Fig. 7-4 for detail dimensions.

In discussing the length of flow in the runner system, we must differentiate between the physical length of flow and the "effective" one. It is the effective length that brings about the actual pressure drop. The factors that increase the effective length are sharp bends, sharp edges in transition section, and similar impediments to smooth flow. When turns in runners are called for, a radius equal to the diameter of the runner is preferred, but even half that amount will prove to be of considerable benefit. Sharp edges in transition sections cause turbulence and also the increase of effective length. The stoning of a sharp edge to a radius of 0.016 R will be of help, although a small radius of 0.016 is still considered visually sharp if needed (Fig. 7-5).

In multicavity runner systems, it is very important that each cavity be filled

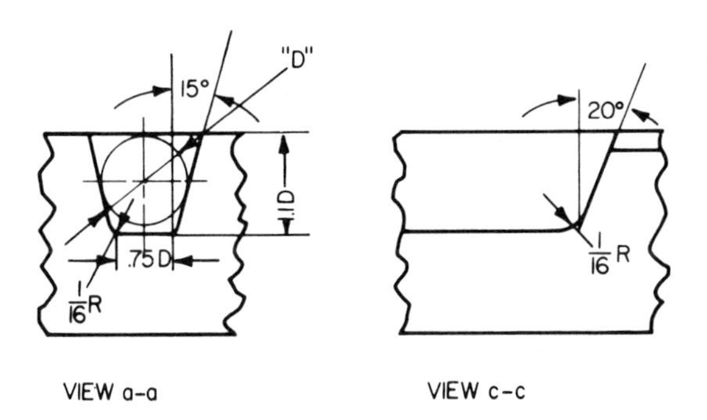

Fig. 7-4. View a-a, round runner of D diameter and equivalent trapezoid. View c-c, transition of trapezoid to gate.

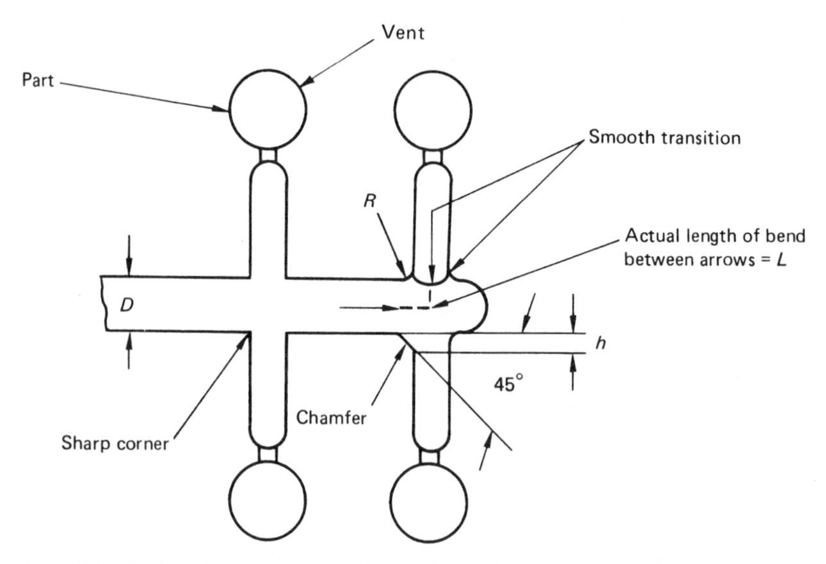

Fig. 7-5. Effective length of runner bends. For R = 1/3 to 1/2 of D, effective length is L; for sharp corners, effective length is 25 L; for chamfer h = 1/3 of D, effective length is 2.5 L.

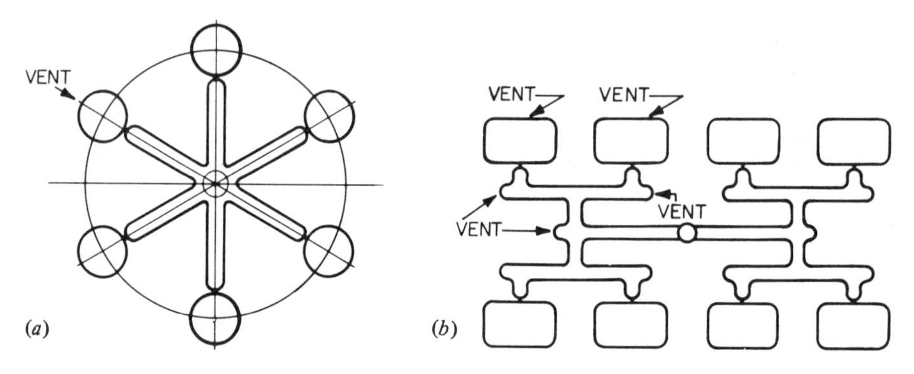

Fig. 7-6. *a*, Balanced spoke runner layout. *b*, Balanced "H" runner layout.

simultaneously, so that not only does each cavity receive material in the same melt condition, but also so that when adjustments in molding conditions become necessary, all cavities will be affected in the same direction. This is called a *balanced system;* it is a must for precision products. Figure 7-6 shows some possible arrangements.

It is good practice to use a runner plate of the same grade of steel as the cavities and that has a surface machined to 50 rms. In some applications, especially in cases of mild usage of a mold, there is a tendency to use the cavity plate for machining the runner in it. If a cavity protrudes on one side above the plate, a runner plate on that side is a must. Runner systems will vary in size and shape, but one thing must be remembered: its pressure drop must be kept to a low value. The runner material is reground for reuse. During this operation, fines are produced, which are a waste; the possibility of contamination is pronounced; inclusion of metal particles is a likelihood. The addition of regrind to virgin material is usually accompanied by the need of adjustment of molding parameters and, in most cases, results in an increase of rejects. For these reasons, the amount of material involved in a runner system should be kept to a very minimum.

It should be emphasized that the round and trapezoidal sections have the least circumferential cooling with respect to the passage area, thus making them the most desirable runner shape. Any other cross section will not perform properly and as a consequence should not be used.

COLD-SLUG WELL

When we consider the heat condition between the nozzle and sprue bushing, we find a nozzle heated to about the same temperature as the front of the cylinder contacting a relatively cool sprue bushing. As a result, we have a temperature at the nozzle tip that is lower than the required melt temperature. There is a gradual rise in heat for about 0.5 to 1 in. depth of the nozzle, at which point the normal melt temperature is existing. The material that is lying in the nozzle zone that is not fully up to heat does not have good flow properties; therefore, if it enters into a cavity, it will produce defective parts. To overcome this, a well is provided as an extension of the sprue, to receive the cool material, thus preventing it from entering into the runner system. The well is equal in diameter to the sprue at the parting line and is about 1 to 1.5 times the diameter in depth. These sizes may vary considerably, but the important thing is to have the inside of the nozzle of such shape and so heated that the volume of cool material is less than the cold-well slug. (See Fig. 7–7.).

In some heat-sensitive materials such as polycarbonate, it is desirable to also have smaller cold-slug wells at the end of the runners or even their branches to prevent some of the runner-cooled material from getting into the cavity. It is especially important for optical products. (See Fig. 7–8).

A cold slug also performs the function of providing the means of extracting the sprue from its bushing, thereby acting as a retainer for sprue with runner on the moving half of the mold. During stripping, a pin, which is attached to the stripper plate and also forms the bottom of the well, moves to eject the sprue with runners from the mold.

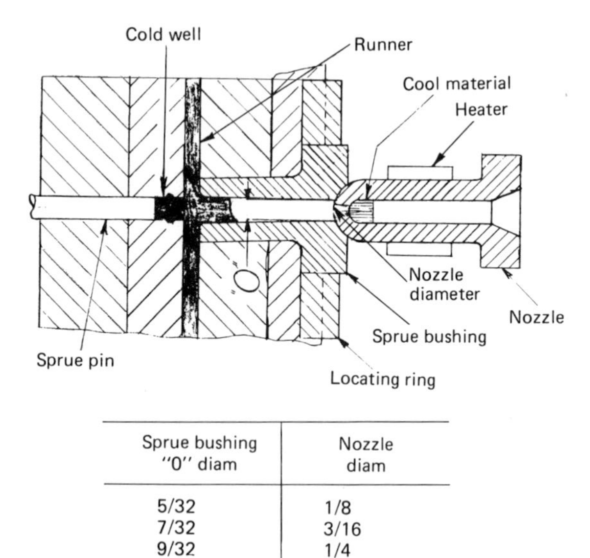

Sprue bushing "0" diam	Nozzle diam
5/32	1/8
7/32	3/16
9/32	1/4
11/32	5/16

Fig. 7–7. Relation of cool material in nozzle to cold well.

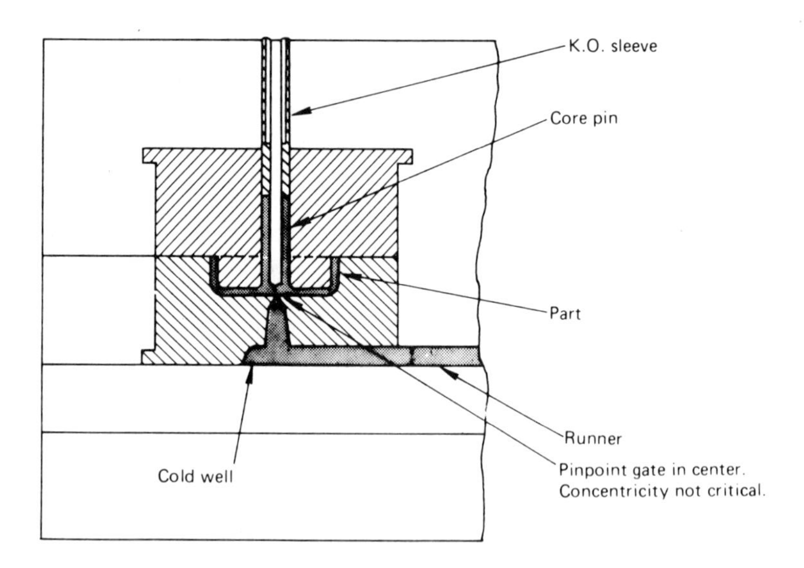

Fig. 7–8. Long cores and material-flow direction.

GATE LOCATION

The gate can have a number of adverse effects on part properties if incorrectly placed. It can be responsible for weak weld lines, gas trapping, and jetting, which results in poor appearance and contributes to an irregular color of the overall product. Since the gate is stressed, it should be placed so that it will not have a bearing on the function of the product. If the part is round and concentricity is important, a center gate is indicated. If the part is oblong and the possibility of distortion exists, the gate should be so placed that the material reaches the outlying points at the same time to avoid variation in cooling rates. This can be accomplished by very slightly varying the runner leading to the farthest point in order to produce a flow that simulates an umbrella-shaped pattern.

Weld-Line Strength

If highest weld-line strength is desired, the gate should be placed close to the area of the weld, so that the material will be at a favorable temperature for self-welding. The strength of a weld line can be determined by analyzing Fig. 4–7. The left-hand illustration shows a condition in which the area A (to the right of the hole) is small and contains material without a weld line; therefore, that area should be about 20% stronger than the left-hand area with its weld line. The area to the left of the hole is about twice the size of that to the right. Using a 10% degradation of weld lines, one would therefore have the equivalent strength of the original material. On the right side of the illustration we have the opposite effect, in which a weld line of cooled material has formed to the right of the hole; its distance is smaller than the undivided section, thereby creating a weld line less strong than the balance of the part. Whenever possible, a gate should be placed in the manner shown in Fig. 4–7, left-hand side, in order to form a part with equal strength throughout its cross section.

It should be remembered that other molding conditions will enhance the strength of the weld lines, including higher mold temperature, fast injection, and vents. If frail cores are present, it is best to have the material flowing parallel to such cores in order to prevent them from bending under the influence of material pressure (see Fig. 7–8 and the section "Gates and Cores," pp. 99–101).

Gas Entrapment

When air entrapment is possible, the gate must be so placed to keep this from occurring. The gas trapping depicted in Fig. 7–9 can occur in almost any molding. It is a byproduct of a nonuniform wall machined into a mold or of the plastic flow pattern in a cavity. These variations cause the flow to bypass the gas

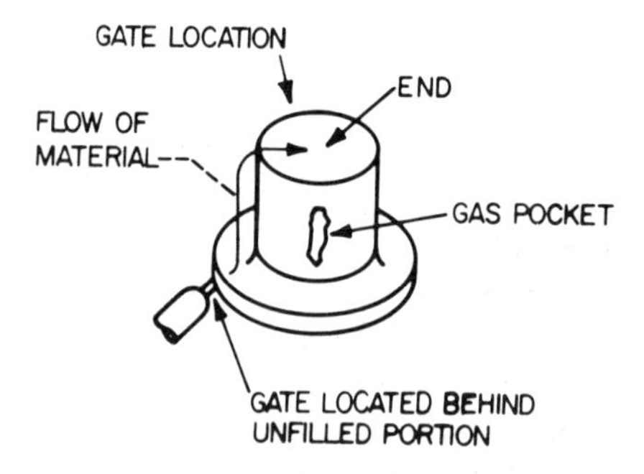

Fig. 7–9. Gate location to avoid gas entrapment.

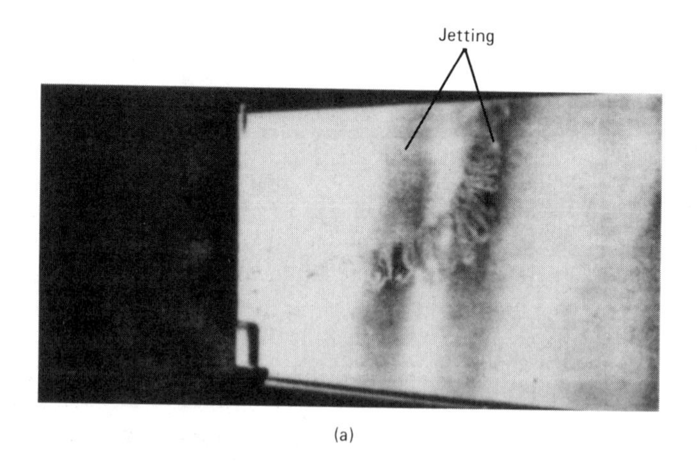

(a)

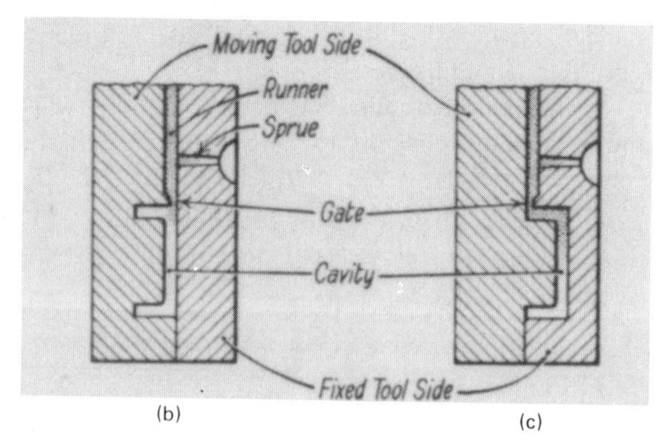

(b) (c)

Fig. 7–10. *a,* A flat-plate mold showing effect of jetting. *b,* Design for possible jetting. *c,* Design to eliminate jetting.

pocket, thus creating a defective product. Defects of this nature can be corrected by polishing the problem area with directional lines to conform to the flow pattern; if that is not sufficient, directional sanding or grinding and polishing of the gas-pocket area should result in an umbrella-shaped flow.

On many occasions the configuration of a product is such that the outlying points to be filled have not been compensated by suitable runners, resulting in formation of gas pockets. Elimination of gas pockets would follow the same directions as just outlined.

As a general rule, the gate should be located on the heaviest portion of a part. Since thicker portions take more time to solidify and have higher rates of shrinkage than thinner portions, placing the gate close to the thickest section allows both the replenishment of voids that might form during pressure holding and the application of full pressure where it is most needed. Another important consideration is to place the gate in such a position that the incoming flow to the cavity will be deflected by a core or mold wall about 1/8 to 3/16 in. from the gate (see Fig. 7–1 for pin and 7–11 for wall).

Jetting

When the incoming flow from a gate is not deflected as stated above, jetting is occurring (see Fig. 7–10). The consequences of jetting are an undesirable appearance and a potential stress condition in the product. Jetting results from the flow characteristics of thermoplastics as explained in melt rheology discussed later in the chapter. The viscoelastic flow of polymers as shown on Fig. 7–25 indicates that the diameter leaving a certain orifice will increase in size. This can be observed in practice by retracting the screw assembly and pushing the polymer through the nozzle. We find that the stream of plastic not only comes out enlarged in diameter in the nozzle, but also maintains its cross-sectional uniformity over a distance of several inches. When a stream of plastic material enters a free and relatively large space from a gate that is smaller in thickness than the cavity, the stream become thicker than the cavity. Portions of the stream at the top and bottom cool faster and freeze before the remainder solidifies. Therefore, we have two rates of freezing in the same stream of material (see Fig. 7–10*a*). This condition shows up as jetting.

The molded product shown in Fig. 7–10*b* and *c* indicates how the same item can be designed to eliminate the possibility of jetting. Jetting can also be eliminated sometimes by enlarging the gate size, which slows down rate at which material enters the cavity.

When designing a mold, the following steps can be used to safeguard against the occurrence of jetting. Use a trapezoidal runner as shown in Fig. 7–11. Here the gate to the cavity is designed so that the material will hit a wall and disperse, thereby preventing a continuous stream from flowing into the cavity.

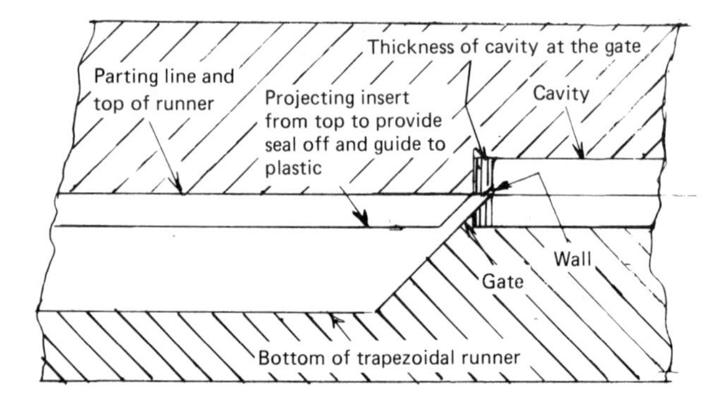

Fig. 7–11. Trapezoidal runner in mold design to avoid jetting.

Another possibility is to have a gate that forms a ribbon-shaped flow; this would eliminate the double freeze-off condition associated with jetting.

Other Considerations

For transparent parts to be used as shields or covers over objects that must be legible or visible underneath, the gate location deserves special attention (See Fig. 7–9). In this case it is important to have the flow directed, so that the last portion of the molded object being filled is not in the area where visibility is required. The location of the gate and the flow pattern of material should conform to this requirement.

In a balanced runner system, uniformity of gate size calls for special attention, so that a uniform rate of filling each cavity is attained.

Last but not least, the location of the gate should be such as to fit the performance and appearance of the completed product and therefore should be approved by the designer engineer involved. In cases where the gates have a sharp edge entering the cavity as a result of machining, this should be corrected by stoning a finish to about .005 up to .020 radius (See Fig. 7–12) in order to facilitate a smooth flow of the polymer.

TRANSITION SHAPE

The transition shape from runner to gate is important because it is essential to keep the material in heated condition prior to entering into the gate. The perfect condition would be to have the full cylinder of the runner come up right against the gate. This is impractical for reasons of tool-life strength. The hemispherical transition from runner to gate as shown in Fig. 7–12 has good strength. Also, because of the small contact surface in relation to volume of material, it loses

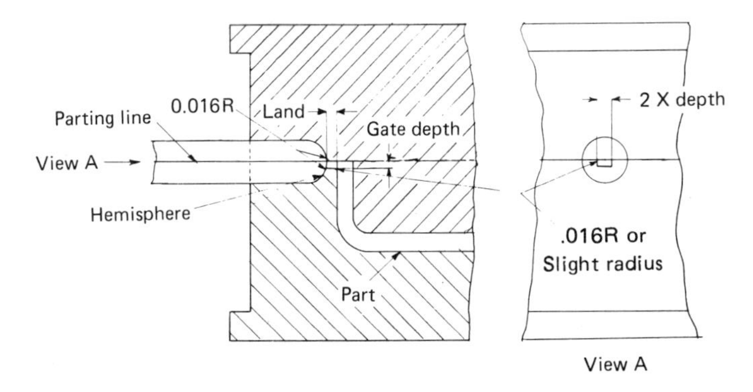

Fig. 7–12. Gate detail requirements.

less heat thereby providing a fully heated source of material supply to the gate. An end taper of the same angle as the side taper of a trapezoidal runner is also satisfactory (see Fig. 7–4). The long tapered transition should be avoided. It exists when the angle of approach is greater than 20° (see Fig. 7–4). In this case, the contact surface is large in relation to volume of material, thus causing the melt to lose more heat.

The same condition of transition shape described above should also prevail on torpedoes or probes that are employed in connection with hot-runner or runnerless molds. By having a form tool for the cavity and probe, the shape can easily be reproduced and the wedge-shaped runner at the front is eliminated; moreover,

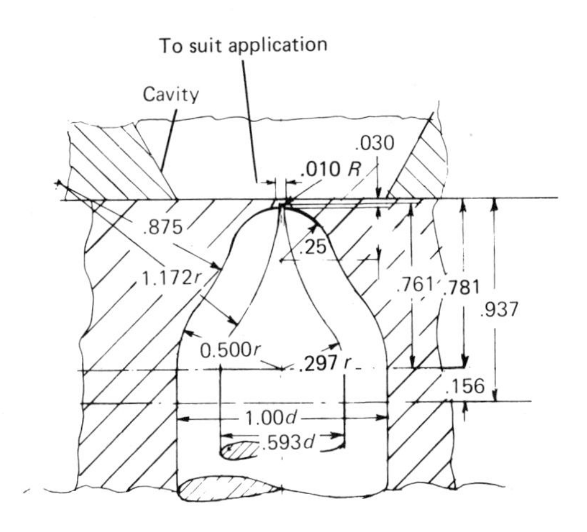

Fig. 7–13. Approach for runnerless probes.

hot material is available at the tip of the torpedo, thereby improving moldability at this point. Figure 7–13 shows the improved shape of an approach for runnerless probes.

On the other hand, the wedge-shaped probe will tend to have a larger freeze-off and thus will offer greater resistance to flow. The net result is a need for more pressure and/or a larger gate. If considerably more pressure is required to supply the melt up to the cavity, there may not be enough pressure left to densify the part in the cavity. This is particularly important in materials that are known to be "sheer rate insensitive." In this specific case, friction as a source of heat and derived from forcing the material through small openings such as gates is not effective in improving flow characteristics of the material. In fact, the danger exists that the molecules might get torn or damaged due to the rearranging force of the pressure and might degrade the properties of the material. This is usually not visible to the eye, but shows up when subjected to accurate tests. The materials affected in the manner described are polycarbonate, polysulfone, modified polyphenylene oxide (noryl), and, to some extent, acetals, linear polyethylene, and polypropylene.

GATE SIZE

The purpose of the gate is to control the flow of fluid plastic into the cavity. It also provides a practical and easy means of separating the parts from the runner system. Furthermore, the gate size is normally a fraction of the part thickness and thereby determines the solidification time of the cavity and, indirectly, the cycle time. Gate sizes vary with each material and its viscosity at the time it passes through the gate. The closer the cylinder nozzle to the cavity, the smaller the gate can be without affecting the moldability of the part. This does not include the shear-rate-insensitive materials, which require a minimum gate size in order to prevent polymer degradation.

The recommended gate sizes as shown in material processing data specify an opening that is a percentage of part thickness with a width of twice the depth of the gate. Round gates about 10% larger in diameter than the specified gate depth should give satisfactory results. Thus, for a part with a wall thickness of 0.090 in. and a gate depth calling for two-thirds of part thickness, there would be a rectangular gate with a depth of 0.060 in. or a round gate with 0.066 in. diam. In addition to gate sizes, gate types demand attention.

The constantly rising costs of molding operations impose upon the mold designer the responsibility of incorporating features in the mold that will lead to negligible or no handling of parts and runners and thus make possible automatic operation of presses. With this in mind, degating in the mold becomes imperative.

The gate size and its location have a decided influence on the quality and appearance of a molded product. The specifications of gate dimensions as stated previously are determined in most cases by suggestions of material suppliers. Most tool designers tend to specify the gate size at the smaller end of specification range in order to facilitate the ease of removing the runner from the molded object. However, if the specifications given in the processing of materials are followed, some of the enumerated difficulties could be eliminated and defects associated with small gates kept to a minimum. According to the flow formula found on page 112, increasing the gate size by only 10% allows the gate to handle about 1-1/2 times the volume of plastic in comparison with the gate as originally provided. This assumes that all the other factors remain unchanged. When gate sizes are smaller than those given in the material processing data, we can expect the following ill effects on the molding:

1. Shots from cycle to cycle can vary in density and strength due to a small change in viscosity of the polymer. As the flow formula indicates, the volume delivered per second (when all the other factors are constant) is inversely proportional to the viscosity; therefore, a change in flow will cause a pressure variation in the cavity and thereby affect consistency in molding.
2. When a shot is completed, some parts of it will display sinks and bubbles due to the freeze-off of the gate before full pressure is applied to the cavity, and solidification begins before the prescribed molding conditions have performed their function. It is important that the gate permit material replenishing from the cushion to the voids created in the molding and that full pressure be applied to the cavity before freeze-off takes place.
3. On multicavity molds where the gates are not balanced to permit the same flow to each cavity, an unequal pressure is applied to each cavity, with the result that each cavity produces a part of uneven properties and appearance.
4. If a gate size is so dimensioned as not to permit the outlying positions of the molding to receive sufficient pressure (cavity pressure of about 2000 to 3000 psi), the product will be lacking in certain properties and have inaccurate dimensions.
5. An undersize gate can cause warpage of the part due to a different cooling rates of material from one end of the cavity to the other.
6. Lamination of the gate onto some materials can also be ascribed to a gate being on the small end of the specified dimensions.
7. Polymer burning at the gate end is due to a large volume of material being pushed through a small gate that by virtue of friction brings about the discoloration of the flowing material. As indicated earlier, increasing gate size by only 10% will enable the gate to handle almost 1-1/2 times the previous volume of material and eliminate the excessive friction.
8. A larger gate size than called for can be responsible for slower freezing and thereby increase the overall cycle time of the product.
9. A system for flattening the gate in the press, so there is no trace of its presence, has been devised by placing a hydraulic plunger where the feeding of

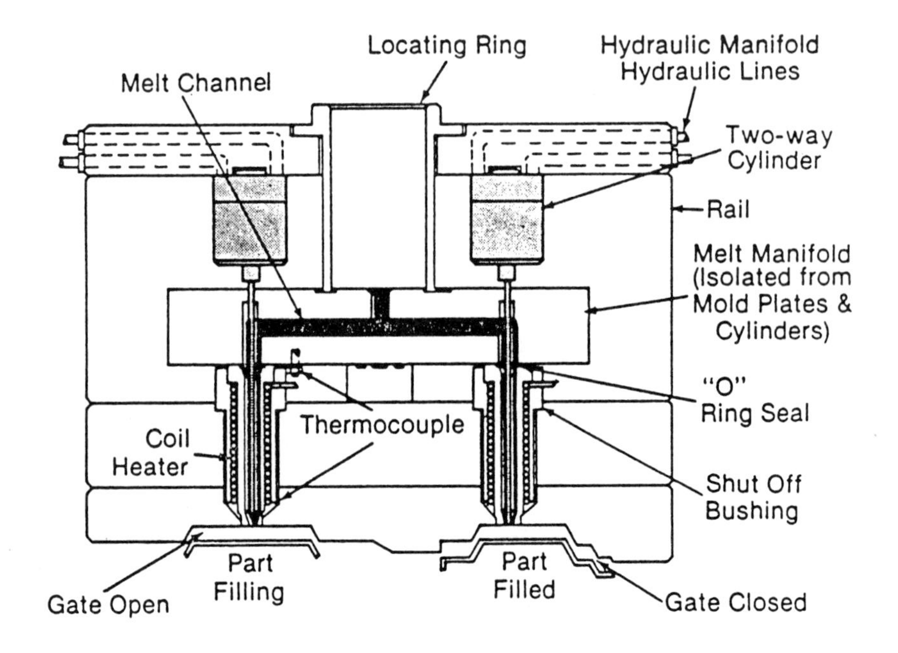

Fig. 7–14. "Zero gate" shut off system. *(Courtesy Incoe Corporation)*

material takes place. In addition, this device is capable of performing other programmed features as claimed by its manufacturer. See Fig. 7–14. Similar gadgets are available from other sources to simply perform flattening of the gate.

Gate Blemishes – Splay and Similar Defects

The principal object of the gate is to admit plastic into cavity. At the same time, however, the dimensions of the gate should be such that it does not differ in appearance from the remaining molding due to its shape of that area. Chapter 17 enumerates the effects of a gate on the quality of products and suggests corrective measures. In this section we will deal with effects of gate configuration that will eliminate the visual change in color due to combining certain gates in relation to cavity configuration.

Splay, smear, blush, and similar defects arise when a plastic material in molten condition contains a considerable amount of gases and some of the gases are trapped between the outside of the gate and cavity. As a result, the gate appears different from the rest of the product. Sometimes an increase in gate size, which allows the gas bubbles to be retained on the inside of the gates, resolves the problem. Otherwise the best solution is to get the gases out of the material before it enters into the gate. For each material and set of operating conditions, the causes for the presence of gases have to be analyzed separately and proper actions taken to eliminate them.

Gates and Cores

Core sizes, can be divided into two categories; (1) up to 1/2 in. diam and (2) those above that size. Let us first treat the smaller-size cores. Cores in a mold are generally provided either to form holes in a product or to support inserts. When the flow of plastic is perpendicular to the cores, the likely action will be to shear them off or to produce a distorted hole. When the gate is in line with a core (as shown in the upper illustration of Fig. 7–15), and the core is

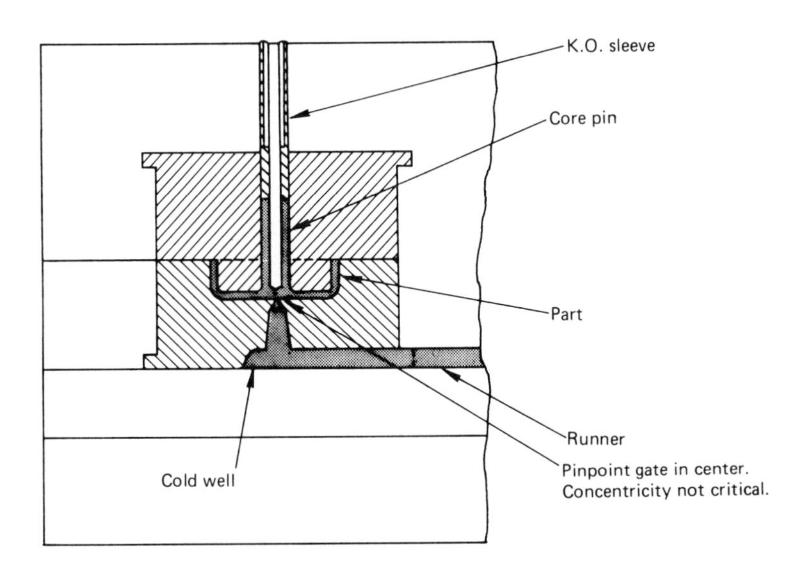

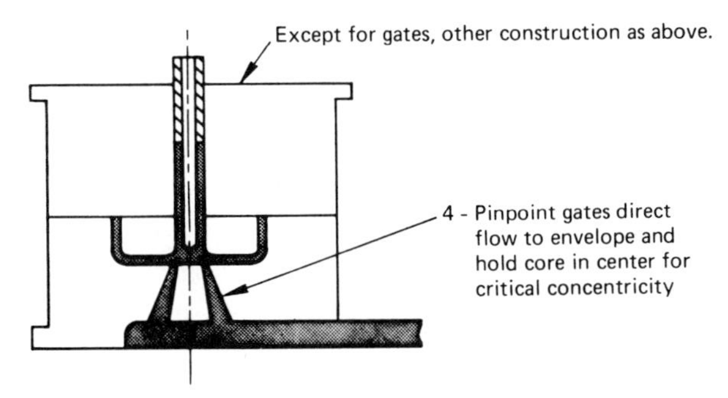

Fig. 7–15. Pinpoint gates in a three-plate mold.

perfectly concentric with the cavity, we could expect a straight hole in the product. In this case the gate would have to be examined under a powerful magnifying glass to insure that the outlet of the gate is perfectly smooth. Preferably the gate should have on its edge a .005 to .016 in. radius to remove irregularities of machining so that the stream of plastic will uniformly cover the core from all directions. For greater precision of a hole or inside diameter, it is best to provide a ring gate out of which three or more individual gates will simultaneously hit the core and thereby retain it in the center position. The separate gates should come down the core and bring about an umbrella-shaped flow which would produce a precision opening. See the lower illustration of Fig. 7–15. All other requirements of gate and concentricity should be same as stated above. Fig. 7–15 shows the core fastened on one end only, but the comments given also apply to cores retained on both ends.

The larger cores, such as seen in medical containers, cups, buckets, and waste baskets, present a story of their own. In this type of container we are dealing with parts that are center-gated at the bottom. The center gate of a molded product has to be located in the geometrical center of the bottom of the cavity and perpendicular to it so there is no irregularity coming from this source. On rectangular or similar objects there will be an auxiliary runner from center to corner so that final result will be a flow of material covering the core in umbrella fashion.

A prerequisite for a desired flow of material is to have a *uniform wall thickness throughout the circumference.* Many products specify minimum and maximum values for wall thickness; for example, an engineering drawing may specify a wall thickness variation of .005. This tolerance could mean that one side of a nominal .80-in. wall could be .075 and the opposite could be 0.85 and still be acceptable from a specification point of view. The difference from one side to the other would amount to about 13% in cross-sectional area. According to the flow formula, this difference would enable one side to handle a volume of material against the 1.6-times-greater volume of the opposite side. These engineering specifications do not emphasize the uniformity of wall thickness regardless of what mold tolerance is being contemplated.

In the absence of uniformity of wall specifications, let us see what manufacturing problems may arise. We will assume that the wall thickness varies but remains within the tolerance limits of the product. A condition is usually created in which one side of the product is larger than the opposite side. According to the flow formula, the thicker side will induce greater flow and be covered first while the opposite is still free of plastic. The pressure of the covered end of the core causes a perpendicular side thrust on the core large enough to eventually cause cracking and finally total loosening of the core from its base. The forces involved can be appreciated from the example shown in Fig. 5–12. The force involved is the projected area times the pressure exerted on it. The projected area in this case is the mean diameter times the length or 6.84 X 21 = 143.64 in. The pressure is taken from the profile of pressure in the cavity at midpoint, which is about 4500 psi. Thus the total force exerted is 143.64 X 4500 $\cong$ 646000 lb or 320 tons. This presents an extreme condition in which the full-thickness tolerance is on one-half of the product.

On the average mold, the side pressure will account for about one-third of the calculated value, which could be approximately 100 tons. Even with this pressure, exerting on every cycle could produce rapid crystallization of the core with respect to the base so that the life of the mold would be limited. When produced with a uniform wall throughout the circumference, this part would not only last at least 10 times longer, but would also operate without gas pockets, and other common difficulties encountered when a core is not concentric with the cavity.

In a product with a large core, uniform wall thickness is the most important specification for a smooth working mold. The total tolerance should be no more than 5% on the wall thickness under consideration. For example, a wall thickness of .080 in. with a tolerance of 5% would give a total variation of .004 in. or a ±.002-in. tolerance of said wall thickness.

On relatively small molds, the precision of concentricity of core to cavity is much easier to maintain; in addition to this, we have smaller projected areas resulting in a much smaller side thrust. Nevertheless, it should be a signal to the moldmaker, mold designer, and molder that uniformity of wall thickness is the most important consideration in a mold that is to have a long life and trouble-free operation. For that reason in practice it is desirable to start out with a mold on the lower specified product thickness, make shots to see how the mold fills, and if need be, make corrections in areas to induce higher flow so that the ultimate result is an umbrella-shaped flow all over the core.

Concluding Note on Gates

Gates are normally machined into a cavity by a grinding wheel or by milling cutters. In each case there is a tendency to have a sharp (razor) edge at the entrance and exit of the gate. This condition is not conducive to the most favorable flowing of a polymer. Fig. 7–12 shows an .016 *R* at each end of the gate; this should be incorporated every place a gate is present. The tendency has been to dimension gates so that gate size is small and the land is long. This is the opposite to the information given in the material processing data (Chapter 18). Plastic material suppliers stress that if proper gate sizes and land – as shown in Chapter 18 – are incorporated, some of these molding problems could be minimized: short shots, sinks, vacuum bubbles, splay, streaking, cloudy appearance, weak parts, jetting, and shrinkage of tolerance.

It should be reemphasized that size of gates and length of land deserve close attention in overall performance of molded parts, and recommendations for these two parameters should be closely observed.

TYPES OF GATE

The following gate types are usually employed, and each has its own advantage for application.

Tunnel gate. The gate in Fig. 7–16 lends itself to automatic operation inasmuch as it separates the runner from the parts during ejection. The gate is limited as to the possible location by appearance of the sheared area and the angular entrance to the part configuration. For multiple cavities, an angular entrance of the gate requires special care in carrying out their machining during moldmaking, in order to insure uniformity of gate opening and consistency in angular approach for a balanced runner system. The angle of approach is determined by the rigidity of material during ejection and the strength of the cavity at the parting line affected by the gate. A flexible material will tolerate a greater angle of entrance than a rigid one. The rigid material may tend to shear off and leave the gate in place, thus defeating the intended performance. On the other hand, the larger angle will give greater strength to the cavity. The smaller angle will give a cleaner shearing surface than may be the case with the larger ones. A tunnel gate for thin or shallow parts is displayed in Fig. 7–17.

For some products a cleaner break between runner and cavity is desirable. This could be accomplished by having a knife action of the gate for the complete cross section of the gate. Normally, the knife effect of the gate occurs only at the center; as we move away from the center, we find the widening of cutting action up to the width of the tool that formed the tunnel gate. A spade shape of the cutting tool would provide full knife action for the width of the gate and thus present a cleaner separation between gate and molding. See Fig. 7–18.

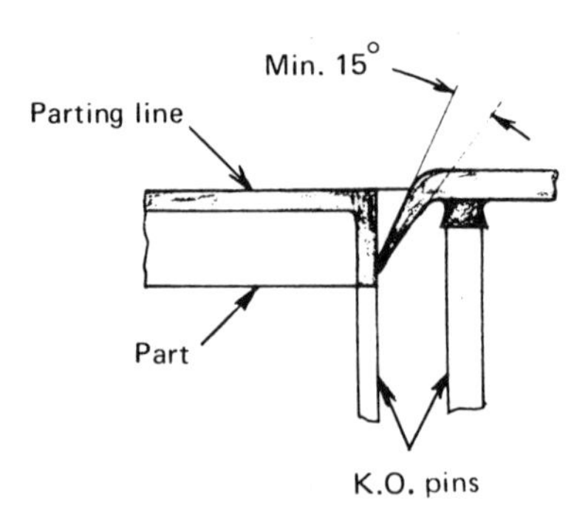

Fig. 7–16. Tunnel gate.

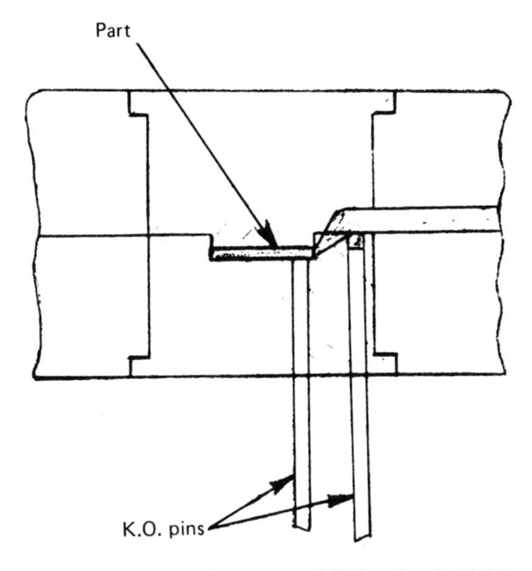

Fig. 7–17. Tunnel gate into parts 0.090 or less in thickness.

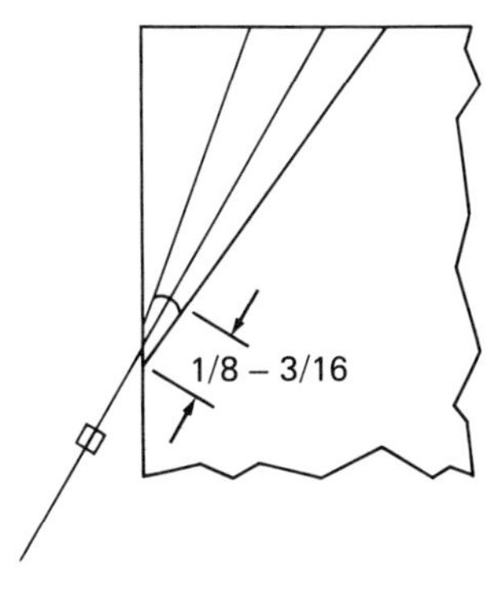

Fig. 7–18. This spade opening can be EDM'd with an electrode having the same tapers from all sides as the body. This will provide a cleaner separation from the molding than a round one.

Tunnel plug gate. Figure 7–19 shows a modification of the straight tunnel gate, and it is used where the sheared surface cannot be tolerated on the outside surface of the product. If the part shown in Fig. 7–16 would not permit a sheared surface, a plug gate could be incorporated. The added plug would have to be removed in a postmolding operation. This plug gate would permit fully

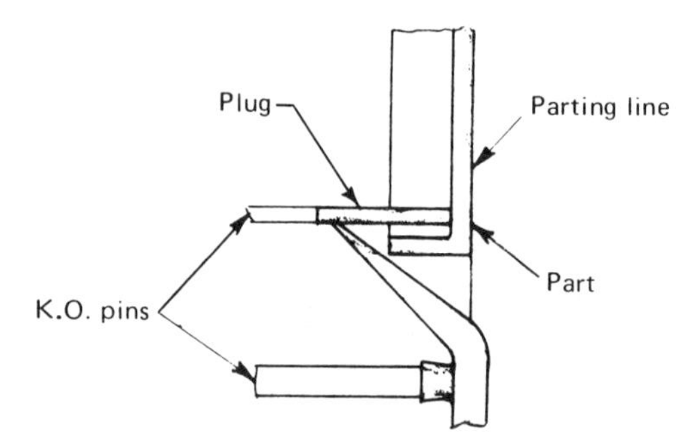

Fig. 7-19. Plug gate.

automatic molding and retain the advantages of consistent repeatability of cycle and, with that, uniform quality and low reject rate.

Pinpoint gate. The pinpoint gate in Fig. 7-15 is generally used in three-plate mold construction. It provides rapid freeze-off and easy separation of runner from part. The size of such gates may go up as high as 1/8 in. provided that the part will not be distorted during gate breaking and separation. A further advantage of pinpoint gating is the ability to provide, with ease, multiple gating to a cavity should such move be desired for symmetry of part or balancing of flow. It also lends itself to automatic press operation if the runner system and parts are arranged for easy drop-off. For a smooth and close break-off, it is best to have the press opening at highest speed at the moment when the plates causing the gate to snap are separating (Fig. 7-15).

Hot runner systems and hot sprue bushings also employ pinpoint gating with favorable results. A word of caution is in order in conjunction with the hot runner gates. The nozzles in the runnerless molding act as do the machine nozzles in conventional molds. The portion of the nozzle that contacts the cavity has a lower temperature than its main portion, therefore, the front part

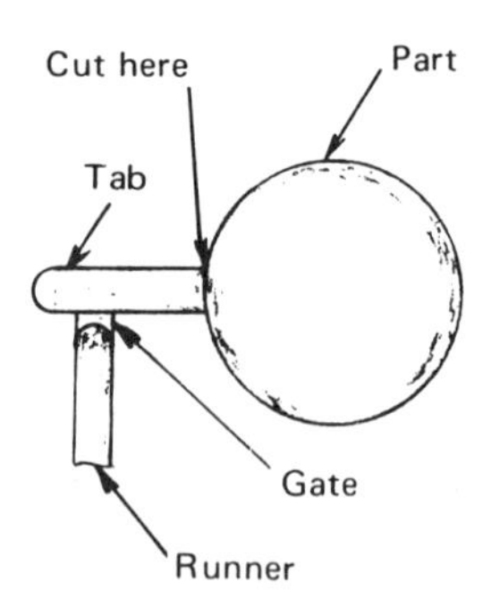

Fig. 7-20. Tab gate.

will contain cooler material entering the cavity. This cooler portion may cause strength or appearance problems, especially for critical parts like gears and cams. To counteract such possibility, the nozzle should be designed so as to have the very minimal amount of cool material, or consideration should be given for a cold-well slug recess. Another possibility is to consider the use of "melt-decompress" action of the machine for its maximum potential pullback of the material from the front of the nozzle, thus keeping the plastic in the well-heated condition. For the melt decompress to be most effective, all passages between gate nozzle and machine nozzle should be free from obstructions to flow and should avoid change in dimensions of passages, so that the suction effect is most productive. It is important in the production of high-strength quality products to make sure that no cooled material of any relative size will enter the molded product.

Tab gate. The tab gate is applied in cases where it is desirable to transfer the stress generated in the gate to an auxiliary tab, which is removed in a post-molding operation. Flat and thin parts require this type of gate (Fig. 7–20).

Edge gating. Edge gating is carried out at the side or by overlapping the part. It is commonly employed for parts that are machine-attended by an operator. It is normally possible to remove the complete shot with one hand and in a rapid manner. The parts are separated from the runner system by hand with the aid of side cutters or, if appearance requirement demands it, by such auxiliary means as sanders, millers, grinders, etc. When degating is performed with the aid of auxiliary equipment, it becomes necessary to construct holding devices that will insure safe, uniform, and satisfactorily appearing gate removal. The edge gating also lends itself to automatic operation, and separating the parts from runners is carried out in a secondary operation. (See Fig. 7–21).

Fin or flash gate. Fin or flash gates are used where the danger of part warpage and dimensional change exists. They are especially suitable for flat parts of considerable areas (over 3 × 3 in.). (See Fig. 7–22).

Direct gate. For single-cavity molds where the sprue feeds material directly into the cavity, a direct gate is applied. Either a standard bushing for an extended nozzle, or a heated bushing, may be used. The type that is selected

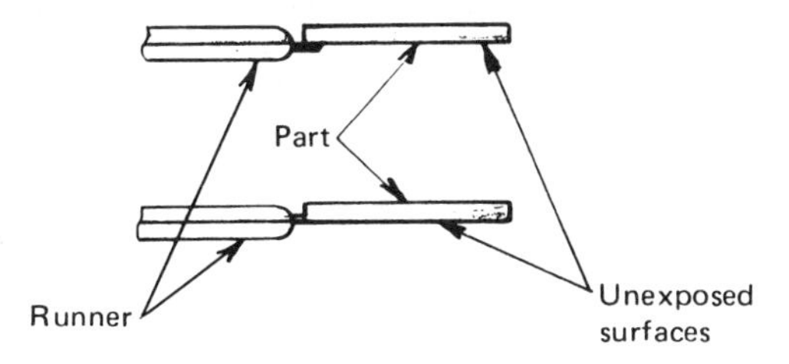

Fig. 7–21. Edge gates.

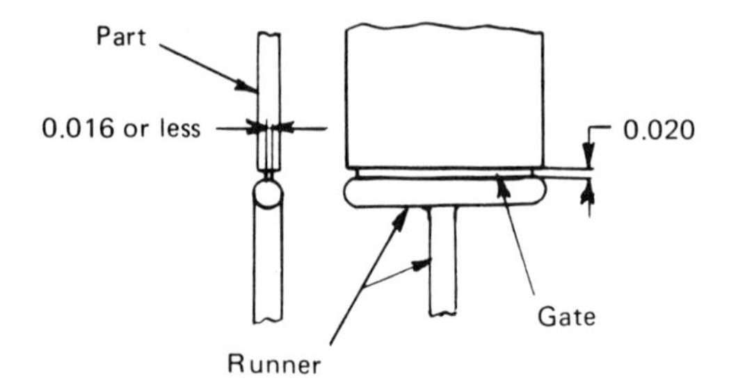

Fig. 7–22. Fin or flash gate.

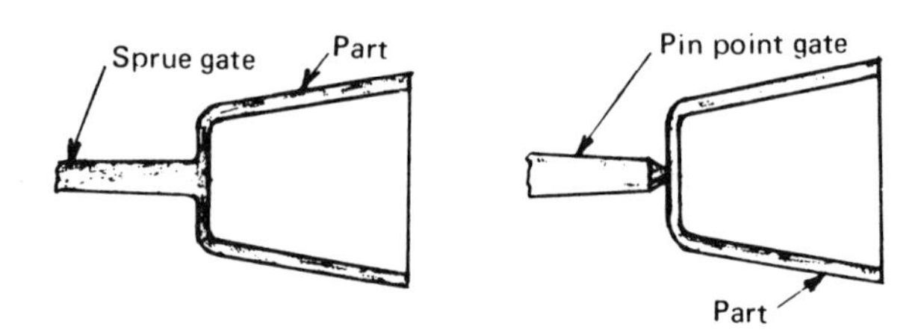

Fig. 7–23. Direct gate.

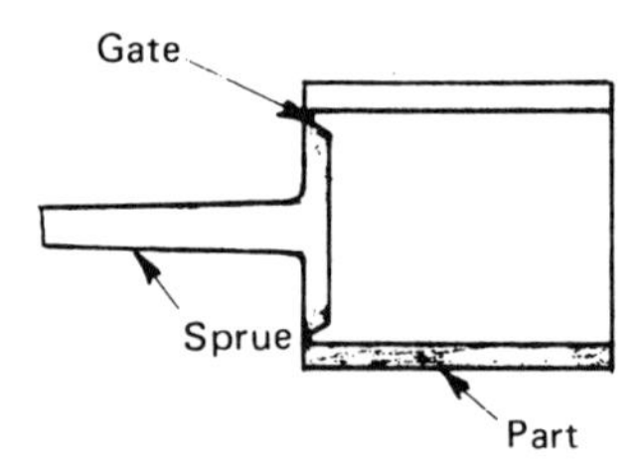

Fig. 7–24. Diaphragm gate.

depends on cost considerations for removing the sprue from the part against the cost of the special bushings. (See Fig. 7–23.)

Diaphragm-and-ring gate. The diaphragm-and-ring gate is mainly applied for cylindrical and round parts in which concentricity is an important dimensional requirement and weld-line presence is objectionable. (See Fig. 7–24.)

LAND OF GATE

Ideally, the length of land of gate should be near zero. However, this is not practical because it would adversely affect the useful life of a cavity. Some length has to be incorporated in each gate. Optimum length is given for each material in the material processing data sheets.

A longer gate than indicated will reduce the pressure available for making the part in the cavity dense and may cause an undesirable flow pattern of the material. Lower pressure in the cavity will cause sinks or voids. A long gate can also cause premature freeze-off. This would interrupt the replenishing with material of any voids formed in the part. Voids will cause dimensional variations and poor properties.

VENTS

Lack of proper venting will cause excessive pressure for injection of the material, short shots, burn spots, poor weld lines, splay marks, and a high degree of internal stresses. Vents not only provide an effective means of displacing the air in a cavity by the plastic, but also permit the escape of gases generated in the act of heating the material. When the pressure on the compressed air in the cavity becomes high, the air temperature rises to a point where the heat will cause burn spots on the plastic. Vents have to be small enough to prevent material from entering them. A cavity can be considered sufficiently vented when the plastic is injected at the highest ram speed without any sign of burns on the part. There are many ways of venting, but in each case one must make sure that, after the vent itself is 0.25 to 0.5 in. long as measured from the inside of the cavity toward the outer edge of cavity block, the remainder of the vents are increased in depth to about 0.010 to 0.015 in. in order to minimize the danger of clogging. Are too many vents dangerous? Yes, if the clamp pressure on the nonvented area of the parting line of the cavity is great enough to cause the metal of cavity to flow or crack.

For steels with a 180,000 psi or more tensile strength, the stress level at the unvented part of the circumference should not exceed 10,000 psi. For materials with lower strengths, the stress should be proportionately decreased.

Aside from the venting of the cavity at the parting line, there are places in which the release of air and gases can be aided (1) by incorporating vents at the extreme "out" positions of the runner system as well as (2) by providing clearance around ejection pins. Of necessity, this clearance can only be large enough to prevent flashing around the pins. The appropriate clearances are indicated on the material processing sheet. In some cases, this clearance could be a substitute for the practice of placing flats on the ejector pins. The venting of runners should be to the full width of the runner and to the depth indicated for each material. The length of the vent should be about 0.25 in., and the remainder should be relieved by a 0.020-in. depth.

A word of caution is in order. When venting parts, the minutest flash may be objectionable—such as may occur with gears. While the depth of venting specified for each material is obtained after extensive testing by suppliers of raw materials, one must remember, in addition to the measured depth, to consider the peaks and valleys from the surface roughness of machining. This roughness measurement plus the "micrometer depth" should be considered as the value indicated in the tabulation. In the case of gears and similar parts, it may be advisable to adopt the following procedure for venting: (1) vent the runner system thoroughly, (2) vent all ejector pins as indicated on "material processing," and (3) water-blast mating surfaces at parting line with 200-grit silicon carbide abrasive.

UNDERCUTS

Functional undercuts are usually either to keep a part in position during ejection or to form a lip to obtain a snug fit of two plastic components. Functional undercuts can be provided in molds such as those (1) shown in Figs. 9–4 and 9–5, (2) with collapsible cores, or (3) in which top and bottom of the mold have protrusions that, when separated, form an undercut in the piece. Many threaded caps are being stripped of the mold without any ill effects on their performance. A close examination of existing thermoplastic closure under proper magnification would reveal that the conditions of the core are such that stripping becomes a successful operation.

The guideline for a successful undercut is that the mold halves be apart so that movement can take place to free the part from a groove. To accomplish this the following conditions should be observed: the undercut should be less than one-third of the material thickness; the product should be ejected when the material is hot enough to permit it to stretch or collapse during ejection; and finally, the steel edge that forms the groove should have a radius against the moving of the part to prevent shearing. These conditions will work in many of the materials that have sufficient elongation and springback after ejection.

MELT RHEOLOGY

Rheology deals with deformation of matter. In the case of polymeric material, when applied to manufacturing, rheology is concerned with one phase of deformation, namely, flow. Of interest to plastic processing is the flow through cylindrical or rectangular ducts. Specifically, we will find out what are the appropriate flow conditions for plastic materials and how these flow conditions contribute to the conversion of the inherent material properties into a finished part. Moreover, we are concerned that the molecular configuration that has been built into the plastic as a result of polymerization not be disturbed—that is, degraded. The flow conditions in a molded product are determined by runners, gates, sprues, cavity, etc.—in short, by every section through which the material has to pass in order to make a product.

Theoretical discussions follow regarding how many cubic inches per second can pass through an opening of definite dimensions when the material is at a certain viscosity and under a certain pressure drop. As these discussions progress, the conclusion might be drawn that a basis is provided for designers to calculate the correct openings for the flow passages. While it is possible to make necessary calculations by making certain assumptions and introducing corrective factors, it is not considered necessary nor practical to use such an approach. The needed information is given in the material data sheet, which was evolved and verified with tests by the suppliers of raw material. However, it is believed that the appreciation of theory would emphasize the need to pay close attention to the data sheet. Theoretical considerations would also discourage the application of passage-dimensions from one material to that of another. For example, successful gates and runners for nylon cannot be favorably considered for polycarbonate.

Plastic materials are called *viscoelastic* in flow characteristic. If passages are not correctly proportioned for these materials, they may cause (1) poorly filled parts, (2) poor surface finish, (3) distortion, (4) voids and bubbles, (5) brittleness, (6) poor welds, (7) poor dimensions, and (8) varying shrinkages. Although it is true that these defects may be caused by other factors, it is important to point out that this one potential source of defects–improperly dimensioned channels–is impressive and should be of major concern to the moldmaker and molder.

Theoretical Considerations

The laws of flow for viscous or laminar fluids, also known as those that follow the Newtonian flow, have been formulated. Plastic materials do not fully conform to the laws of Newtonian flow, so the established formulas have to be modified to make them applicable to plastics.

To understand how plastic materials flow, let us examine a simple rheometer (Fig. 7–25). In practice, a rheometer is used for measuring the rate of flow under

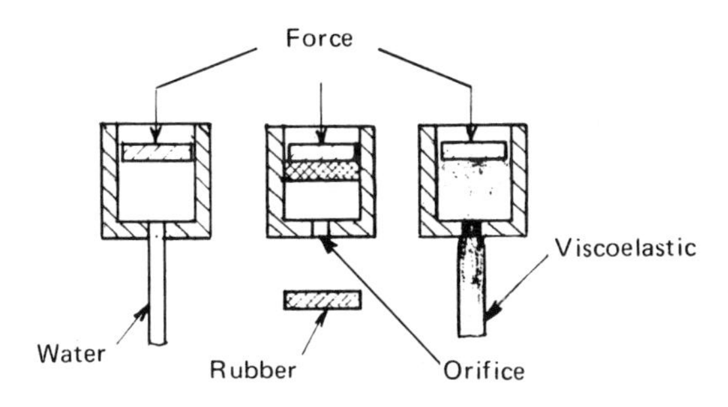

Fig. 7–25. Rheometer.

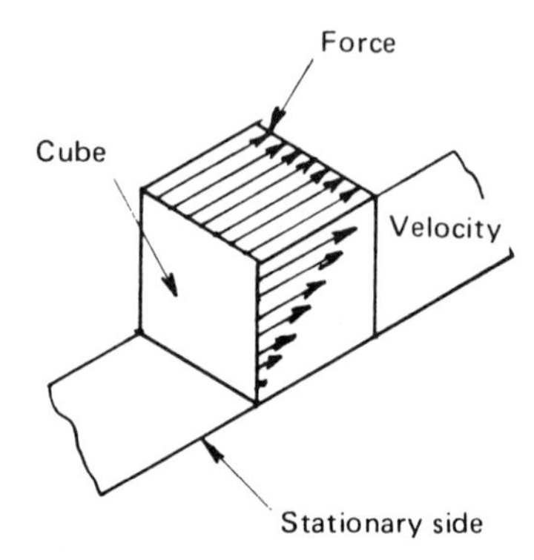

Fig. 7–26. Graphic concept of viscosity.

conditions of changing pressures and temperatures, and establishes melt index information for polymer suppliers.

Referring to Fig. 7–25, we find that if a fluid such as water is forced through an orifice, the stream coming through is of the same size as the opening. If a block of vulcanized rubber is pressured through the same rheometer, the material that comes out assumes its original shape. This is due to the full springback ability of the rubber, i.e., memory. If a molten plastic is forced through the same rheometer, the flow on the outside of the orifice is larger in diameter than at the exit opening. This behavior, known as *viscoelastic,* means that the molten plastic acts partially like rubber and partially like a fluid.

If all conditions on the rheometer remain the same, the amount or volume of molten plastic that comes through in 1 sec will depend on the internal resistance to flow of the liquid. This resistance is known as the *viscosity.* Viscosity is a property that must be controlled so that the flow along the passages will be such as to protect the polymer properties from degradation.

Viscosity can be explained with the aid of Newton's hypothesis. Newton conceptualized viscous materials as consisting of minute parallel layers known as laminar layers. (See Fig. 7–26.)

Graphic Presentation and Description of Viscosity

Viscosity can be visualized with the aid of the following: When a force of 1 lb is applied to a square inch of the top layer of a cube, shearing will take place, resulting in the movement of 1 in. forward by this layer. The velocity or shear rate will be 1 in./sec with respect to the wall of the duct.

When a viscous material flows through a tube, the layer adjacent to the wall sticks to it and does not move (Fig. 7–27). The next layer moves and slides over the wall-adhering layer. The remaining layers move with respect to each other

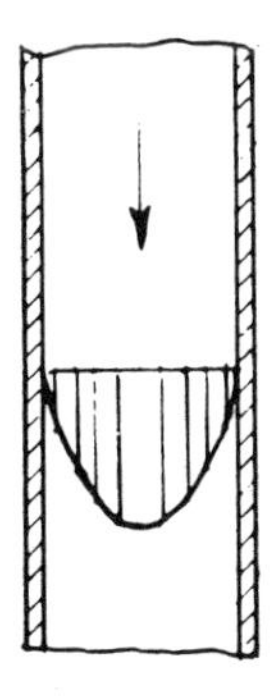

Fig. 7–27. Graphic concept of laminar flow.

at an increasing rate as the distance from wall to center increases. This imaginary layer movement is known as *shearing.* This type of shearing takes place in the injection or extrusion cylinder, and in the nozzle, sprue, runners and gates, and cavities. The unit pressure on the fluid that is subjected to the action of shearing is the *shear stress;* the speed of the movement of layers with respect to each other is the *shear rate.* Now, we can state Newton's hypothesis in equation form:

$$\text{Viscosity} = \frac{\text{Shear stress}}{\text{Shear rate}}$$

That is, viscosity is the proportionality factor between the shearing stress of adjacent layers of a fluid and the rate of shear within the fluid. The viscosity relationship can be expressed mathematically.

Substituting for shear stress, the shear force exerted over the area that resists the flow, we have shear stress

$$= \frac{\text{Unit pressure} \times \text{Area}}{\text{Area resisting flow}} = \frac{P \times R^2\pi}{2\pi R \times L} = \frac{PR}{2L}$$

if Q = material (in.^{3}/sec)

R = radius of a cylinder through which it flows, in.

L = length of cylinder, in.

P = pressure, psi

μ = viscosity, lb. sec/in.^{2} (reyn)

h = height of a rectangular duct, in.

w = width of a rectangular duct, in.; mostly = 2 h

The inside circumference of a cylinder ($2\pi R$) and its length (L) is the area that offers resistance to flow. The shear rate is

$$\text{Shear rate} = \frac{4Q}{\pi R^3} \text{ for cylindrical shapes}$$

$$= \frac{6Q}{wh^2} \text{ for rectangular shapes}$$

$$\text{The viscosity, } \mu = \frac{PR}{2L} : \frac{4Q}{\pi R^3} = \frac{\pi P R^4}{8QL}$$

$$Q = \frac{\pi P R^4}{8 \mu L} \text{ for cylindrical shapes}$$

$$Q = \frac{P h^4}{9 \mu L} \text{ with } w = 2h \text{ for rectangular shapes.}$$

Thus, for fluids behaving as Newtonian (i.e., where the stress change has a corresponding shear rate), the amount of material Q that will flow per second will depend upon the unit pressure P and the geometry of the channel R, and will decrease with increase of length of channel L and increase of viscosity. The viscoelastic materials do not behave in this predictable manner. Their viscosity changes with change of shear stress, and the relationship between shear stress and shear rate does not follow a law as is the case with viscous flow. Therefore, if a value for viscosity is to be established, a specific temperature, a definite shear stress, and a shear rate at that particular shear stress must be determined.

Furthermore, various plastic materials react differently to applied shear stresses due to the configuration of molecules from which they are constituted. The term applied to such differences in reaction is "shear-rate sensitive" or "shear-rate insensitive."

The shear-rate sensitive materials respond to the pressure by having their molecules readily shifted and aligned with the direction of flow. Their viscosity decreases, and, as a result, flow is easier. They are normally of the lower molecular weight, which is the shorter molecular chain; usually possess lower properties; have a different rate of shrinkage in the direction of flow and a different rate to the direction of flow; and have a tendency to orient with the direction of flow. Thus, differences are created in strength between a test of the material in the direction of flow and a test at 90° to that direction. On the other hand, the shear-rate insensitive materials consist of long chain molecules running from hundreds of feet to miles long. These long molecules are so intertwined that the application of shear stresses only causes greater entanglement. The net result is that the viscosity does not change with increased application of force. In addition, a danger of affecting properties exists because the entanglement formed in the polymerization may be disturbed.

In order to better understand what occurs during the flow of viscoelastic shear-rate sensitive and insensitive materials, an analogy is appropriate. Let us imagine a cable made of fine wire, having a very large number of strands parallel to each other. If we try to force this cable through an opening of dimensions that are the same as those of the cable, the wires near the wall of the opening will have a tendency to stick while the others move in the direction of the application of force. The moving wires will vary in speed and will be at a maximum toward the center. Once the wires close to the center go through the opening, the others follow and thus flow takes place.

This arrangement corresponds to the shear-rate sensitive polymers. For shear-rate insensitive polymers, let us visualize a cable made of fine wire similar to the one just described, except that the wires are twisted into a sharp spiral. Again, we want to force the cable through an opening of the same size as the cable. When the force is applied, the wires get more twisted and entangled than when they were originally formed. A continued increase of force damages or even breaks some strands without improving the flowability of the cable. This is similar to the condition of shear-rate insensitive polymers, where forcing a plastic through an opening of incorrect size will cause degradation of properties.

Getting back to the practical side of rheology, it can be readily recognized from these descriptions that putting down a value for the viscosity of a material is not a simple matter, and it is wise to let the researchers of plastic materials convert their findings into practical information usable by the molder.

Taking a closer look at the flow formula of viscous fluids, which includes oils, we do not find a factor pertaining to the roughness of the cylinder through which the material is flowing. Actual experience, however, proves that the inner surface of a cylinder has an effect on the pressure drop and indirectly on the flow. Research pioneers in the field of hydraulic fluids have developed an empirical formula for the pressure drop, which shows the relationship to all the factors involved:

$$P_d = f \frac{L}{d^4} q^2 \times 0.0123$$

where

P_d = pressure loss, psi

f = friction factor, dimensionless

d = cylinder inside diameter, in.

L = length of cylinder, ft

q = flow, gpm

For drawn smooth steel tubing, f is approximately 1/15 of that of wrought tubing. This indicates the value of smooth passages in relation to pressure drop for plastic material flow.

The question arises: how do these theoretical concepts and formulas relate to manufacture of products? The flow law,

$$Q = \frac{\pi P R^4}{8 \mu L} \text{ for round ducts}$$

can be said to apply to viscoelastic materials except that the viscosity is not constant as with the Newtonian behavior. In the molding operation, however, there is a relatively simple way of determining the viscosity, namely, by delivering a certain volume of material to the mold within a specified time. As long as this is repeated from cycle to cycle, we know that the material is at a constant viscosity. Some process controls for viscosity are based on the principle of measuring the time for the distance of a screw travel to a very precise value and signaling an order for change if time is not within prescribed limits. We therefore measure the viscosity or ability to flow through certain established openings at a prearranged pressure.

Q, the volume of material in in.^{3}/sec, is governed by an oil-flow variable control valve that determines the gpm of oil delivered to the hydraulic cylinders, which actuate the movement of material through the heating chamber. This movement may take from 1 to 10 sec depending on the setting of the control valve. The recommended speed of material movement is, in most cases, the maximum the machine will produce, i.e., in the range of 1 to 2 sec. This means that the whole cavity is filled almost simultaneously. For most materials, this is desirable since there would be no variation in solidification time from one area of the cavity to another. A uniform material temperature throughout the cavity will minimize stresses and provide favorable conditions for good weld-line strength.

The next element of concern in the equation is the gate radius or gate depth. Since this is the smallest passage point and the viscosity is at the desirable working condition, the need may arise for a larger opening than the one selected from the material processing sheet. If a correction in size is considered, we must recognize that for a round opening, a 10% increase in diameter or, for a rectangular gate, a 10% increase in depth, will provide the ability to handle 47% more in.^{3}/sec of the flowing material. A gate that is too small can be detected by blemishes around the gate area and surface imperfections on the part. These defects, when checked in the troubleshooting information of the material processing data, will normally call for gate increase.

The next item in the equation that is of interest is P, the psi pressure that provides the energy for shearing action. It represents the work input from screw rotation, pressure in moving the material forward, back pressure against which the screw rotates, and all the pressure drops encountered in the molding of parts. In selecting these pressure components, care must be exerted that the sum of heat from the pressures plus the plasticating heat of the cylinder provides all the needed energy for the desired viscosity. Too high a work input may show a higher temperature on the pyrometer than the one for which the setting was made. Special attention has to be given to shear-rate insensitive materials, so that the work input and heat derived from it are a small percentage of the total plasticating heat required. This means keeping the back pressure low and screw rotation at a slow rpm.

The final factor in the equation is the length of the passages through which flow takes place. When we speak of length, we have in mind the effective length and not just the physical dimensions of each length component. Each time the flow makes a turn, the effective length becomes greater depending on the shape of the turn. From the empirical formula on pressure drops, we find that it is affected by the surface roughness of the passage and it thereby also contributes to an increase of the effective length. Figure 7-5 shows how each shape of a bend changes the effective length. A surface roughness of 50 rms would be most favorable for flow and is the finish indicated for sprues, runners, gates, and cavities.

These discussions point out that the theory of rheology when translated into molding parameters and features incorporated in moldmaking will lead to parts with desired properties. The molding parameters and mold features are detailed in the material processing data. Familiarity with and close attention to these data will lead to satisfying performances.

8 Details of Molds

A number of mold components are made by mold-base and molding-aids manufacturers as off-the-shelf items. These items perform a variety of functions and, because of frequent usage, they have become "standards." In this category we find such components as heated sprue bushings, collapsible cores for molding undercuts and female threads, early ejector return systems, and cam slide retainers. These components continue to grow in number and so does their application. It is wise to carefully review those readily available components, their intended performance, and their cost. They can be evaluated against those in-house designed parts for comparable performance and price.

Certain details in mold design require attention in order to insure satisfactory mold performance.

EJECTION PINS

To be most effective, ejection pins should act against walls, ribs, bosses, or any area of a part that will present the most rigidity. When pins are made to act against large open areas, they tend to cause pulling of such areas; this, in turn, brings about tightening of the sides against the core, thus making ejection more difficult. In addition, an excessive impression of the knockout pin will also occur.

Ejection pins should be of the largest size the part will allow, but hardly ever less than 1/16 in. diam. Smaller pins embed themselves deeper into the plastic and sometimes even pierce the surface that they are pressing on while ejecting the part. Occasionally, buckling of pins occurs when unusual part sticking takes place.

We can determine how a 1/16-in.-diam pin will perform when the molded parts require the full stripping force to remove the parts. It is assumed that each mold pin will absorb its share of stripping force, and this is predicated on the thought that the pins are strategically placed wherever sticking tendency will exist.

Let us select a mold with 40 K.O. pins of these sizes: 10 of 1/4 in. diam, 20 of 3/16 in. diam, 6 of 1/8 in. diam, and 4 of 1/16 in. diam. The mold will operate, for example, in a 200-ton, 14-oz press with a stripping force of 10 ton or 20,000 lb.

$$P \text{ (force)} = A \text{ (area)} \times S \text{ (stress)}$$

$$\text{or } \frac{P}{A} = \text{Stress}$$

The area is composed of the areas of all the pins; thus, we will have 10 X 0.0491, 20 X 0.0276, 6 X 0.0123, and 4 X 0.0031,

or

$$\begin{array}{l} 0.491 \\ 0.552 \\ 0.074 \\ 0.012 \\ \hline 1.129 \text{ in.}^2 \text{ total} \end{array}$$

$$\frac{P}{A} = \frac{20{,}000}{1.129} = 17{,}750 \text{ psi}$$

The force exerted on the 1/16 in.-diam pin will be

$$\begin{aligned} P_1 &= \text{Area of 1/16-in. pin} \times \text{Stress} \\ &= 0.0031 \times 17{,}750 = 55 \text{ lb} \end{aligned}$$

The compressive force of 55 lb is exerted on a slender item. In this case, failure will occur because of buckling rather than compression. The critical force that will cause buckling can be calculated according to the *Machinery's Handbook* ("Euler Formula for Columns"):

$$P_{cr} = \frac{S_y A r^2}{Q}$$

P_{cr} = critical load that would cause buckling

S_y = yield strength of the pin in psi = 250,000

A = area of pin in in.^{2} = 0.003

r = radius of gyration $= \dfrac{0.0625}{4} = 0.016 = \left(\dfrac{d}{4}\right)$

$$Q = \frac{S_y l^2}{n\pi^2 E}$$

l = length of pin in inches = 6

n = 2 for the column fixed at one end but guided and free to float at opposite end. An ejection pin is fixed in the stripper plate and guided, free to float in the cavity.

E = modulus of elasticity = 30×10^6 psi

Substituting the values in the formula we have,

$$P_{cr} = \frac{250{,}000 \times 0.0031 \times 0.016^2}{Q = 0.016} = 11.65 \text{ lb}$$

$$Q = \frac{250{,}000 \times 6^2}{2 \times 9.8 \times 30 \times 10^6} = 0.016$$

If the full force of stripping is required to remove the parts, the 1/16-in. pin will buckle under the load of 55 lb because its resistance to buckling is about 11 lb.

We can figure how long the 1/16 pin can be in order to withstand the 55-lb force. The length appears in the Q formula, and we therefore have to find out what Q will be when the critical force is 55 lb.

$$55 = \frac{250{,}000 \times 0.0031 \times 0.016^2}{Q}$$

$$Q = \frac{250{,}000 \times 0.0031 \times 0.016^2}{55} = 0.00338$$

$$0.00338 = \frac{250{,}000 \times 1^2}{2 \times 9.8 \times 30 \times 10^6} \quad \text{or}$$

$$1^2 = \frac{0.00338 \times 2 \times 9.8 \times 30 \times 10^6}{250{,}000} = 7.95$$

$$1 = 2.82 \text{ in.}$$

This calculation shows that the 1/16 in. diam can only be about 2-3/4 in. long, and the remaining 3-1/4-in. has to be larger in diameter, about 1/4 in. to resist buckling. This would become a shoulder pin. On molds with ejection strokes of several inches (e.g., over 4 in.), even larger diameter pins of considerable length should be checked out along with the method outlined.

Another case where small ejection pins may be involved is a thin-walled box or cover. In this application, it is possible to use larger-size pins, letting their segment engage the wall of the box. The segment should have a large enough area so that embedding in the wall will not be perceptible. An area equivalent to 1/8 in. diam should accomplish the result. This can be determined by a layout or calculation. If the wall thickness of the part is 0.065 in., we can figure the desirable segment area and corresponding larger pin size.

From tables in the *Machinery's Handbook* ("Segment, Circular for 1″ Radius"), we can establish the segment areas that correspond to pins with radii of 0.125, 0.156, and 0.1875. The value h in the table, when multiplied by the pin radius, will give the actual segment height of 0.065. This gives an equation:

$$\text{Pin radius} \times \text{Segment height} = 0.065$$

or

$$\text{Segment height} = \frac{0.065}{\text{Pin radius}}$$

$$0.125 \times h = 0.065; 0.156 \times h = 0.065; 0.1875 \times h = 0.065$$

$$h = \frac{0.065}{0.125} = 0.52; h = \frac{0.065}{0.156} = 0.417; h = \frac{0.065}{0.1875} = 0.346$$

In the tables, there is a corresponding "area of segment A" value to these "segment heights - h" in the tables. They are

$$A_1 = 0.6406 \qquad A_2 = 0.467 \qquad A_3 = 0.3701$$

When these areas are multiplied by the square of pin radii, we obtain the segment area of the 0.065-in. wall.

$$S_1 = 0.6406 \times 0.125^2 = 0.010 \text{ in.}^2 \text{ or } 3\text{-}1/2 \times \text{Area of } 1/16\text{-in.-diam pin}$$

$$S_2 = 0.467 \times 0.156^2 = 0.0113 \text{ in.}^2 \text{ or } 3\text{-}5/8 \times \text{Area of } 1/16\text{-in.-diam pin}$$

$$S_3 = 0.370 \times 0.1875^2 = 0.013 \text{ in.}^2 \text{ or } 4\text{-}1/4 \times \text{Area of } 1/16\text{-in.-diam pin}$$

From these figures, it appears that 1/4-in. diam pins would satisfy the needs for a hard material like polycarbonate, whereas a low-density polyethylene would call for a 3/8-in.-diam pin. Should it become necessary to go to 1/2-in.-diam pin in this type of arrangement, it will still be more economical than a stripper plate.

Commercial ejection pins are purchased in standard lengths nearest to the length required for a mold. This necessitates cutting the pins to the required length. By so doing, a razor edge is created, which may tend to shear the plastic that is subjected to ejection. The cutoff edge should be stoned lightly (not to exceed a radius of 0.005) to eliminate the ill effects from a razor-sharp edge. Care should be exerted not to break the corner to a point that it will allow plastic to flow in.

Most engineering specifications call for the ejection marking to be recessed by 0.002 to 0.005 in. The stoned radius of 0.005 in. would also aid in releasing the part from the pins.

Ejector pins are manufactured to a tolerance of +0.000 and -0.001 in. The reamers for the holes in which the pins move have a tolerance of +0.0001 to +0.0004 in. This would conceivably produce a maximum clearance of +0.0014 in. between pin and hole. This clearance would not cause the material to flow. There are occasions when the clearance between pin and hole is used for venting air and gases from areas where entrapment of these would produce a porous or otherwise defective surface. If normal clearance between pin and its opening is maintained, venting means are provided by placing a flat, 0.001 in. deep, on the diameter of the pin in two, three, or four places of the circumference.

In the majority of the molds, the ejector pins do a satisfactory job of removing parts. There are some shapes with relatively thin walls in which stripping can be accomplished by engaging the whole circumference of the part, causing it to move from its core. This is done by a so-called stripper plate, which has a limited movement over the leader pins.

CAVITIES AND CORES

On multiple-cavity molds, each cavity and core should be identified by its own number on the parts and placed so that there is no possible interference in end use. This is done in order to tie down a potential problem to a specific cavity. In addition, the outside of cavity and core should be marked with the grade of steel and hardness. This is important for cases of repair or replacement. The designations should be according to SAE numbers and separated by a dash from the hardness number. Thus, H13–54 would indicate a grade H-13 tool steel heat-treated to 54 RC hardness. Furthermore, the surface finish in rms, surface treatments, if any, and any other surface requirement should be indicated on cavity specifications.

The finish in cavities should be such that it will facilitate stress-free removal of parts. There should be directional finish lines perpendicular to the parting line. Finish in the mold should be designated in microinch surface roughness numbers, which may vary from area to area depending on the need of a particular surface. It is to be remembered that too rough a surface will present ejection problems.

When cavities and cores are mounted in plates with a machined opening in the picture-frame style, they should be provided with shoulders so that they will not move with respect to the plates in any direction. In blind pockets, the cavities are retained by means of cap bolts. In both styles of mounting, the cavities and cores are locked in position by means of keys or dowel pins so that they are in proper registry with each other at all times.

MOLDING INSERTS

Some products incorporate metal inserts that are molded in the plastic part. Whenever other means of placing the metal insert into the product are feasible, they should be seriously considered. Metal inserts can be satisfactorily attached by ultrasonic insertion or by molding the space for the insert in the product and spinning a retaining wall over the insert, etc.

Molded-in inserts are a potential source of mold damage and in many cases contribute to an irregular cycle. If they must be used, they should be placed on the stationary half of the mold to avoid loosening during ram movement. They should be supported so that there is no chance of hydraulic pressure from the plastic moving them out of position. If at all possible, a holding pin from the opposite half of the mold should be provided to hold the insert in position during the molding operation. A heavy spring loading for the holding pin should be arranged in order to compensate for variation in insert.

Some products are so shaped that the only economical way they can be made is by having a mold insert that has to be removed from the part outside of the press. These inserts are usually placed on the ejection side so that they can be removed with the parts. Such inserts should have a detent that will not permit loosening and yet be ejectable. Its socket arrangement should allow for only one-way placement. In multicavity molds, the inserts should be perfectly interchangeable to avoid possible damage.

MOVABLE MOLD PLATES

Molds that are constructed with movable plates, such as three-plate molds or stripper plates, govern the distance that plates travel by means of shoulder bolts or similar arrangements. The movements of the plates have to be parallel, or else binding takes place between the leader pins and their bushings. To maintain parallelism, the shoulder pins should be exactly the same length from under the head to the shoulder; the counterbores for the head should also be of the exact same depth. To prevent easy distortion, the moving plates should preferably be of a tougher steel. The lack of parallelism in moving plates is a frequent cause for improper functioning of a mold, and any factor that can contribute to it should be guarded against.

PROTRUDING CORES

Slender cores that protrude deep into a cavity present not only a temperature control problem, but also a danger of being bent under the hydraulic pressure of the flowing plastic. Such products as medicinal vials and pen-and-pencil barrels are typical examples. The flow of the plastic must hit the core at the top and envelop it simultaneously all around so that it will be properly centered. Theoretically, this should be easily accomplished, but in practice it has been found that there is enough variation in gate location, etc., to make it difficult to attain. It is more practical to assure concentricity of the core by making four small gates evenly placed on the circumference of the part. With the four flows hitting the core at the same time, no perceptible movement in the core takes place.

If at the end of the part an opening exists through which a pilot could be incorporated, which would enter into a suitable hole in the opposite half of the mold, this would provide an additional means of core support and aid in maintaining its straightness. When long cores are piloted or even touch any section of the cavity, the leader pins should engage their bushings before contact is made between core and cavity. This will insure that the job of aligning the halves is performed by the leader pins and not by frail cores.

STRESS LEVEL IN STEEL

When we examine the great variety and complexity of plastic parts, we realize that the molds in which they are produced are even more complex. There are a great many factors to keep in mind when a design layout for mold is made, but none is more important than maintaining a low stress level in the steel of all the components of cavity and core. Highly stressed parts mean short tool life. The cost of making a mold can run into tens of thousands of dollars. If a mold fails before producing the expected quantity, the designer's must examine every phase of moldmaking to make certain that the design is conducive to low stress level in fabrication and to specify stress relieving after every major step of moldmaking. The main problems are outlined in the following paragraphs.

Heavy and high-speed cuts during metal removal, severe grinding action, and electric discharge machining all produce stresses to a different degree in various tool steels. Stress relieving will minimize the danger of failure.

Molds that are built for long life and high activity are heat-treated either initially or whenever the intermediate hardness of the cavities begins to unfavorably affect the quality of the product. From a heat-treatment standpoint, the tool designer should be on the lookout for the following:

1. The parts should be so shaped that they will heat and cool as uniformly as possible. A part that may heat so that a temperature difference exists between two points, will produce a harmful strain when quenched.

2. A balanced section will heat and cool more uniformly, thus guaranteeing a much lower stress level (Fig. 8-1).

3. Addition of holes that would reduce the mass of metal in one area to offset the lower mass in an adjacent point (Fig. 8-2).

4. Sharp angles and corners are a most common error, which, with a little effort, could be minimized. Sharp corners and angles are points of high stress concentration (Fig. 8-3A).

When a rectangular insert is being made for the cavity, the sharp corner in the

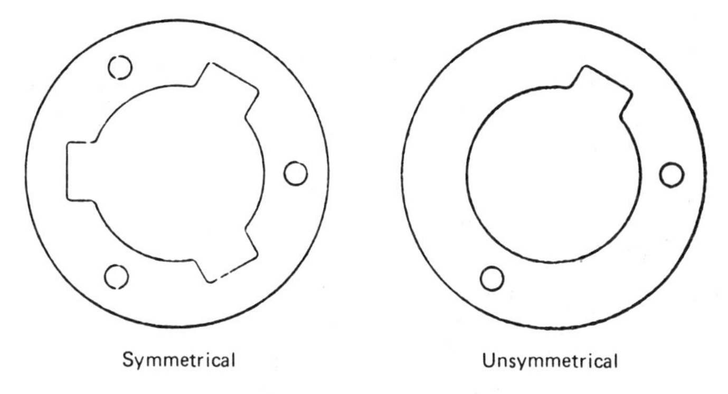

Fig. 8-1. Symmetrical design for uniform heating and cooling.

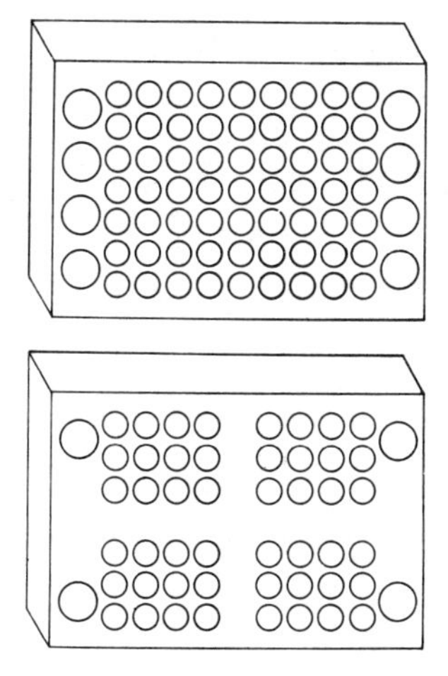

Fig. 8-2. Additional holes in top block lead to part with low heat-treating stresses.

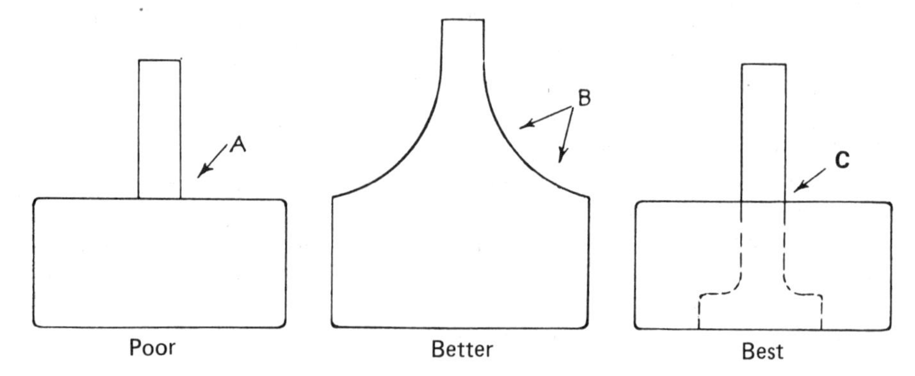

Fig. 8-3. Stress concentration in sharp corners.

plastic will most likely tolerate a radius of 0.020 in. If this is not permissible, the insert portion that is even with the cavity can be made larger and can have a generous radius; the portion that is molding can be of whatever shape is required (Fig. 8-4).

5. The thin section will cool faster than the thick one during quenching and will set up stresses. A larger radius or even taper in the transition area will minimize stresses (Fig. 8-3B).

6. For whatever purpose they are intended, blind holes should be eliminated. The through-hole makes for greater uniformity in cooling and eliminates the stress concentration from the sharp corner at the bottom of the hole. Junction of holes such as may be planned for fluid circulation should be avoided in favor of drilled-through holes since the intersection of holes will act as a stress raiser.

The designer should be aware that the best choice of material coupled with the best effort of the heat treater cannot overcome faulty design. When layouts for the cavity and core are made, the outline of components should be presented to a heat treater for a recommendation of design modification that will lead to parts with low-level stresses. It is a matter that deserves serious consideration. An order to the heat treater should specify: "To be stress-relieved if heat-treating steps will not accomplish it."

DRILLING HOLES FOR COOLANT

Deep hole drilling (over 12 in. in length) is being carried out by specialty shops that are equipped with appropriate facilities to do the job in an economical and technically correct manner. Drilling holes in cavity blocks is normally done by the moldmaker, and in many cases the result is short of being satisfactory. One can experience runout of the drill with the consequence that the distance of

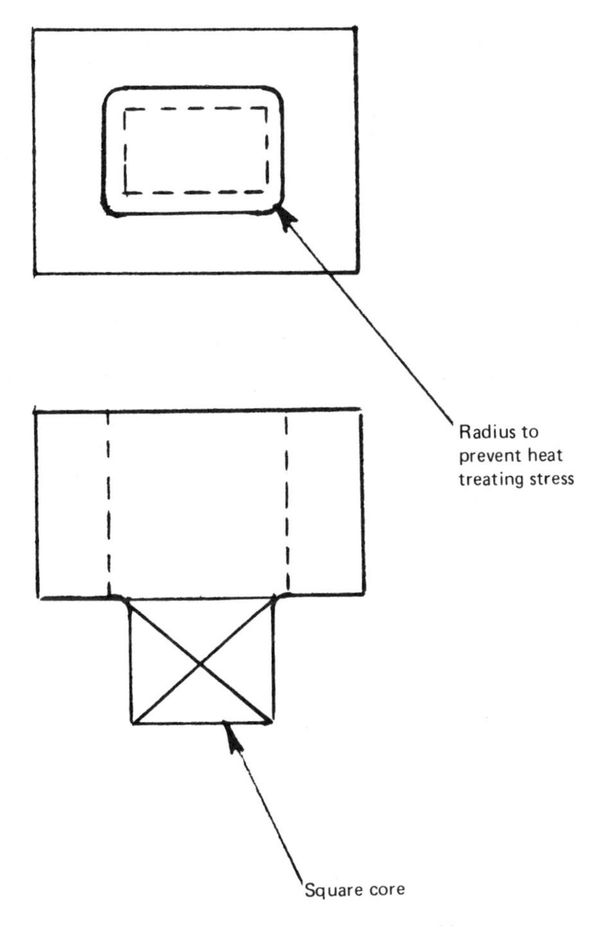

Fig. 8–4. Two-piece construction to avoid stresses.

hole from cavity varies. In some cases, the hole is so close to the cavity that, after a relatively short operation, a crack develops in a cavity because there is insufficient wall to resist repeated flexing and thermal shock. When special facilities for hole drilling are lacking, a drill grind should aid in accomplishing the desired result.

Figure 8-5 shows a drill ground with a notch in each lip, which are staggered with respect to one other. When power-fed at recommended feeds and speeds, this type of drill has produced straight holes with a hole accuracy greater than that generated by drills without notches. The advantages of this grind are that it permits free and efficient circulation of coolant around the cutting edges, and produces a continuous chip flowing up the flutes. This chip is flowing up quite

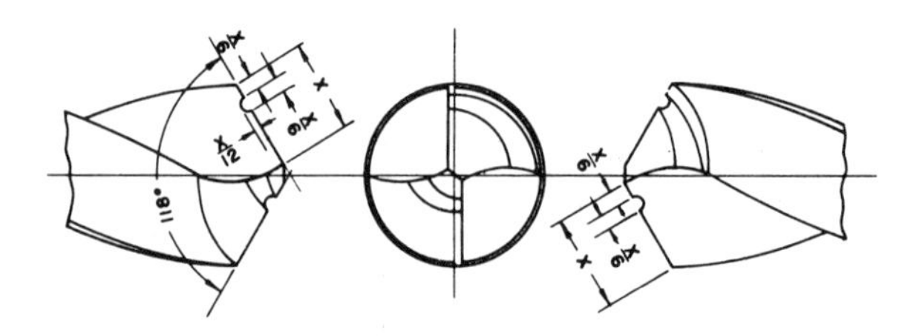

Fig. 8-5. Drill grind for straighter holes.

rapidly and has to be cut frequently to prevent entanglements and possible hazard. This grind eliminates the need for backing out of the holes for chip clearance. The important thing is to carry out the grind so that it will be sharp at the notches; the width of the notch should increase toward the back as shown in the figure and should have clearance at the heel so that it will cut freely.

The drill itself should be correctly ground, in line with the best recommended practices. It should have correct heel clearance; the length and angle of both cutting edges should be the same; the position of the center of the drill should coincide with the axis of the drill; and the web thickness at the point should be about half of that in the body of the drill.

A trial run in a scrap piece of a steel should provide the necessary confidence for proceeding with the drilling in a cavity or plate. The feed and speed for alloy steel recommended by drill manufacturers are: for 7/16 in. diam, 0.009 in. feed and 437 rpm, and for 37/64 in. drill diam, 0.010 in. feed and 359 rpm.

The operation of drilling, when properly carried out as far as quality of grind, feed, and speed are concerned, can prove to be a valuable yet inexpensive aid for the moldmaker and can safeguard the proper distance from cavity to circulating hole.

DIMENSIONING MOLD COMPONENTS

Any manufactured part should be dimensioned so that it will be in harmony with the equipment on which it will be produced. It should not be necessary for the machine operator to add, subtract, or convert the drawing specification into figures needed to manipulate the machine tool.

Once the general outline of the mold design has been made, the designer should invite shop supervision to help decide on the method of fabrication that will be used for the best overall economy.

With the method of fabrication for each component agreed upon, the mold component details can be drawn, and the dimensions as well as tolerances can be specified in line with the selected method of fabrication. On cavities and cores, a vital starting point—with product performance in mind—should be jointly established by product and mold designers from which all dimensions and pertinent notes radiate.

The next step is to determine what parts may later need replacement. Such parts and the spaces into which they fit should be dimensioned so that interchangeability is assured. If fitting in some area is unavoidable, then the dimensions of fit related to a specific cavity and core should be recorded by the toolmaker and incorporated as a permanent record of the tool drawing. Shrinkage dimensions should be included in the detail and carefully analyzed as indicated in Chapter 7.

Cores that shut off against solid surfaces should be dimensioned for an interference of 0.001 and so designed that, if necessary, additional interference can be added by shimming, etc. After a mold is in operation for a number of hours, a certain "settling" takes place, which may alter some interference dimension. On the other hand, too much interference plus possible expansion due to heat can cause stressing and ultimate failure.

Shutoff at an irregular parting line should be dimensioned similar to the cores except that the fitting of the parting line should be arranged by an insert so that it can be readily examined by visual observation rather than by indirect judgment of bluing, etc. In most cases, there exist similar designs that could be used as a guide, provided that the mold made from such a design performs well and any modifications after delivery were properly indicated.

Leader pins should not terminate in blind holes. Such holes invariably collect plastic pieces, which interfere with correct functioning of the mold. Whenever there is a chance of plastic material flowing in between mold components, the parts involved should be dimensioned for interlocking surfaces by means of tapers or have just enough clearance so that a particular material will not enter (see material processing data, Chapter 18). "O" rings, when called for, should be dimensioned according to the supplier's engineering specifications.

Cavity and core blocks should be dimensioned for a light press fit into their plates. This means that the dimension of block and plate are the same. The transition from shoulder to body of the blocks should have a slight undercut with a radiused tool to prevent sharp-corner-stress concentration in heat-treating.

The dimensioning of relatively deep cores, when center-gated, deserves special attention. All precautions should be exercised to insure that concentricity between cavity and core are safe-guarded. If the flow of the plastic is not symmetrical, we have transverse loading, causing a shift in the core, uneven thickness, and strains on the aligning and guiding means of the mold. The result is difficult mold opening, interruption of production, and ultimately damage to the alignment of halves.

The ultimate proof of good dimensioning can be found in the answer to this question: How would you make the mold, and what information would you need to accomplish this goal?

MOLD PROTECTION

A tool that has received all the necessary attention and care from the designer and moldmaker should be handled with extreme care so that the expanded effort is fully protected. Any protruding parts should be protected against damage in transfer. The mold surfaces, especially cavities and cores, should be covered with a protective coating against surface corrosion. The coating should be easily removable before the molding operation starts. The protection of mold surface applies equally to the time after a run, when the mold is ready to be removed from the press and to be stored for the next run. In some areas where the atmosphere is highly corrosive, the mold must be protected while in the press for anticipated operation. This is especially important over a long holiday weekend of 72 hours or more. Commercial coatings are available for this purpose; before being used, however, they should be carefully evaluated for their ability to protect the area involved.

In any event, molders should realize that a very expensive tool is placed in their trust and it should be treated with care.

9
Types of Bases and Molds

The mold base performs certain important functions in the molding operation. Thus it should be selected with care and attention to such requirements for each job as strength for rigid mounting of cavities and cores; adequate provision for incorporating cooling passages; properly guided stripper plates; accurately ground component plates; ease of disassembling and self-aligning means for reassembling of components; overall parallelism of complete base; and centering means in the press. The base provides a housing for cavity and core that will withstand the forces encountered in the operation. It incorporates accurate guiding means for the alignment of molded halves. It facilitates easy and rigid mounting in the press and adaptation to the injection cylinder for receiving of the plastic. When the product is properly shaped, the base has the means for easy removal of the parts from the mold. The mold bases are composed of standard interchangeable components; this is a valuable feature whenever the need for replacements arises.

MOST COMMON TYPE OF MOLD BASE

The most common type of mold base is shown in Fig. 9-1. Mold bases come in a large variety of sizes of width times length. From all indications, there is a standard width times length available to fit almost any job and any press, beginning with a 3-1/2 X 3-1/2 and ending with a 23-3/4 X 35-1/2.

Any one length times width is available with numerous thicknesses of A and B plates from 7/8 to 5-7/8 in steps of 0.5 in. Mold bases also have a variable for the space in which the stripper plate moves. Any one of the bases may be equipped with horns to actuate cams that will form openings or shapes at a right angle to the press movement. Hydraulic or air cylinders as well as springs may also bring about the right-angle movements of cores. The cylinders must be coordinated electromechanically with the press movement to both insure proper functioning and avoid damage to the mold. This coordination simply involves consideration of following:

At what point of press opening does core withdrawal take place?
At what open distance of the mold does stripping begin?

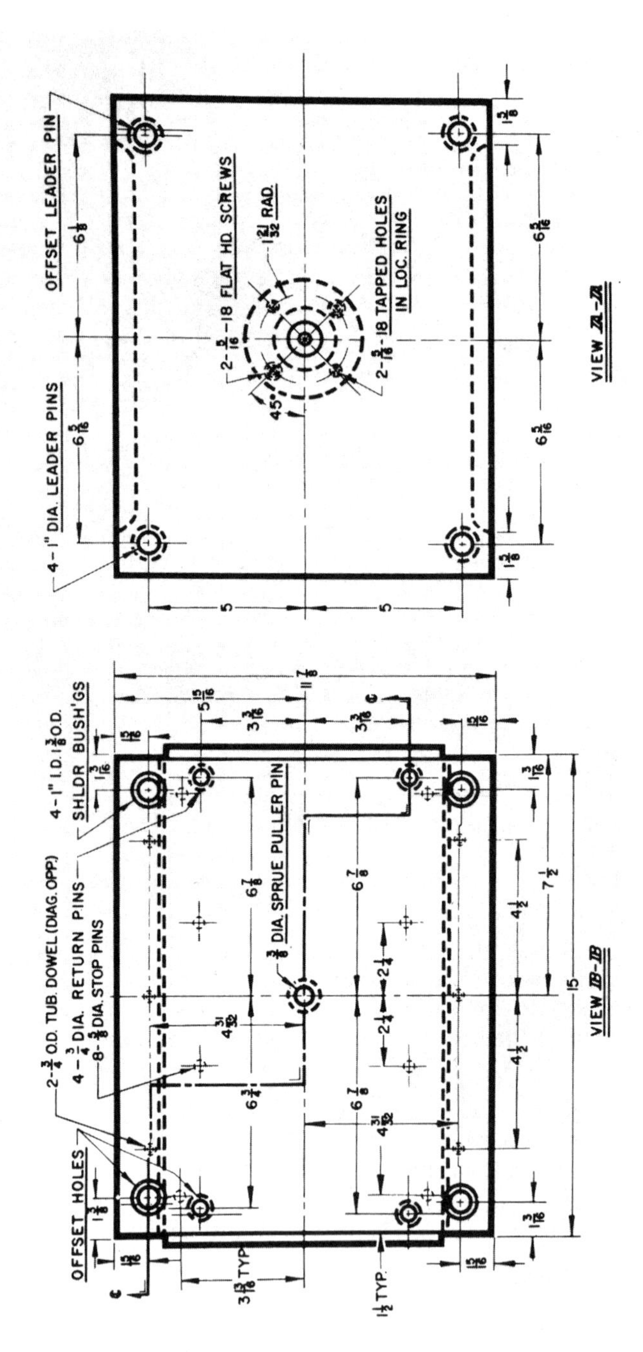

Fig. 9-1. Standard mold base. (*Courtesy of D-M-E Co.*)

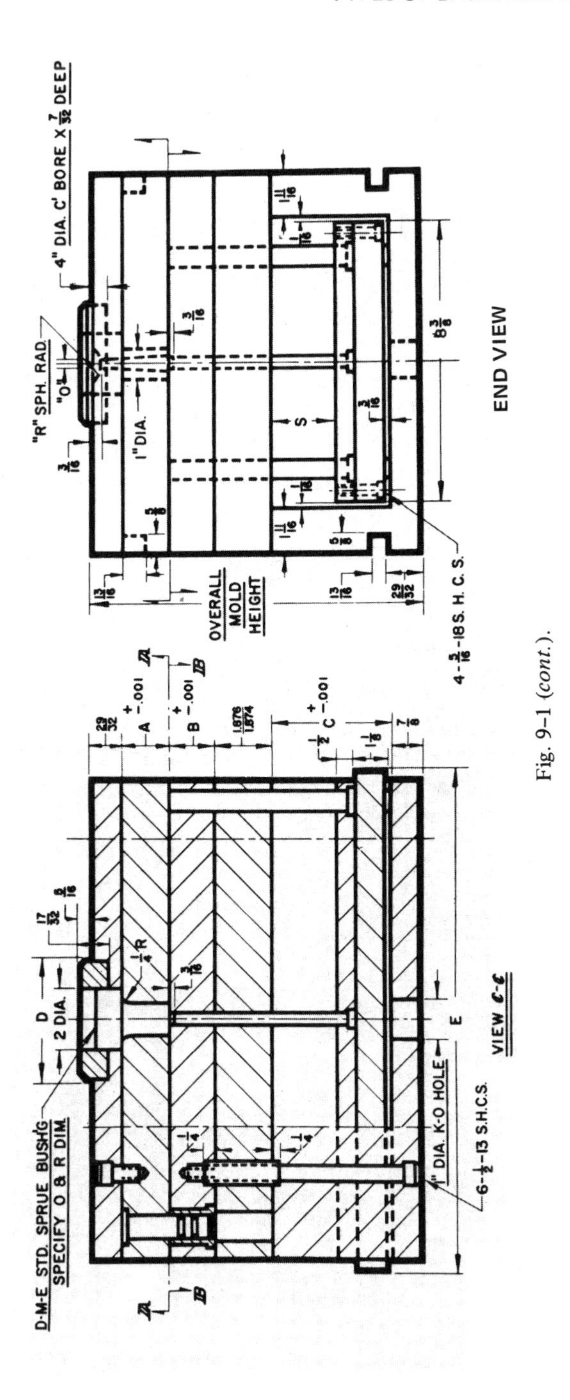
D-M-E STD. SPRUE BUSH'G
SPECIFY O & R DIM.
2 DIA.
D
VIEW C-C
1" DIA. K-O HOLE
6-1/2-13 S.H.C.S.
E
"R" SPH. RAD.
4" DIA. C' BORE X 7/32 DEEP
1" DIA.
S
OVERALL
MOLD
HEIGHT
4-5/16-18 S. H. C. S.
END VIEW

Fig. 9–1 (*cont.*).

When do the stripper pins return to zero position? When does the core engage the mold for the injection position?

When is every moving component in proper position ready for final mold closing?

Some machines are equipped with hydraulic stripping cylinders as well as with hydraulic core pulling circuits that make it easy to synchronize all the movements for safe mold operation. Whenever these features are absent, it may be necessary to provide heavy springs over the free portion of return pins. This brings about early return of the stripper plate, and with limit switches strategically placed, they are interlocked with press movement so that they will bring about the desired action.

Figure 9-2 shows a cam arrangement actuated by a horn. The angle of the

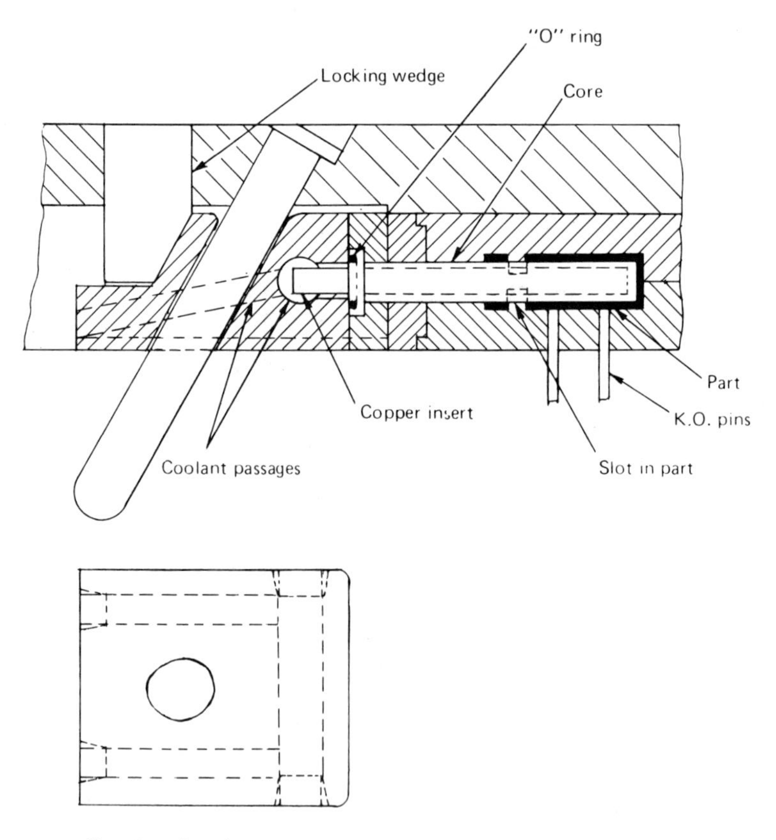

Fig. 9-2. Cam arrangement.

horn should be 30° or less. A 45° angle can be used whenever the pulling force of the cam is very light (10 lb or less). If for any reason it becomes desirable to delay the cam movement until the mold opens up to an inch, then an offset rectangular horn as shown in Fig. 9-3 can be applied. The straight portion of the horn governs the delaying action.

The sliding element in the cam is subjected to a downward pressure of considerable force during mold closing. It should, therefore, be well lubricated and its forward corner rounded to prevent accidental gouging.

As far as horn strength is concerned, there are two factors to keep in mind. The pulling force of the slide will tend to (1) deflect the horn and (2) create a stress at the fixed point of support. From *Machinery's Handbook* ("Beams Fixed at One End and the Load at the Other End"), we find the maximum deflection at the end to be:

$$\text{Deflection} = \frac{Wl^3}{3EI}$$

in which W = load

l = length of beam

E = modulus of elasticity

I = moment of inertia and for round sections = $0.049\ d^4$

This formula tells us that the length of the horn should be kept to a minimum, since the deflection increases with the third power of the length, while the

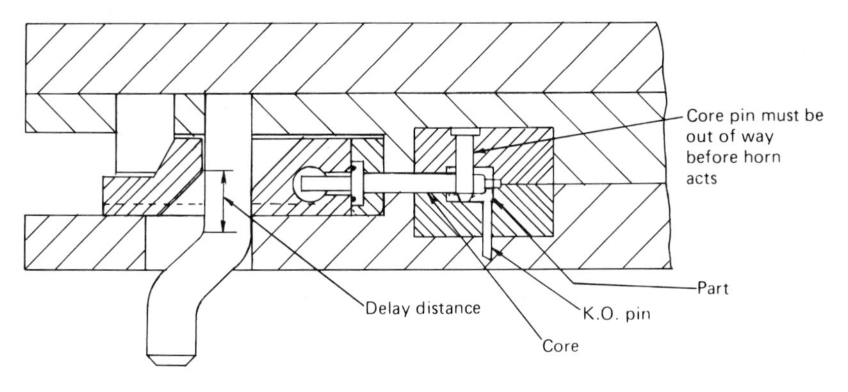

Fig. 9-3. Delayed cam action.

diameter of the horn on the higher side will cause a decrease of the deflection with its fourth power. The stress at point of support is

$$S = \frac{Wl}{Z}$$

$$Z = \text{section modulus for a round bar} = 0.098d^3$$

This formula points out that the stress will increase in proportion to the length but decrease with the third power of horn diameter. In both cases, it is important to have the horn diameter on the large size in order to keep the deflection and stress at support at low values.

If the force W could be established with any degree of certainty, it would be rather simple to determine the diameter of the horn. One way of approximating the value of the force is to calculate in a manner as shown in the following. The part will be as per Fig. 9-2 and is made of polycarbonate.

A stress is generated in the tube because it wants to shrink and is prevented from so doing by the core. The circumference on the inside of the tube when removed from the mold is 1.96. When the part is on the core, its circumference will be greater by 0.012 due to the shrinkage allowance of 0.006 in./in. This 0.012 is the amount of elongation of tube circumference while on the core. From the *Machinery's Handbook*, the elongation

$$e = \frac{PL}{AE}$$

where $P = W$ and $\frac{P}{A} = \frac{W}{A}$ = stress resulting from shrinkage

E = modulus of elasticity. For polycarbonate $= 3.45 \times 10^5$

L = length of the circumference after removal from core.

The thickness of the tube from Fig. 9-2 is 0.07.

Substituting these values in the formula, we have

$$W = \frac{0.012 \times 0.07 \times 2 \times 3.45 \times 10^5}{1.96}$$

$$= 296 \text{ lb}$$

From the *Machinery's Handbook,* the stress in a beam fixed at one end and the load at another

$$S = \frac{Wl}{Z}$$

where l = length of horn = 5

Z = section modulus = $0.098\,d^3$
(resistance of section to flexing)

The safe stress for a leader pin is 60,000 psi.

Thus,
$$60{,}000 = \frac{296 \times l}{0.098\,d^3} \text{ or } 0.098\,d^3 = \frac{296 \times 5}{60{,}000}$$

or
$$d^3 = \frac{296 \times 5}{0.098 \times 60{,}000} = 0.252$$

$$d = 0.632 = \text{approx. } 5/8 \text{ in. diam}$$

The choice of a 3/4-in.-diam horn would reduce the stress to about 35,000, giving it a higher safety factor.

The preceding calculation leads to a practical size and may be followed if circumstances of molded shape make it desirable. It is to be recognized that in the above calculations, several factors are neglected such as coefficient of friction, taper, and the temperature of the plastic at which withdrawal takes place. Nevertheless, the calculated size may represent a least favorable condition or the larger size horn that might be needed for core removal. A rule of thumb is to have the horn size 1/8 in. smaller in diameter than the leader pin for application where the length of horn does not exceed 5 in. and the angle is 30° or less.

The sliding or moving part of the core holder should have a detent in the "out" position to insure safe entrance of the horn during mold closing. The holder should be locked in position of molding by means of a wedge supported by the horn side of the mold.

It is interesting to note that the hydraulic force from the injecting material in the preceding example would be

$$F = \text{Exposed projected area} \times \text{Injection pressure}$$

$$= (0.625 + 2 \times .07)^2 \frac{\pi}{4} \times 20{,}000 \text{ psi.}$$

$$= 9120 \text{ lb}$$

This sizable force has to be contained by the locking wedge to prevent change in size of the part.

The entrance side for the horn on the core holder should be well rounded to allow easy engagement.

From the calculations, we find that considerable force is exerted by the plastic onto the core with the result that very good contact exists between the plastic and metal core. This tight contact is responsible for most of the heat from the plastic being transferred onto the core. Under these conditions, the conductivity of the heat from core to mold base will not reach a point of balance; however, in all probability, for a polycarbonate part, the core temperature could well be 70° to 100°F higher than the surrounding base. The temperature increase will cause lengthening of the core and a corresponding change in thickness of the bottom. What will this length increase be?

$$e = \text{Length} \times \text{Coefficient of expansion} \times \text{Temperature rise (°F)}$$

$$= 5 \times 6.33 \times 10^{-6} \times 100 = 0.00316$$

Since the bottom has a tolerance of +0.001, steps have to be taken to control the temperature of the core so that it will approximate that of the base. There is one other requirement that has to be met, namely, the core has to be completely out of the cavity before stripping begins.

Let us now assume that the part we have been working on calls for outside surfaces free of any knockout pin marking; it then becomes necessary to mold it in a design with wedge type splits as shown in Fig. 9-4. This arrangement calls for the splits to be guided at the end for a movement that will separate them as they rise upward, and keep them parallel by means of pins that are parallel to the face of the B plate. In this application, the mold is to have two stripper plates; one to raise and separate the splits until there is adequate space for the ejecting sleeve that removes the pieces from the mold, the second one to actuate the sleeves that strip the product. The splits should be hardened to at least 54 to 56 RC; the rails (mounted in the base) for riding of the splits should also be heat-treated to the same hardness. The stripper bars that raise the splits should also be heat-treated for a hardness of at least 54 to 56 RC.

This type of action derived from the splits may require a smaller press daylight but, on the other hand, may present more difficulty in providing good temperature control of cavities. Another type of cam action is the "tulip," which is basically the same as the wedge-split type except that it applies to round parts with undercuts. The tulip design is shown in Fig. 5-1.

It is also possible to provide undercuts by means of shifting offset pins or sliding core sections actuated by the stripper pins. Some of the more common ways of arranging cam actions that provide capability to mold undercuts in a product are shown in Figs. 5-1, 9-4, and 9-5.

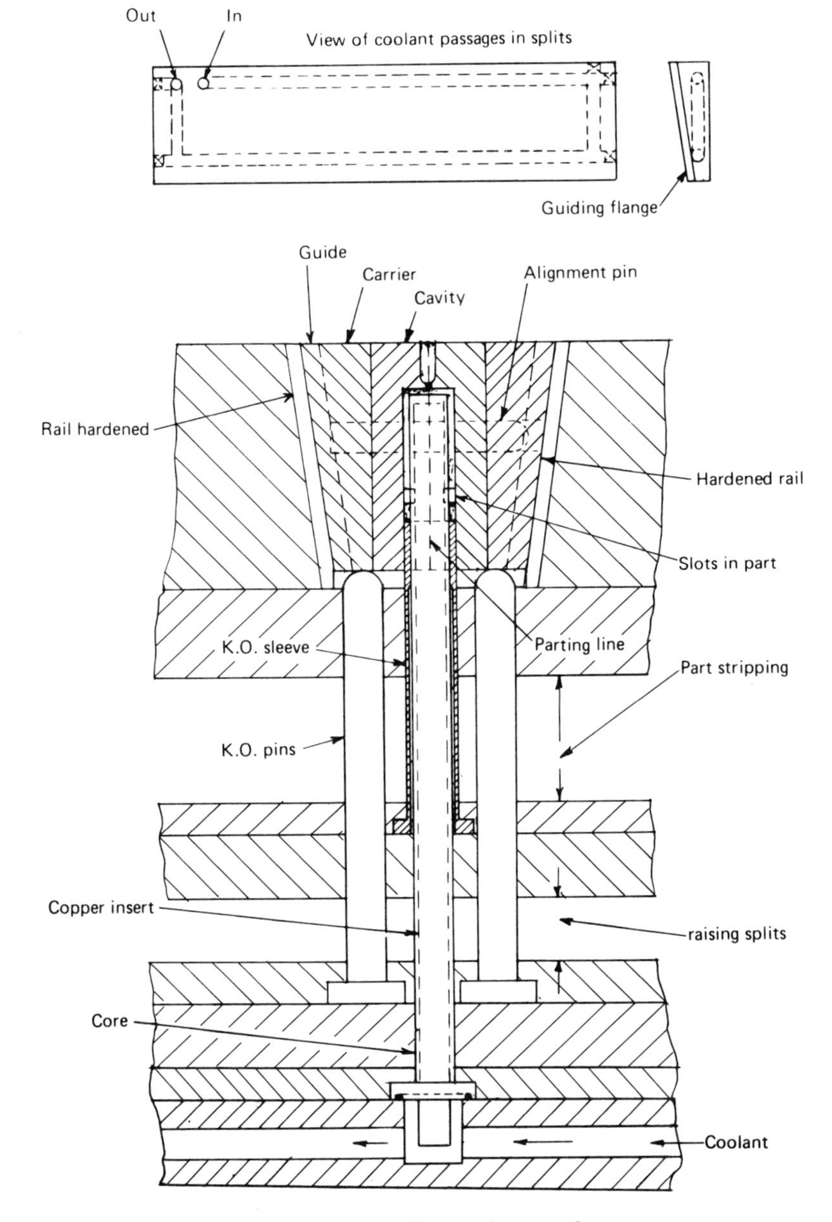

Fig. 9-4. Mold with tapered cam wedges.

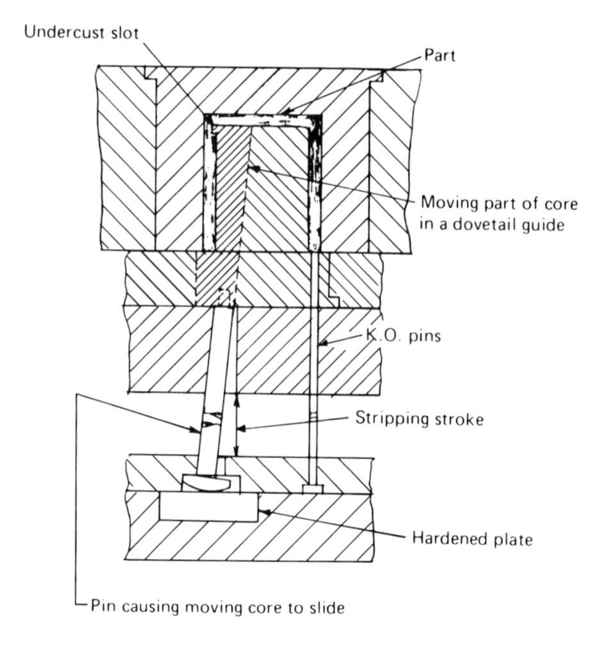

Fig. 9-5. Mold for inside undercut.

MOLD BASE WITHOUT BACKUP PLATES

Mold bases without backup plates are used not only for low-usage parts but also whenever the A and B plates are 2-7/8 in. or thicker, in which case it is possible to include all the needed base features. These features are liquid passages, clamping slots, blind pockets, adequate strength, and allied requirements. This mold base is usually less expensive than the one called "standard."

THREE-PLATE MOLD BASE

When a third plate of the floating type is added to the stationary half of the mold, it is designated as a *three-plate mold base.* This type is used for multi-cavity parts whenever a center gate or multiple gating near the center becomes desirable in order to have closer control of dimensions. It lends itself to automatic degating and operation (Fig. 5-2).

The three-plate base operates as follows:

Springs are placed between the A cavity plate and the intermediate plate. During mold opening, they cause the A cavity plate to move and remain in contact with the B plate at the same time they retain the intermediate plate in closed position. This action causes the gates to break. Mold opening continues, while the press opens. The moving mold half is connected to the A plate by means of threaded rods, and through this connection it moves the A plate until it reaches the limit as established by the shoulder bolts. At this point, the stopping shoulder bolt transmits the pulling action onto the intermediate plate. The movement of the intermediate plate (about 3/8 in.) brings about the stripping of the runner from holding pins, which in turn causes the runner system to drop out of the mold. The intermediate plate has its travel limited by shoulder bolts attached to the clamping plate. When the mold is fully opened, there is a space, between the cavity plate and intermediate plate, which is governed by the length of the cavity-plate shoulder bolts. The space is wide enough to permit free falling of the runner system.

The leader pins have to be long enough to firmly support the sliding plate in its outer position. Should the plate become heavy (e.g., over 250 lb), consideration should be given to larger-size leader pins and, preferably, the plate should be equipped with ball bushings.

Let us calculate the deflection of 7/8-in.-diam leader pins in a mold with a floating plate of 250 lb and where the length of the pins to the center of the plate thickness is 5 in. We shall assume that the weight is equally supported by all pins. From *Machinery's Handbook* ("Beams Fixed at One End"), we find the maximum deflection at one end to be:

$$\text{Deflection} = \frac{Wl^3}{3EI}$$

where $W = 0.25$ of 250 lb = 62.5

$l = 5$ in.

$E = 30 \times 10^6$

I for 7/8 in. diam = 0.029 (from *Machinery's Handbook*)

$$\text{Deflection} = \frac{62.5 \times 125}{3 \times 30 \times 10^6 \times 0.029} = 0.003$$

This is an excessive deflection, and an increase in leader pin size is indicated. The next size standard leader pin is 1 in. diam. Since deflection changes in reverse proportion to the moment of inertia, which in turn changes as the fourth power of the diameter, we can take $(1/7/8)^4 = 2.9$ and divide it into 0.003 deflection to obtain a value of 0.001. This deflection can be viewed as tolerable and therefore acceptable.

The remaining items to be calculated for strength are the pulling rods and shoulder straps. Again, we will use the 200-ton, 14-oz press as the example in which the mold will be run. In this press, the stripping force is 10 ton or 20,000 lb. The mold is equipped with four pulling rods and four shoulder bolts at each separating point. The allowable stress of rods and shoulder bolts is 40,000 psi, with a safety factor of 4.5, meaning that the steel from which the parts are made should have a tensile strength of 180,000 psi.

From *Machinery's Handbook,*

$$S = \frac{P}{A}$$

where S = allowable stress = 40,000 psi

P = force = 20,000 lb (stripping)

A = area at root of threads

$A = \frac{P}{S} = \frac{20{,}000}{40{,}000} = 0.5 \text{ in.}^2$ for four pulling rods or eight shoulder bolts

The area of each at root of the rod = 0.5/4 = 0.125 in.^{2} The area of each shoulder bolt thread = 0.5/8 = 0.0625. The nearest thread size for the root diameter area of 0.125 is 1/2–13, and for the 0.0625 area is 3/8–16. The rods will be 0.5 in. diam with 13 thread/in., and the shoulder bolts will have a threaded end of 3/8–16 with a body diameter of 0.5 in.

STRIPPER PLATE BASE

Another type of a mold base is one in which stripping is performed by means of a moving plate. In this case, the leader pins are mounted on the B side of the moving side of the press. Plate stripping is usually done when knockout pin marks on the surface of a part are objectionable or when it is difficult to remove the part from a core by means of pins due to its geometry.

The question of floating stripping plate weight and size of leader pins should be treated in the same manner as outlined in the preceding section, "Three-Plate Mold Base."

The movement of the stripper plate should be limited by shoulder bolts or other suitable means. The pins that are used for moving the stripper plate should be 3/8 in. diam or larger. There are combinations of any of the previously described bases, such as the three-plate base with stripper plate, but there is also a number of special arrangements that as yet do not lend themselves to standardization. Included in the variety are runnerless molds.

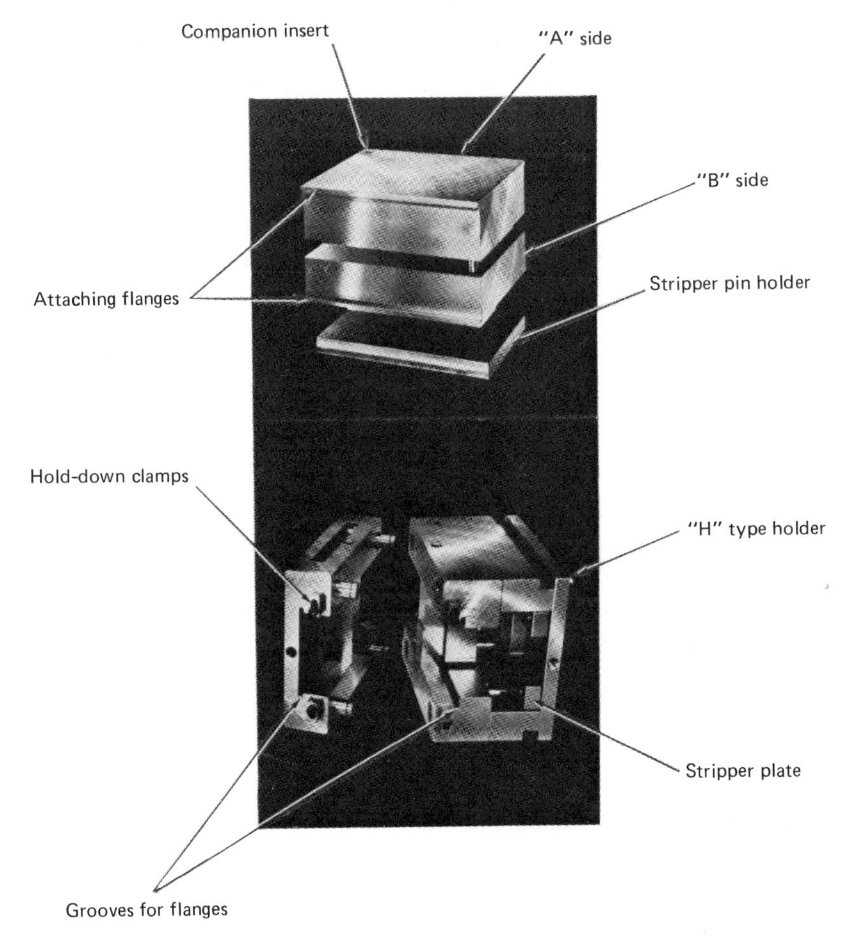

Fig. 9–6. Holder for companion insert. (*Courtesy of Master Unit Die Products Inc.*)

BASES FOR INTERCHANGEABLE CAVITIES

We normally think of molding as a high-activity operation involving multicavity costly molds. As in all activities, there are exceptions to the rule. There is a need for an inexpensive single-cavity mold for low-activity parts–less than 1,000 per year–or prototype pieces, which can be tested and evaluated prior to the making of a production mold.

Some molding plants have their own approach to the solution of the low-activity mold problem. The majority of processors use one or more of the following.

1. Commercially produced mold-base holders and companion inserts such as

manufactured by Master Unit Die Products Inc., Greenville, Michigan. In this system, the mold base contains the common components such as locating ring, sprue bushing, leader pins and bushings, and stripper plates with return pins. It also is machined to receive the companion inserts for easy and fast placement, locking them in place without disturbing the holder in the clamped position of the press. Cooling channels may also be added to the frame if deemed necessary.

It takes four to six insert cavities to offset the cost of the frame against the price of complete molds for same cavities, however, after that number, the savings become of sizable magnitude.

The companion inserts consist of A and B halves (which correspond functionally to the A and B plates of standard bases) include K.O. pin holders, and are made in various steel grades. The most common type holder is of the "H" design, which is capable of holding two sets of companion inserts for molding parts at the same time, as long as the plastic material is the same. There is a great variety of holders for special machines and special shapes available.

The companion inserts can be advantageously utilized for multicavity parts of small and relatively simple shapes with considerable savings in mold cost without loss of productivity (Fig. 9-6).

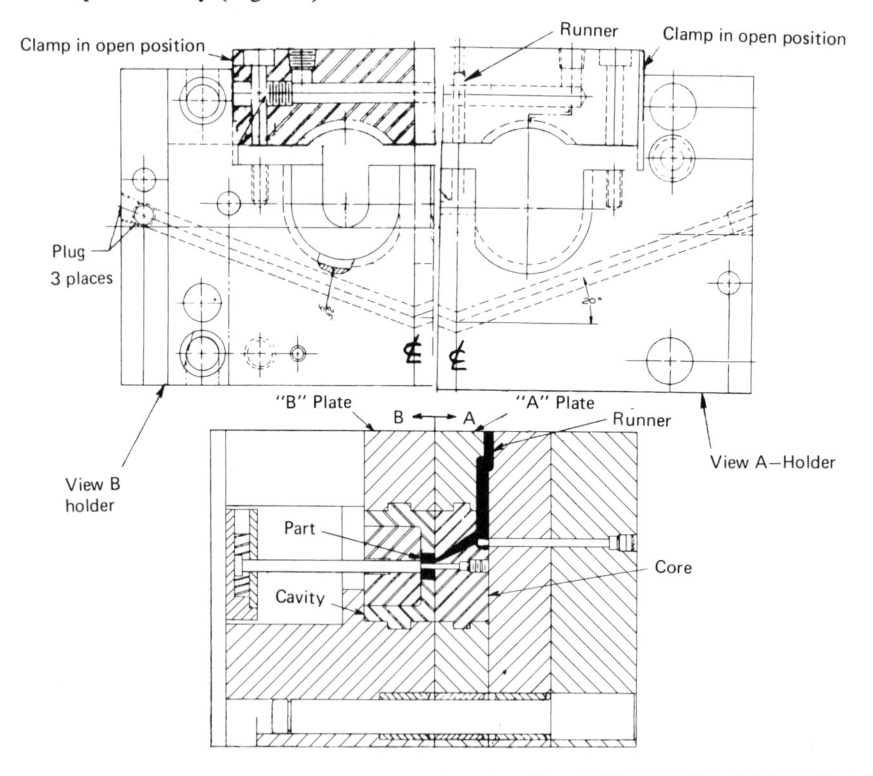

Fig. 9-7. Universal holder—three plate. (*Courtesy of Rockwell International Corp.*)

2. Another type of universal holder was designed by the author. It was built principally for low-activity gears. The inserts for the holder are round as are the cavities and cores for parts. Each insert has the gate and runner incorporated in it as well as means of centering the A and B inserts by the conical male and female arrangement. The holder is of the three-plate design and fits the Newbury Hornet 25 or similar machines. The cavities are cooled by means of passages incorporated in the holder—parts that are in close contact with the inserts. The cost of inserts compared with a single-cavity mold is about one third for cavity inserts against a complete single-cavity mold (Fig. 9-7).

3. Another type of universal holder consists of machining a standard mold base to receive round cavity inserts of a specific size, e.g., 4 in. diam. The inserts are also arranged for rapid interchangeability and in most cases are sprue-gated. With constantly rising costs of tool-making, the mold designer may view the described mold-base arrangements as sources of potential cost reduction.

10
Cavity Materials

When a decision is made to produce a plastic part, all those connected with the project are aware of the large expenditures that will be involved in making a mold to manufacture it. In comparison with such production tools as drill jigs or punch press dies, molds are very expensive. A mold usually costs many thousands of dollars, and, when the parts to be made are complex, requiring multiple cavities, the cost figure runs into tens of thousands. Moldmaking is a one-shot deal; it has to be done correctly the first time around. A heavy responsibility rests on the shoulders of those who control the materials that go into the making of a mold, for the mold must perform well and meet expectations. Every known precaution has to be exercised so that the work of skilled toolmakers is safeguarded and so that the tool steel used is economical in quality and quantity.

Whenever applicable, calculations should be made to insure that the mold is not overdesigned and thus wasteful, nor underdesigned and thus destined to bring about failure. Analysis of wear conditions, operating condition, etc., should have a bearing on cavity material selection.

STRENGTH OF MATERIALS

When we apply materials in the design of any product, the object is to use the least amount that will meet the needs of an anticipated job and at the same time have a built-in margin of safety. The amount and type of material used in a design can be optimized by application of formulas found in technical textbooks or handbooks under "strength of materials." The mechanical handbooks list all the necessary formulas, fortified with examples of how they are to be applied. A correct application of a formula becomes much simpler when there is an understanding of the principles underlying it.

When we discuss the strength of materials, we are confronted with external forces tending to exert their influence upon bodies. Whenever an elastic body is deformed as a result of an external force, a resistance is set up to counteract such deformation. This resistance is called a *stress.* There are five kinds of stresses: tension, compression, bending, twisting, and shear. The questions then arise:

1. What type of load and resulting stress are applied to a part under consideration?

2. What is the environment-temperature and chemical exposure?
3. What materials are best and most economical for the purpose, and what sections are needed for proper functions?

All five stresses can be mathematically expressed in one simple formula, namely: the stress is equal to the force and its manner of application, divided by the cross-sectional property that totalizes the stress.

Let S = stress

F = force in whatever manner it is applied

C = cross-sectional property that totalizes the stress

$$S = \frac{F}{C}$$

In tension, compression, and shear, the formula will be

$$S = \frac{P}{A}$$

Where $F = P$ = a force in pounds perpendicular to the cross section of a body when in tension or compression but parallel with the cross section when in shear,

$C = A$ = a cross section of a body in square inches that totalizes stress

There is an exception to the compression formula, namely, when the proportion of length to diameter of the loaded element is 25 or greater. This proportion is called a *slenderness ratio.*

It is true that an ideal column—perfectly straight, with perfectly homogeneous material, entirely free from any flaws of manufacture, and loaded in the perfect axis—would fail by direct crushing the same way as a short block would. In actual practice, this kind of column is nonexistent, and when a long column is loaded, it will fail by buckling. The critical force that would cause buckling in a machine column that is unbraced, according to *Machinery's Handbook* ("Euler's Formula for Machine Elements in Compression"), is

$$P_{cr} = \frac{S_y A r^2}{Q}$$

$$Q = \frac{S_y l^2}{n\pi^2 E}; r = \frac{d}{4}$$

where

P_{cr} = critical load in pounds that would result in failure

A = cross-sectional area, in.^{2}

S_y = yield point of material, lb/in.^{2}

r = least radius of gyration of cross section, in.

E = modulus of elasticity, lb/in.^{2}

l = column length in inches

n = coefficient for end conditions; for one end fixed and the other free but guided, $n = 2$.

This formula is applicable to ejector pins, and an example is calculated in Chapter 8 for such pins.

In bending, which is a combination of compression and tension with respect to the neutral plane, the formula is

$$S = \frac{F}{C} = \frac{M}{Z}$$

where $F = M$, the bending moment, in.-lb

$C = Z$, the section modulus, a property of cross section that totalizes stress, in.^{3}

In torsion, the formula is

$$S = \frac{F}{C} = \frac{T}{Z_p}$$

Where $F = T$ torsional or twisting moment, in.-lb

$C = Z_p$, polar section-modulus, a property of cross section that totalizes stress, in.^{3}

The section modulus in bending and torsion can also be defined as a measure of stiffness or resistance to bending or twisting as the case may apply.

Now it is necessary to identify the different stresses that will be considered during the design process. We may have in mind a safe operating stress, a yield

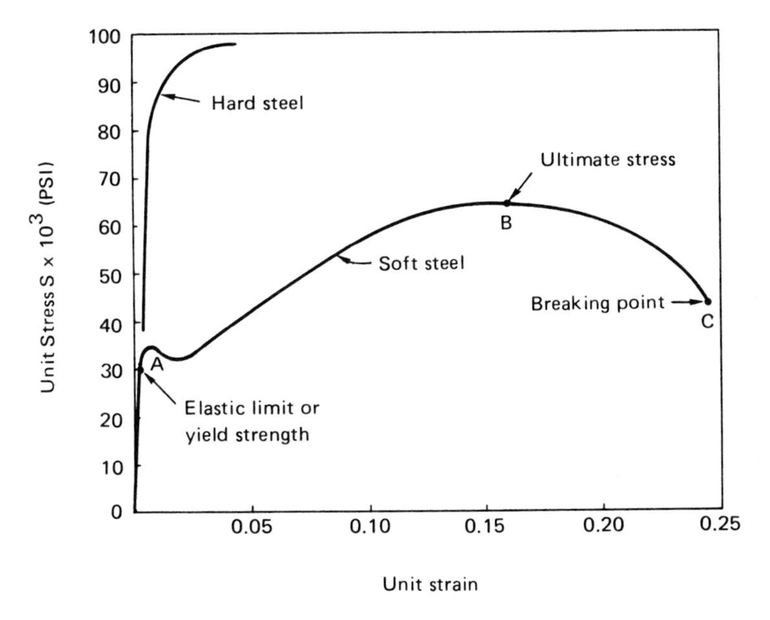

Fig. 10–1. Stress-strain curve.

stress, an ultimate stress, or breaking stress. The data for all these stresses are established by subjecting a test bar to increasing load increments and recording the strain produced in the test bar. This information generates a stress-strain curve from which a considerable amount of essential information is developed.

On Fig. 10–1, we see a straight portion of the curve up to the point *A*. According to Hooke's law, the deformation or strain is proportional to the stresses within limits established by test for each material type. Mathematically, this is expressed as

$$\frac{\text{Stress}}{\text{Strain}} = \text{Constant } K = \text{Modulus of elasticity } E$$

The strain can be also shown as a proportion of elongation to the original length L of test bar:

$$\text{Strain} = \frac{\text{Elongation}}{\text{Original length}} = \frac{e}{L}$$

Substituting in the modulus formula e/L for strain, and the force divided by area or P/A for stress, we have

$$\frac{\text{Stress}}{\text{Strain}} = \frac{P}{A} : \frac{e}{L} = \frac{PL}{Ae} = E \text{ in psi}$$

$$= \frac{PL\ (\text{lb} \times \text{in.})}{Ae\ (\text{in.}^2 \times \text{in.})} \text{ or lb/in.}^2$$

Thus, we see a relationship between modulus of elasticity, force, area, elongation, and original length. This relationship is used when lengthening or shortening of members causes stresses.

Point A on the curve (Fig. 10-1) is called the *proportionality part* or *elastic limit.* When parts are strained at values below this limit, they will return to the prestrained condition upon removal of the stress or load. Therefore, the parts do not suffer permanent change.

Point A is the beginning of *yield strength.* An increase in force above this point will have a certain corresponding amount of permanent deformation. Thus, a yield strength at 0.2% "offset" is that value of stress which would cause a permanent deformation of 0.2% or 0.002 in./in. when the load is removed (Fig. 10-2). The yield point above A is the transition point from the elastic stage of a material to the plastic stage. The plastic stage is a condition of deformation—under a heavy load—that is permanent. At this stage, creep flow becomes pronounced.

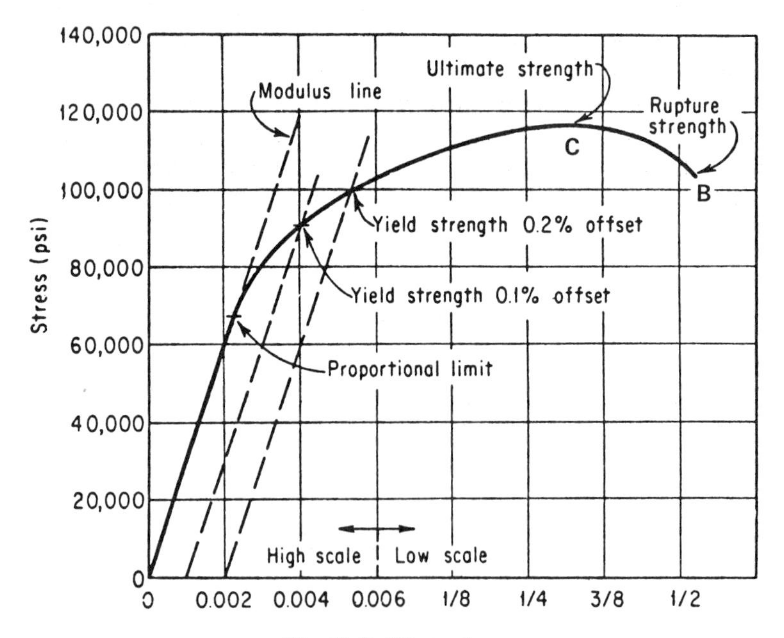

Fig. 10-2. Offset values.

The next point of interest is the *highest stress point B,* which is called the *ultimate strength* or *tensile strength.* The information shown on the stress-strain curve is some of the important data found in tables on metals. Continued elongation of a test bar results in rupture, and this point is called the *rupture strength C.*

Suppliers of metals usually give the tensile strength, the yield strength, elongation at the strength, and the amount of offset. With that much information, it is a simple matter to construct a curve up to the yield point. For practical mold construction purposes, the strength of interest is the elastic limit and the values below it.

In many instances, it is taken for granted that a hardened steel may offer greater rigidity or less deformation than the untreated steel. The curve in Fig. 10–1 shows that within the elastic limit of the soft steel, the hard steel behaves exactly the same way. However, the hardened steel has a much higher tensile strength and will have its own proportionality limit extended to stresses several times that of the soft steel. On the other hand, the ultimate or tensile strength of the hard steel is very close to that of its yield strength. As far as the modulus of elasticity is concerned, all steels, regardless of alloying elements or heat treatment, have the same modulus–about 28 to 30×10^6 psi.

The tests and curves discussed were carried out on steel at room temperature. Plastic molds are subjected to temperatures considerably higher than room temperature. Even when we consider the full range of temperatures at which molds operate, we find that the recommended tool steels retain all the significant properties at those elevated temperatures.

While the description of the modulus of elasticity is still fresh in our minds, we want to look at the modulus of shearing, which applies to stresses and strains generated by torque. Here the twist in a shaft is the strain (Fig. 10–3).

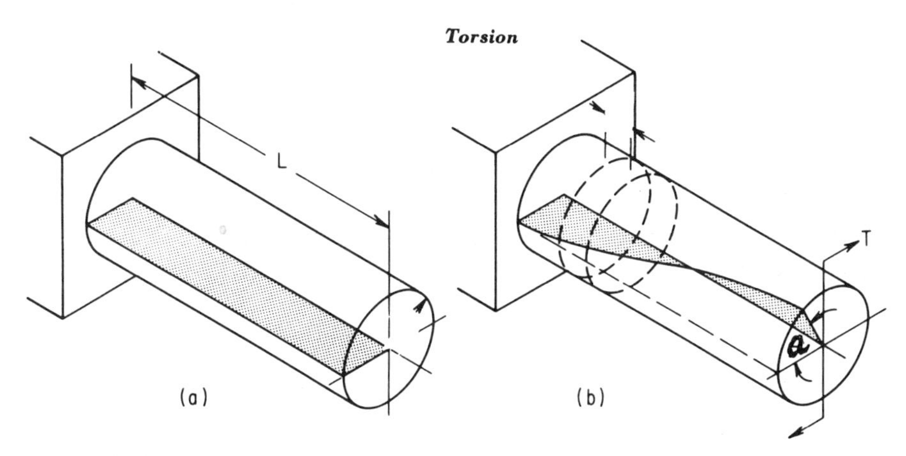

Fig. 10–3. Shaft in torsion.

$$\frac{\text{Stress}}{\text{Strain}} = \text{Constant} = G \text{ Modulus of shearing} = 0.4E$$

or

$$12 \times 10^6 \text{ psi}$$

Through several mathematical steps, we arrive at a practical formula for angular distortion in a shaft. The deflection in degrees is:

$$a = \frac{584\ TL}{D^4 G}$$

in which

a = angular deflection of shaft, in degrees

T = torsional or twisting moment, in.-lb

L = length of shaft, in.

D = diameter of shaft, in.

G = torsional modulus of elasticity = 12×10^6 psi for steel

If the horsepower and rpm of a driving motor are known, the torsional moment T can be expressed in this form

$$T = \frac{63{,}000\ P}{N}$$

in which

P = horsepower

N = rpm of shaft.

With the aid of additional mathematical steps, we obtain a formula for shaft-diameter determination,

$$D = \sqrt[3]{\frac{16T}{\pi S}}$$

where S = allowable shearing stress.

Let us now apply the strength-of-material information to examples encountered in moldmaking. In Chapters 2 through 9 inclusive, several calculations were made, in which stress values were selected that were considered to be practical and to have an ample safety margin for specific load conditions.

EXAMPLES

1. *Tension.* A three-plate mold uses four rods for actuating the movable plates. The stripping force per rod is 5000 lb. Since the rods are threaded, we will use the root diameter of the thread as the area that will resist the external force.

$$S = \frac{P}{A}$$

The steel we will use is a low-carbon steel with a tensile strength of 70,000 psi. The safety factor we will choose is 4 or $S = 17{,}500$ psi. Thus,

$$S = \frac{P}{A} = \frac{5000}{17{,}500} = 0.286 \text{ in.}^2$$

A root diameter that corresponds to this area is 19/32 in., and the nearest thread size would be 3/4 in.-10 as found in the *Machinery's Handbook* ("U.S. Standard Bolts"). This size is out of proportion to the other mold components. From the stress-strain curve, we conclude that a harder steel will have a much higher tensile strength. The 4100 series steel with a 320 Bhn has a tensile strength of 160,000. Using again a safety factor of 4, we will have an allowable stress of 40,000 psi

$$S = \frac{5000}{40{,}000} = 0.125 \text{ in.}^2 \text{ at the root}$$

and this corresponds to a 1/2 in.-13 thread size. This is in line with requirements for strength and proportion.

2. *Compression.* The formula in compression is the same as in tension, namely,

$$S = \frac{P}{A}$$

A mold was being repaired, and it was found that one of the causes for malfunction was an indentation in the backup plate by the supporting pillar. This indicated that the compressive stress in the plate reached the yield-strength value and brought about permanent displacement of metal. According to the preceding formula, the problem could be corrected by either increasing the area

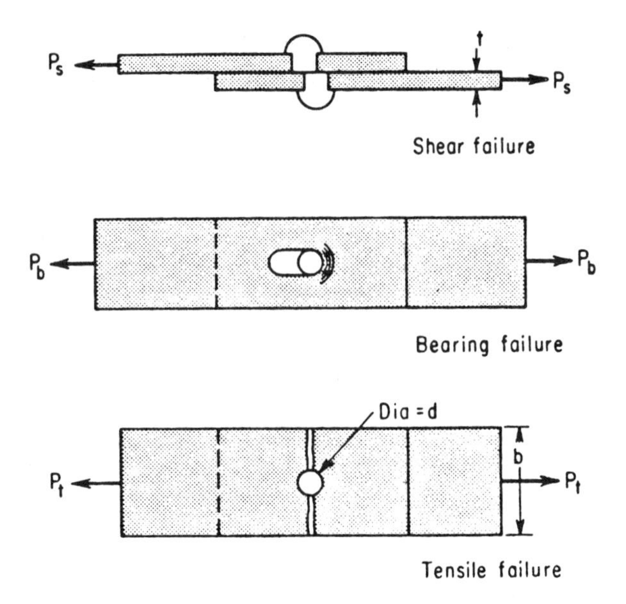

Fig. 10–4. Shear, bearing, and tensile stresses.

of support pillars or hardening the plate so that the yield strength is raised. It is neither practical to change the area of existing support pillars nor possible to add another pillar to reduce the stress level, due to the location of stripper pins. The only solution left is to harden the existing plate or to substitute one with a higher tensile strength and correspondingly higher yield strength.

3. *Shear.* The shear strength for steels is 0.75 of the tensile strength. The formula is the same as in tension, i.e.,

$$S = \frac{P}{A}$$

Figure 10–4 shows two plates riveted together by a 1/4-in.-diameter rivet and subjected to a force of 400 lb. This assembly has three possible chances of failure: shear failure of rivets, bearing failure which would cause the plate to distort at point of riveting, and tensile failure which could bring about the breaking of the plate at the rivet hole.

Figuring the three stress values, we have

$$S_s = \text{Shearing stress} = \frac{P}{A} = \frac{P}{0.7854\, d^2} = \frac{400}{0.049} = 8170 \text{ psi}$$

$$S_b = \text{Bearing stress} = \frac{P}{A} = \frac{P}{d \times t} = \frac{400}{0.25 \times 0.125} = 12{,}800 \text{ psi}$$

$$S_t = \text{Tension stress}\ \frac{P}{A} = \frac{P}{(b \times t) - (d \times t)} = \frac{P}{t(b-d)} = \frac{400}{0.125 \times 1.25} = 2560$$

These results show that the bearing stress is the highest, but within a safe limit.

4. *Bending.* In the molding operation, even though the stress of bending may be in the low range, the deflection associated with bending is of prime concern. Whether the deflection relates to cavities built out of separate sections, deep cavities, or leader pins supporting heavy plates, it has to be limited in magnitude in order to safeguard the proper functioning of a mold. In Chapter 5, calculations were made for several conditions, and appropriate deflection formulas were applied. The important part in selecting a deflection formula from the *Handbook* is to recognize the condition in the mold as compared to the type of beam listed.

Let us analyze the deflection formula for a beam fixed at both ends and loaded in the center.

$$\text{Deflection } d = \frac{Wl^3}{192\,EI}$$

W = load in the center and is obtained by multiplying the injection psi with the molding area that will act like the beam described.

l = length of the molding area

E = modulus of elasticity for steel 30×10^6 psi

I = moment of inertia or the property of the cross section that resists bending.

If we look at Fig. 5–8 and apply the deflection formula to this construction, we have a moment of inertia:

$$I = \frac{bd^3}{12}$$

In this formula, d represents the combined thickness of cavity side and plate width as shown in the figure. Nowhere in the formula is there a factor related to stress and indirectly related to hardness. The formula simply tells us that to keep deflection to a controlled value, the wall thickness d must be of calculated magnitude. The depth b is assumed to be fixed by the design of the part. Hardness of the steel has nothing to do with resistance to beam deflection; the only thing that matters is the thickness of wall d, which increases to the third power its resistance against deflection. The same general reasoning applies to any mold component subjected to deflection. The side effects of cavity deflection include parts out of shape and parts out of dimensional tolerance; it also contributes to difficulty of mold opening and inconsistency of cycle. The same force that causes the wall of a cavity to deflect will, upon solidification of the plastic (decay of inside pressure), bring about an equal force from the springback of the steel and exert a pressure on the wall of the plastic, making mold opening difficult.

When leader pins are under a load and deflect a few thousandths of an inch, the result will be misalignment and possible damage to the mold. An increase in diameter to a higher standard size will correct the difficulty by reducing the deflection to tolerable values.

Where the deflection is a problem, the only solution is to increase the moment of inertia (i.e., by increasing the factors in the formula, mainly the dimension that is in line with the direction of the force) and not to substitute steel grades.

5. *Torsion.* A motor with an output of 1.5 hp at the shaft running 50 rpm is to drive an unscrewing device that is required to be synchronized with all cavities. The driven shaft is 16 in. long and 1.25 in. diam. How much will the shaft distort? There are four cavities, each with equal torque requirements. The angular distortion,

$$a = \frac{584\ TL}{D^4 G}$$

where

$$T = \text{torque} = \frac{63{,}000\ P}{N} \quad \text{and } P = 1.5 \text{ hp and } N = 50 \text{ rpm}$$

$$= \frac{63{,}000}{50} \times 1.5 = 1890 \text{ in.-lb at driving shaft}$$

$$L = \text{length of shaft} = 16 \text{ in.}$$

$$D = \text{diameter of shaft} = 1.25$$

$$G = 12 \times 10^6 \text{ psi}$$

The distortion at 4, 8, 12, and 16 in., with each station taking 25% of the total torque, will be

$$a = \frac{584\ TL}{D^4 G}\ ;\ D^4 = 2.43,\ T_1 = \frac{1890}{4} = 472$$

$$a_4 = \frac{584 \times 472 \times 4}{2.43 \times 12 \times 10^6} = 0.0378$$

$$a_8 = 2 \times 0.0378 = 0.0756$$

$$a_{12} = 3 \times 0.0378 = 0.1134$$

$$a_{16} = 4 \times 0.0378 = 0.1512$$

This amount of distortion is more than the job can tolerate. In the general formula for distortion, the only variable is the shaft diameter, and nowhere is there mention of material strength. Therefore, the only way to minimize distortion is to increase the shaft diameter.

The job can tolerate an angular distortion of 8 min expressed as a decimal of a degree = 0.1333. Thus, at 4 in. (based on the previous figures), it should be

$$\frac{0.1333}{4} = 0.033$$

$$0.033 = \frac{584 \times 472 \times 4}{D^4 \times 12 \times 10^6} \text{ or}$$

$$D^4 = \frac{584 \times 472 \times 4}{0.033 \times 12 \times 10^6} = 2.78 \text{ or } D = 1.28 \text{ or } 1\text{-}5/16$$

In conclusion, it is to be emphasized again that when a formula does not call for "stress" or "strength," the only way to improve the resistance to deformation is by increasing the dimensions of the cross section.

HARDNESS DETERMINATION AND ALLIED PROPERTIES

After the stresses have been analyzed, there are other material characteristics that play a vital part in the successful performance of a tool or mold and should be kept in mind.

In essence, the required characteristics are hardness, toughness, resistance to wear, and behavior at elevated temperature. In addition, there are some related to those mentioned:

1. *Hardness* is a property very difficult to define. It means many things in many applications. Hardness in metals can be interpreted as
   a. Cutting hardness, meaning resistance to machining
   b. Abrasive hardness, meaning resistance to wear when in rotating or sliding motion
   c. Tensile hardness, meaning the strength at elastic limit and ultimate strength.
   d. Indentation hardness, meaning resistance to indentation or a measure of plasticity (peenability) and density

e. Deformation hardness, meaning resistance to distortion especially in sheet steel.

We can gain a better understanding by learning the methods used in making hardness determination.

Resistance to indentation is actually the hardness that is measured in making hardness readings. In the Brinell hardness test, a load of 3000 kg is applied for 10 sec by a hand-operated hydraulic press through a hard, 1-cm-diam steel ball (Fig. 10-5). The diameter *d* of spherical impression made with the steel ball is measured, and the depressed surface is calculated. The load of 3000 kg divided by the calculated surface of spherical impression gives a Brinell hardness number in units of kilogram/square millimeter (kg/mm^2).

There are some limitations to the Brinell hardness test. For hardness above 480 Bhn, a tungsten carbide ball is used; for softer nonferrous metal, the load is much lower than 3000 kg. Care also has to be exerted that the thickness to be tested is at least 0.1 in. and the test is undertaken away from the edge to avoid bulging.

The Brinell hardness test actually is indicative of the strength of the material, and it has been found that there is a correlation between the Bhn and the tensile strength of iron-based metals.

Each two Bhn points correspond approximately to 1000 psi. Thus, a steel with a 300 Bhn will have an approximate tensile strength of 150,000 psi. This Brinell hardness test actually indicates the compressive strength and permanent deformation of a material.

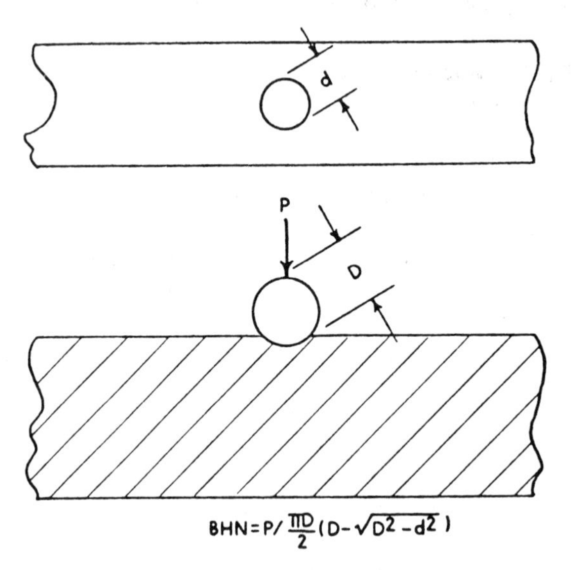

Fig. 10-5. Brinell hardness test.

In the Rockwell hardness test, a predetermined standard minor load is applied to the surface being tested and then followed by a standard major load, which is released by a loaded lever system (see Fig. 10-6). Only the major load indentation is used for reading on a calibrated scale. For most hardened metals, the C scale with a diamond indenter called a *brale* is used. A load of 150 kg is applied. The indenter has a diamond cone with a spherical apex. The dial or hardness number increases with the hardness. The steps taken to arrive at a reading are:

a. Apply minor load. Dial set at zero.
b. Apply major load and minor load. Dial reads "penetration."
c. Release major load. Dial reads "hardness number."
d. Minor load withdrawn. Dial is idle.

By implication, we have indicated that hardnesses tested would extend throughout the material. There are, however, many cases in which only a very thin skin of an object is to be examined for hardness. In that event, the "superficial Rockwell hardness," "Vickers," "Tukon," or "Knoop" tester is employed. The principal use of these testers is for very shallow penetration (0.005 in. or less) of an impression, such as with very thin material or shallow surface-heat-treatment like nitriding.

2. *Toughness* is another characteristic that has a different meaning to differ-

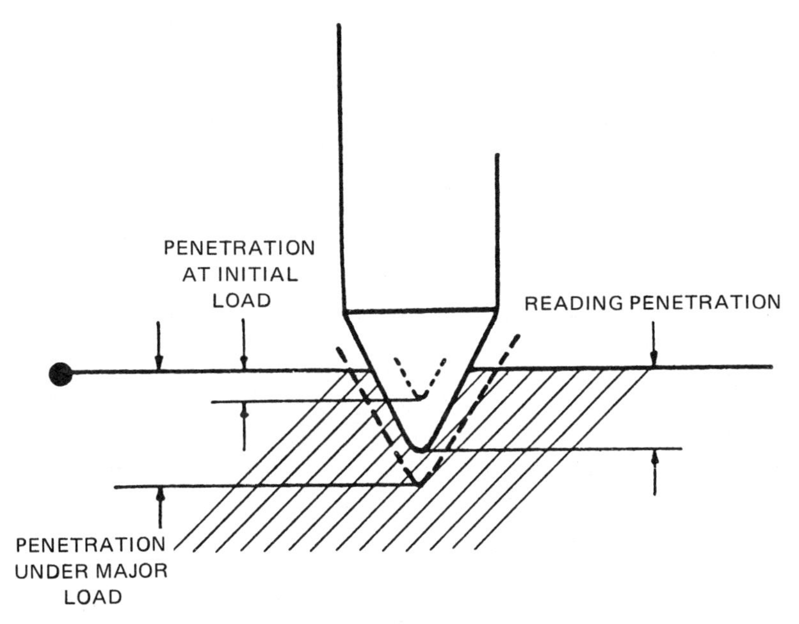

Fig. 10-6. Rockwell hardness test.

ent people. Toughness can be defined as the *resistance of a cubic inch of metal to the work that will cause its fracture.* Toughness is equal to the area under the stress-strain diagram shown in Fig. 10–1. From the diagram, it can be seen that hard steel does not have the toughness of its annealed counterpart. It can also be stated that toughness represents the work-absorbing capacity both in terms of elastic and plastic deformation. Toughness shows some relationship to impact strength.

WEAR RESISTANCE

Resistance to wear is an important characteristic of contacting metal surfaces that are moving against each other. Wear may be caused by:

1. Friction, which causes small particles to be torn away from the surface; it depends on the coefficient of friction between materials and the perpendicular force—normal force—to the plane of friction.
2. Cracks resulting from highly stressed surfaces, thus creating an uneven area over which sliding takes place.
3. Galling, which is a local removal of surface material. When the removed metal particles are caught between the sliding parts and gather, they may cause seizure of the surfaces.

Wear of metal parts moving against each other can be minimized by certain precautionary steps. Some of the steps are:

1. Using hard materials to reduce penetration of the surface or materials with a high compressive strength to avoid deformation.
2. Keeping the forces normal to sliding surfaces at low value. This will keep the friction force also low.
3. Using tough materials to prevent breaking of small particles. Parts must be designed so that chances of edges breaking are eliminated.
4. Providing smooth surfaces, preferably with machining lines of both sliding surfaces parallel to the direction of movement.
5. Using dissimilar materials to prevent galling. In galling, the tearing of metal pieces may be due to self-welding tendencies of spots or sawtooth-shaped surface finish. The surface may be improved against galling by stoning or waterblasting with abrasives. Materials with a lower tendency to indentation will also be more resistant to galling.
6. Providing lubrication.

The practical application of the preceding information is illustrated by the following cases.

In the applications for wear plates on plastic molds, the tendency is to use heat-treated tool steel. Looking at the compressive properties of cast iron, bronze, and even brass, we find them to be adequate for the purpose. Cast iron has a compressive strength of 85,000 psi, bronze 120,000 psi, and brass 80,000 to 90,000 psi. In addition to those properties, the dissimilarity of sliding materials provides an added benefit because of nonadhering tendencies.

Another case is behavior at temperatures above normal–i.e., at elevated temperatures, problems of wear can be much more pronounced because practically all of the properties of metal are affected by the increase of temperature. The most important consideration, however, is the coefficient of expansion and making proper allowance for it. Other important features are increased plastic flow of the steel, decreased strength, etc. While the mold steels such as type H-13 are known as "hot work-tool steels" and are formulated to retain good tensile strength, hardness, and toughness at elevated temperatures, there are steels involved in the base, etc., that require attention when used at higher molding temperatures.

The lack of toughness of some steel parts may be overcome by eliminating sharp corners in areas where danger of chipping may exist. In any case, paying close attention to the stress level involved may result in the application of materials that are of low cost, easy to machine, and do not require heat-treating.

TOOL STEELS

The term *tool steel* is applied to a special quality steel that is electric-furnace-melted. The processing and conditioning of tool steel are carried out with great care during the many mill operations from ingot to the salable product. Rigid physical and chemical standards are maintained to insure consistent high quality. To the end user or toolmaker, a uniformity of quality is extremely important because of the amount of toolmaker labor involved in each piece of steel. Plastic moldmakers are considered to be the most highly skilled toolmakers, and are therefore the highest paid. Any loss of a toolmaker's time due to inconsistent tool-steel quality is of major concern to supervisory personnel.

Following are the seven basic grades of tool steels, which are listed in order of relative alloying complexity.

1. *SAE "W" Water-Hardening Carbon Tool Steel.* In these steels, carbon is the principal control element. When hardened, the surface is very hard and provides good wear qualities. A significant characteristic is that this steel, when hardened, will have an outer layer or "case" of 65 to 67 RC hardness and a core of 40 to 45 RC. A slight variation in alloying or melting practice can bring about variation in depth of the "case." Thus, the steels are graded into shallow, medium, medium deep, and deep hardenable types. Tool steels that have a

tendency to distort appreciably in heat treatment have to be ground to required finished sizes. If grinding is required after hardening, an appropriate depth of case should be specified so that after grinding there is still enough case left to do the expected job. This type of steel is commonly used for cold chisels, hand punches, cold header dies, cold forming tools, shears, and wood chisels. For maximum hardening of these steels, a fast quench in water is required. There is a tendency to distortion and dimensional changes in the heat treatment. The analysis of W-2 steel is:

Carbon	0.93
Manganese	0.30
Silicon	0.25
Chromium	0.05
Vanadium	0.18

There are formulations from W-1 to W-5:

2. *SAE "O" Oil-Hardening Tool Steel.* The addition of an appreciable amount of manganese and some chromium and tungsten to the "W" grade permits this steel to be hardened in oil. It hardens uniformly throughout its section. Thus, it allows more grinding and sharpening. It also enables the making of very complex punch press dies and similar tools that require frequent sharpening and dressing. When heat-treated, it will distort less than carbon steel. These features make it a useful tool steel for the same applications as carbon steels but also for gauges and master hobs.

The analysis of 0-1 steel is:

Carbon	0.94
Manganese	1.20
Silicon	0.30
Chromium	0.50
Tungsten	0.50

The range of formulations is 0-1 to 0-7.

3. *SAE "S" Shock-Resisting Tool Steels.* A number of tool applications—such as cold cutters, screw and nut drivers, concrete breakers, shear blades, header dies, rivet sets, chipping chisels, forming dies, and swaging dies—require extreme strength to resist shock. Thus, a higher toughness value is called for. This can be enhanced by reducing the carbon content and adding more of the elements such as managanese, chromium, molybdenum, or tungsten. These elements improve the ability to absorb repeated stresses and offset some of the loss of hardness due to reduction of carbon content.

The analysis of S-1 steel is:

Carbon	0.55
Manganese	0.25
Silicon	0.25
Chromium	1.35
Vanadium	0.25
Tungsten	2.0

The range of formulations is S-1 to S-7.

4. *SAE "A" and "D" Air-Hardening Tool Steel.* While oil-hardening tool steels represent an improvement with respect to distortion and size change, air-hardening tool steels carry this feature a decided step forward. The slower cooling rate in the hardening results in less strain, and thus lower distortion. The important element that classifies these steels as the air-hardening type is molybdenum. By uniting with carbon and chromium, the molybdenum produces hard carbides, which give these steels high abrasion resistance. This abrasion resistance makes it a preferred tool steel for blanking and forming dies, rolls, cams, and similar long-running applications. The higher alloy content calls for a higher hardening temperature. Vanadium is introduced to prevent grain coarsening.

The analysis of A-2 and D-2 steels is:

	A-2	D-2
Carbon	1.00	1.50
Manganese	0.70	0.50
Silicon	0.30	0.30
Chromium	5.25	12.0
Vanadium	0.25	0.90
Tungsten	1.10	0.75

The steels are available in formulations A-2 to A-9 and D-1 to D-7.

5. *SAE "H" Hot-Work Steels.* As the name implies, these steels are used in operations where the temperature of the tools employing the steels may reach 1200°F. Typical applications are forging dies, hot extrusion, hot shearing, die casting, plastic molding, etc. The principal requirements in these steels is resistance to softening within the range of operating temperatures, resistance to abrasion, and ability to withstand high pressure and intermittent shock. The carbon content is lower than in any previous steel discussed. When the percentage of chromium predominates over that of tungsten, that steel will be more resistant to repeated stresses and will also better resist heat checking from rapid temperature changes.

When the tungsten percentage predominates over that of chromium, the steel will have a higher heat and abrasion resistance.

The analysis of H-13 steel is:

Carbon	0.40
Manganese	0.40
Silicon	1.00
Chromium	5.25
Vanadium	1.0
Molybdenum	1.20

The steels are available in formulations H-10 to H-43.

6. *SAE "T" and "M" High-Speed Steels.* These steels are the most alloyed and are principally used as cutting tools in the metal-cutting industry. They have the ability to maintain a hard cutting edge even when the cutting tool becomes a dull red, a characteristic known as the "red hardness." Tungsten and molybdenum are the predominating elements and may be used singly or in combination with each other. There are many varieties of compositions in which these two elements change with corresponding variations in performance.

The analysis of M-1 and T-1 steels is:

	M-1	T-1
Carbon	0.80	0.75
Chromium	4.0	4.10
Vanadium	1.15	1.10
Tungsten	1.5	18.0
Molybdenum	8.50	0.70

The steels are available in formulations M-1 to M-56 and T-1 to T-15.

STEELS FOR PLASTICS

In selecting a tool steel, the most important consideration is that it maintains integrity in shape and size over prolonged usage. We must always keep in mind the costly moldmaking labor and do everything to protect it against loss due to poor steel selection or improper treatment. It should therefore possess outstanding abrasion resistance, particularly when the molded part is made of either glass-filled thermoplastic or thermosets. It should be tough so that it will withstand extreme conditions of molding—for example, so that when a plastic piece is caught at the parting line, the steel will neither deform nor crack. Its properties at molding temperatures should not change to any noticeable degree. The dimensional change after heat-treating should be negligible. It should have good corrosion resistance. The polishability of the steel after heat-treating to the desired hardness should be such as to attain a good mold finish with ease. A good machinability index is an important factor especially when configurations demand the application of small cutters.

Each grade of steel and each size of bar has a rolling or forging surface that has to be machined off so that the remaining steel to be used for the mold will have the specified composition and will respond to the needed heat treatments as indicated in the supplier's literature. The amount to be machined off is indicated by the supplier and this amount should be added to the dimension of the needed size.

Although the H-13 steel was discussed on p. 161, it is repeated here because of its importance in the plastic field. The following are the most popular grades used for plastic molds and their AISI designations.

1. H-13 is a 5% chrome, high-alloy steel with all the properties required in plastic cavities. It possesses excellent abrasion resistance and toughness, is not affected by temperatures encountered in molding, and is easily polished. It can be heat-treated to 54–56 RC hardness. Its change in dimensions and shape after heat-treating is insignificant. When an unusual cross section is heat-treated where a danger of distortion may exist, an annealing temperature should be specified within which the least change would take place. (See Fig. 10-7.)

H-13, which is manufactured under a variety of trade names, is a grade suitable for cavities where machining does not involve very delicate sections. This type of

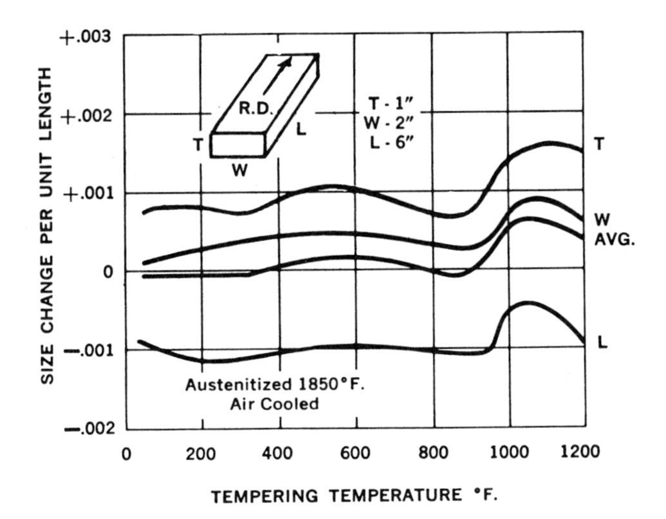

Fig. 10-7. Size change during heat treatment.

steel heat-treated to 54–56 RC should be applied for thermosets, glass-filled thermoplastics, and other materials that have an abrasive action during flow. The resistance to abrasion is enhanced by hard chrome-plating. It is standard practice to plate thermoset molds with hard-chrome, especially for the production rates of 10,000 or more per year per cavity.

The free machining grade has a hardness of 20 RC, and the composition is the same as H-13 except for added sulfides for freer machining. It is a mold steel with machining properties comparable to free-machining screw stock. It is also heat-treatable to 54–56 RC. Where extremely high polishes are required, the standard H-13 grade is preferable but machinability is sacrificed.

When the sulfide-modified H-13 is prehardened to about 44 RC, its resistance to abrasion is comparable to hardened H-13. Its machinability is tough. It is suitable for unfilled thermoplastics.

The modified H-13, 20-RC grade suggests itself for applications where close tolerance dimensions have to be machined with delicate cutting tools and where a high polish is not a significant factor. In general, it could be said when frail sections are machined for precise finished shapes, this steel will be of considerable help in reaching the anticipated result.

The modified H-13, 44-RC prehardened grade is used where the minutest distortion in heat-treating cannot be tolerated and where it might be necessary to correct sizes by machining after a trial run has been made.

Before proceeding to other grades of steel, it is to be emphasized that if a mold steel is to be hardened and there is concern for dimensions, distortion of fits, or brittleness, the grades H-11, H-12, or H-13 are a must. There have been too many unfavorable experiences in which a better wearing steel, under conditions of molding operations, has cracked, making it necessary to replace a costly mold component with a steel that is tough, although the wear resistance of the substitute may not be equal to the one that failed. It is safer to sacrifice one outstanding characteristic of a steel, such as exceptional abrasion resistance, and to use a mold material with the many proven advantages of the H-11, H-12, and H-13 groups.

Some cavities requiring chrome-plating for additional wear resistance may be so irregular in shape that they become very difficult and costly to plate. In this event, remember that the H group can be nitrided, thus providing an excellent abrasion-resistant surface such as exists on ejection pins.

Precautions should be taken that the mold has been fully debugged in production, and all engineering requirements should be thoroughly checked out prior to nitriding. The depth of the case is a function of time exposed to the nitriding gas, and the required depth should be indicated to the heat treater. A case of 0.005 is more than adequate in most applications. When very thin sections are involved in the parts contemplated for nitriding, they should be a danger signal because they may become very brittle and not able to function under conditions of molding. The familiarity of the heat treater with nitriding the H steels should be verified prior to proceeding with such a move.

2. P-20 is a chrome-molybdenum low-alloy steel mostly in prehardened state to 300 Bhn. It is suitable in this hardness for injection materials of the unfilled grades and where the tendency toward stringing and associated indentation is at a minimum. This tool steel can be carburized to a depth of about 1/16 in. with a resulting surface of 54–61 RC; the hardness of the remaining depth would be 41–45 RC. For parts with generous dimensional tolerances, P-20 can be a suitable material for thermosets or abrasive materials. The composition is carbon, 0.30; chromium, 0.75; and molybdenum, 0.25.

3. P-21 is a nickel-aluminum, precipitation-hardening, prehardened tool steel for cut cavities. It is available in hardnesses of 300–330 Bhn. It is also suitable for unfilled thermoplastics, when the tendency toward stringing is not pronounced. Additional wear characteristics can be obtained by nitriding, which results in a very high skin hardness 15N-90-92 Rockwell and a core of 39–40 RC. Since the nitriding is done at relatively high temperatures, the consideration of dimensional and distortion changes should be kept in mind. The composition is:

Carbon	0.20
Silicon	0.30
Manganese	0.30
Chromium	0.25
Nickel	4.10
Vanadium	0.20
Aluminum	1.20

4. The next categories are hobbing steels; the favorable conditions under which they are to be applied are as follows:

- When more than two cavities are needed.
- When it is important to have cavities duplicated with respect to each other.
- When prospects of additional cavities for future requirements are favorable.
- Where cost of polishing would be an appreciable factor in the overall cost of a cavity.

The drawbacks are the cost of making the hobs. Also, hobbed cavities are mostly heat-treated by carburizing with accompanying potential of distortion and core softness.

Like other processes, it has its place, advantages, and limitations, and each case has to be analyzed on its own merits.

The hobbing steels are:

• P-1 is a low-carbon hobbing steel. It is capable of easy hobbing, thereby reproducing maximum detail of design. After carburizing for surface hardness, the core hardness requires careful attention. The possibility of spot depression while in use should be analyzed, particularly while determining wall thickness of cavities. The greater the thickness of the soft core material between the hardened surfaces, the more susceptible an area will be to possible cave-in. This grade is for use when molding pressures are on the order of 2000 psi. Composition: carbon, 0.10.

• P-2 and P-3 are chrome, nickel, and molybdenum alloy steels. They are somewhat difficult to hob, but provide a good core hardness after carburizing to resist sinking in use. The core hardness of 15–20 RC is usually obtained after carburizing. The compositions are:

	P-2	P-3
Carbon	0.07	0.10
Manganese	0.30	-
Silicon	0.15	-
Chromium	1.0	0.60
Molybdenum	0.25	-
Nickel	-	1.25

• P-4 is a 5% chromium air-hardened steel. It possesses very low distortion in heat treatment and develops a high core hardness. It is suitable for shallow cavities and those subjected to high pressures and higher mold temperatures. Composition: carbon, 0.07; manganese, 0.40; silicon, 0.25; chromium, 4.50; and molybdenum, 0.45. The core hardness is about 30–38 RC.

• P-5 is a chrome-nickel-alloy steel. It represents a compromise between ease of hobbing and core hardness. It is easier to hob than P-4, but the core hardness is not as good. The core hardness ranges from 15–25 RC. The composition is:

Carbon	0.08
Manganese	0.4
Silicon	0.3
Chromium	2.3

• P-6 is a chrome-nickel high-strength alloy in which hobbing in conjunction with machining can be used to good advantage. Shallow impressions can be made by hobbing some of the configurations. The remaining required shapes are machined. Good for high-strength cavities subjected to large molding forces. The core hardness is 37–38 RC. The composition is:

Carbon	0.10
Manganese	0.50
Silicon	0.30
Chromium	1.50
Nickel	3.50

Some shops prefer to work with a A-2 or S-7 steel cavities and cores. Actual experience shows that their toughness is not quite up to that of the H-13. For some cavity shapes, if the steels are heat-treated according to the exact specifications of the steelmakers, they will perform well.

5. Beryllium-copper cavities are made by the "high-pressure cast" process in which a steel hob is used. Molten metal is poured around the hob, and heavy pressure is applied to make a dense casting. If the parts are properly annealed, machined, and heat-treated, the end result is a casting with a hardness of 40–48 RC. If correct steps are followed, the end product is a most satisfactory cavity for injection-molded parts. If the finished cavity is properly chrome-plated, it is also adequate for filled thermoplastics.

The steel hob, usually made of H-13 steel, should be larger by 0.004 in. to allow for shrinkage of the beryllium-copper casting. When the castings are to be cores (male), the shrinkage allowance is 0.008 in./in.

Nickel-plated beryllium copper when heat-treated will give a hardened surface comparable to nitrided steel. There is a shrinkage problem in connection with this heat treatment. When heat-treated, the plated and unplated beryllium copper shrink differently. Therefore, at the time of product design, it has to be decided whether the beryllium copper will be heat-treated with nickel-plating so that the proper shrinkage allowance can be accounted for in making the hob. The heat conductivity is four times better than tool steel, and its coefficient of friction against steel is quite low.

When any of the listed material grades are being considered for actual use, the supplier should be contacted for literature in which all the pertinent characteristics are described in detail. Each AISI designation covers a type within a certain range of composition. One steelmaker may prefer one range of composition while another may choose a range quite different, and yet both will fall within the same AISI number. The variation in composition, however small, will lead to different surface-hardening methods, core hardness, and properties, which may influence a specific application. Therefore, the mold designer and molding supervision should discuss and analyze the grade of steel being considered so that the one with the most favorable properties will be applied. Each steel supplier has a comparison chart of the trade names against the AISI designation, which, in effect, indicates all the suppliers for any one grade of steel. It is a useful source of information when comparative evaluation is undertaken.

A listing of the steels used in plastic molds with trade names and suppliers is seen in Table 10–1.

HEAT-TREATING

Most mold steels employed are subjected to a heat treatment of some sort in order to obtain the desired performance in use. It therefore appears appropriate to have an explanation of the heat-treating processes and their effect on tool quality.

The basic steps of heat-treating tool steels are heating, quenching, and tampering. Frequently, heating is done in two steps–i.e., a preheat and high heat. The controlling factors in determining the magnitude of these steps are configuration and mass of material.

The hardening temperatures of tool steels are at such levels that reactions take place between the surface of the steel and the heating atmosphere. The reactions may cause a loss of carbon from the surface. This loss may reach a depth of 0.010 to 0.015 in., thus creating a layer of a working face that is poor in carbon. Underneath the decarburized layer, the composition contains the correct proportion of elements, but carbon is missing where it is needed. This condition can be responsible for an inferior working surface and thus defeat the purpose of heat-treating. Carbon is an important element in tool steels; therefore, decarburization should be prevented by heat-treating in the presence of carbon-rich gases, which is referred to as "controlled-atmosphere heat-treating."

The first step in heat-treating is to preheat the steel to a comparatively lower temperature–between 1000° and 1600°F. This serves to (1) minimize decarburization, (2) avoid heavy scaling, (3) relieve cold-work stress, and (4) cause uniform and gradual heating.

In some cases, two preheating steps are indicated, especially when the objects to be hardened are relatively large.

The second and most important step in heat-treating is the high-heat cycle. Close attention to time and temperature is imperative in order to control the "grain structure" that will produce desired properties. Frequently, a "soak" is called for at the high heat, which means that steel should be held at quenching temperature for a specified duration. This is done because (1) it assures that the entire section is thoroughly heated and (2) it overcomes the sluggishness with which the high-alloy steels respond to appropriate conditioning.

The third step, the quench, is performed at the end of the high-heat cycle. During this step, the actual hardening takes place. The change in grain structure progresses as the temperature decreases to within a range of 700° to 400°F on down to 150°F. At this point, the reaction was found to be practically complete. Each alloy requires a different cooling rate to assure full hardness. Different cooling media are used, and their function is to reduce the temperature

Table 10–1. Comparison of Mold Steels.

AISI	BETHLEHEM STEEL	BRAEBURN ALLOY STEEL	CARPENTER TECHNOLOGY	CRUCIBLE STEEL DIVISION COLT INDUSTRIES	D-M-E	LATROBE STEEL	UNIVERSAL-CYCLOPS	TELEDYNE VASCO
H-13	Cromo-High-V	Pressurdie 3	No. 883	Nu-Die V	No. 5	VDC Viscount 20 Viscount 44	Thermold H-13	Hotform V
A-2	A-H5	Airque	No. 484	Airkool		Select B FM	Sparta	Air Hard
D-2	Lehigh H	Superior 3	No. 610	Airdie 150		Olympic FM	Ultradie 3	Ohio Die
P-1			Mirromold					
P-2	Duramold B							
P-4	Duramold A		Super Sampson					
P-5			Sampson Extra					Vasco Chromold VM
P-6	Duramold N		No. 158					
P-20				CSM No. 2	No. 3		Thermold Z	
P-21						Cascade		
T-420					No. 6	420 Stainless		

quickly at first, followed by a slower temperature reduction so that hardening stresses can be accommodated.

After hardening by quenching is completed, the next step is tempering. Tempering is reheating, after hardening, to an intermediate temperature below the range of hardening, and then cooling to room temperature. The primary purpose is to toughen the steel, which is accomplished by relieving stresses that were set up in hardening as well as by conditioning the changed structure of the steel particles that have developed during hardening. Even though it is a simple step in the heat-treating operation, it may be considered of utmost importance. Control of the time between quench and temper is vital. Since actual hardening takes place gradually during quenching, it is important that the process not be interrupted by premature tempering. On the other hand, excessive time between quenching and tempering may cause the hardening stresses to build up and bring about rupture. The standing rule is that tempering be initiated as soon as the quenched material can be held with the bare hand.

The tempering range is between 300° and 1200°F depending on the grade of steel and hardness desired. The time ranges between 1 and 6 hr. In some steels, double tempering is called for in order to achieve best results.

This heat-treating information is intended to give the reader a general concept of the operation. The instructions for heat-treating a steel are much more elaborate, as will be seen from the following example of instructions for H-13 steel.

An atmosphere-controlled furnace carefully regulated to assure a "neutral" atmosphere is recommended for the 1400°F preheat and the 1850°F high-heat operations. The generous use of thermocouples to assure accurate temperature control is desirable.

Preheat large H-13 dies at 1150°F, and hold at this temperature for 3 to 6 hours after uniform heating. Then raise the preheat temperature to about 1400°F, and hold another 3 to 6 hours after uniform heating to this higher temperature.

Then raise to an 1850°F austenitizing temperature, and hold the die at this temperature for ½ hour per inch of thickness *after* it is uniformly heated to 1850°F. Then air-cool in still air to a temperature of less than 125°F. To minimize warping and bulging of large dies, air-cool from the hardening furnace to a temperature of approximately 700°F. Then place in a tempering furnace controlling at 700°F, and allow the die to cool to approximately 200°F in the furnace. This furnace cooling time should be adjusted to approximately 6 to 8 hours. At about 200°F, remove from the furnace, and complete the quench in still air.

After the die has cooled to less than 125°F, place it in a tempering furnace controlling at approximately 200° to 300°F. Slowly raise the temperature to the desired tempering level, and hold at this temperature for a minimum time of 2 hours per inch of thickness. Air-cool to room temperature. Retemper using the previously described procedure.

Table 10-2. Heat-Treating for Three Mold Steels.

AISI Type	HARDENING (°F)		QUENCH	TEMPERING (°F)	ANNEAL (°F) (1 hr./in. thickness)
	PREHEAT	HIGH HEAT			
H-13	1400–1500	1800–1850	In air-controlled atmosphere	950–1100	1550–1600
P-20	Stress relief	1550–1575	In oil to 1250°F	Double temper 1050–1150	1425–1450
A-2	1200–1400	1725–1750	Air cool	Double temper 300–900	1550

Exceptional wear resistance of mold surfaces can be accomplished by carburizing or nitriding the steel, which is usually called *case hardened.* The depth of penetration of these treatments may vary from a few thousandths up to 1/16 in.

Steels to be nitridable should contain in their composition either aluminum, chromium, or molybdenum.

Carburizing is accomplished by heating steel between 1600° and 1850°F in the presence of a solid carbonaceous material, a carbon-rich atmosphere, or liquid salt.

Nitriding consists of subjecting parts to the action of ammonia gas at temperatures of 950° to 1000°F in order to impregnate the surface with nitrogen.

These treatments can produce skin hardnesses considerably above the maximum hardness obtainable in heat-treated tool steels and provide excellent resistance to abrasion.

Table 10-2 gives a general outline of heat-treating information on the types of steels mentioned for mold use. Needless to say, specific heat-treating data must be obtained from the supplier of a particular grade of steel.

Anyone dealing with mold steels should be aware of potential sources of stresses that can be detrimental to the life and usefulness of the finished tool. These sources of stresses may be a result of tool design, heat-treating, machining, grinding, electric discharge machining, welding or brazing, or any other causes that contribute to heating of the steel in a nonuniform manner.

The problem of stresses is treated in Chapter 11 under each fabrication method and in Chapter 8.

11
Cavity Fabrication

When a decision for making a mold is made, the cost is predicated on producing a specified quantity of parts without additional tooling expenditure. Sometimes, the anticipated quantities are exceeded; other times, they fall short of requirements, and costly repairs become necessary in order to supply the needs.

In the making of cavities by machining, grinding, or electric discharge machining, there is constant drive to improve the rate of metal removal. Cutting tools as well as machine tools are developed for heavier and faster cuts; grinding wheels are tailor-made for special steels to allow deeper cuts per pass; and EDM machines (pp. 179–181) are revamped to burn the metal at an accelerated pace. It is fully appreciated that faster metal-removal rate leads to more economical manufacture, but at the same time it must be recognized that the newer cavity fabrication is associated with generation of more heat and indirectly with higher stresses that if not relieved can cause premature failure.

Suppliers of tool steel caution the user against fabricating stresses and strongly advise a stress-relieving operation. When a steel is to be heat-treated and a preheat cycle is part of the heat-treating specification, then the metal-removal stresses will be eliminated.

A great number of cavities are made of prehardened steel, and therefore would not be heat-treated. For those cavities, a stress-relieving operation should be carried out immediately after fabrication. The stress-relieving temperature as a rule is about 100°F below the tempering heat and is held for 30 min. for each inch of steel thickness. It is best to check the stress-relieving heat and time with the maker of the steel.

The information about fabricating stresses has always been emphasized by the steelmakers, but for some reason it has not been given the attention it deserves. Since a tool drawing should cover all the requirements of a tool element, it would be the appropriate place for a note such as the following:

Note: For heat-treated steel: "*Note:* Use preheat and harden to RC ____."
Note: For prehardened steel: "*Note:* Stress relieve @ ____ °F for ____ hours per ____ inch of thickness."

Every effort should be made to eliminate the invisible source of problems, namely, fabricating stresses.

Mold cavities can be produced by a variety of processes. The process to be used is determined first of all by the lowest cost at which the cavity can be produced for the desired end result. Other factors include precision of reproduction from cavity to cavity, quality of surface finish, durability, and repairability. Frequently, a combination of processes is employed in order to meet all the specified requirements. The most common processes are discussed in the following sections.

MACHINING

Machining processes are those that employ standard machine tools such as lathes, milling machines, grinders, etc. Modern shops are equipped with duplication or copying attachments to machines that will reproduce complicated contours from a model at economical rates. With a precise model as a master, in which dimensions are modified to allow for shrinkages and the outside contours are to required specifications, it is a relatively easy procedure for a mold shop to cut cavities on duplicating facilities. The problems arise when a uniform wall thickness is to be maintained on the product. This phase usually requires a certain amount of toolmaker planning, fitting, and juggling in order to produce the expected uniformity of wall. In any event, where the inside of a product is used for mounting or assembling of parts, such as may be the case with portable tools, etc., the planning for core and cavity machining should be coordinated so that possible discrepancies are avoided.

In this case, where the inside dimensions are exact and to close tolerances and the inside contours have to conform to parts that will fit into them, it would be preferable to make the core first. After the core is made to finished dimensions, a sheet of formable material in proper thickness, such as lead, could be added and formed to the core so that a casting could be made from this assembly. This casting could now be compared with the model and checked out to determine if any thin or thick spots in the wall or other discrepancies are uncovered. Since the outside in most applications is for appearance purposes, it can be modified with relative ease. The inside, however, has set dimensions that are governed by parts that fit there; for this reason, adjustments to the inside shape become a formidable task.

With the high cost of plastic materials, the luxury of making up discrepancies with higher material usage is nonexistent. The purpose of these comments is to call attention to existing problems in tool fabrication, so that the moldmaker may be cautioned against the possible pitfalls. Many products are dimensioned so that regular geometrical contours are involved, in which case straightforward machining practices can be employed for producing cavities and cores. When the intended machining is to be carried out with heavy-metal removal rates and the parts have intricate shapes and precise sizes, it is best to first rough the machine within 1/16 in. of finished dimensions, stress relieve, and follow with finish machining. Otherwise, the machining stresses added to the heat-treating stresses can cause warpage or even cracking.

Grinding, if done so that considerable heat is generated between the wheel and the steel, will produce stresses. If the grinding temperature is above the critical temperature, rehardening may take place. If such rehardening is not followed by retempering, the result may be failure of the component. If the grinding temperature is below the critical temperature, a soft surface may be produced, which would affect the wearing qualities of the steel. Grinding stresses can be minimized by using a wheel that is "soft" with open pores, employing a good coolant, and taking small cuts per pass (approximately 0.0001 in.). There are special wheels for the different grades of tool steels formulated to produce little grinding heat.

Tool designers are familiar with machine tools, their feeds, and speeds as well with the variety of types of cutting tools. All these are well described in mechanical handbooks, and thus there will be no additional information here on the subject. Other cavity-making processes not commonly known will be treated in a more elaborate manner in order to enable the mold designer to make comparative evaluations.

CASTING

Specifically, investment casting may be considered for applications where the number of cavities is greater than six and tolerances of dimensions are in the range of ±0.005. It is particularly adaptable to complex shapes and unusual configurations as well as for surfaces that are highly decorative and difficult to obtain by conventional processes. These decorative surfaces may have a wood grain, leather grain, or textured surfaces suitable for handle grips, etc.

Almost any alloy of steel or beryllium copper alloys can be cast to size and heat-treated to any desired metal hardness that is within the range of the alloy being cast. A comparative cost evaluation will in many cases favor the investment process. Cavities from investment cast tooling when produced by qualified people can be of the same quality as those machined from bar stock, i.e., they can be free of porosity, proper hardness, uniform with respect to each other, and–where the time element is a factor–can be produced in days instead of weeks. In this process, cavities have been made that weigh as much as 750 lb.

The investment casting method calls for a model made of a low-melt material such as wax, plastic, or frozen mercury. The model is a reproduction of the desired cavity block and, when cast, is ready for mounting in the base. It incorporates shrinkage allowances as well as a gating system for metal pouring. The complete model is dipped in a slurry of fine refractory material and then encased in the investment material, which may be plaster of paris or mixtures of ceramic materials with high refractory properties. With the encased investment fully set up, the model is removed from the mold by heating in an oven to liquefy the meltable material and cause it to run out. The molten material is reclaimed for further use. The mold or investment casing is fully dried out during the heating. After these steps, the investment is preheated to 1000° to 2000°F in preparation for the pouring of the metal. The preheat temperature is governed by the type of metal to be poured. When pouring is completed and solidification of the metal has taken place, the investment material is broken away to free the casting for removal of the gates and cleaning.

The making of the model for cavity and core blocks of meltable material is an intermediate step. These model blocks are cast in molds that are the starting point for the process. The starting-point mold consists of the part cavity or core where the parting line width as well as block portion for mounting, etc., are built around the part cavity and core, and thus form the shape needed as the complete block.

The investment-casting process was developed commercially to a high degree of precision and quality during World War II for the manufacture of aviation gas-turbine blades. The turbine blades were made of alloys, which were difficult or impossible to be forged. Subsequently, refinements have been developed in the investment-casting process that are especially valuable to the moldmaking field. Most of these improvements are in the area of investment materials for the purpose of maintaining closer tolerances on the castings. Some mold shops have equipped themselves with the ability to produce investment castings alongside their regular fabrication facilities.

HOBBING

Cavities made by a process of pressing a master into a metal block to obtain a desired shape are known as *hobbed cavities.* Application of the hobbing process has decreased to a large degree, not only because of the advantages and economics of other processes, but also because it is not generally understood. Thus, potential favorable applications are overlooked or at least not investigated. It is difficult to prescribe a set of conditions that would dictate the application of this process over others because there are certain considerations that can influence a decision. These pertain to shape (contour) of part, depth to be formed, details of sections, pressures involved in molding, etc. It can be stated, however, that when four or more cavities are being analyzed, the part under investigation should be considered for hobbing if the accessibility for polishing cavities is difficult, and thus the cost of polishing would be high; if the exactness of shape reproduction from cavity to cavity is essential; or if dimensional tolerances (±0.005) are not very close. Examples include handles for cooking utensils and knobs for ranges and radios.

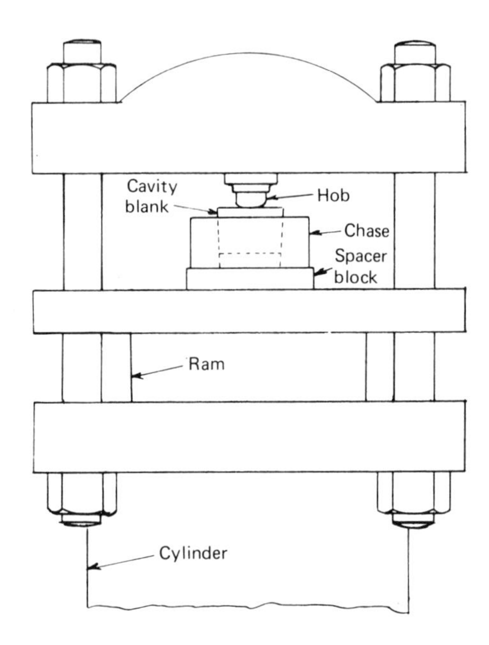

Fig. 11-1. Hob and hobbing press.

The hobbing process consists of pressing a master made of hardened tool steel into an annealed steel blank by a slow-acting hydraulic press. (See Fig. 11-1 for hobbing press and for hobbing components in press.) The pressures necessary for driving the hob into the steel blank range from 40,000 to 200,000 psi. They are determined by the yield strength of the material. From Fig. 10-1, it is evident that there is considerable elongation of the steel at yield stress levels; thus, the steel will flow without appreciable increase in force. An increase in force above yield strength reaches the ultimate strength of the material, causing rupture or tearing of the steel being hobbed.

With the knowledge of the yield strength of the steel that is to be hobbed and the projected area of the cavity, the necessary force can be calculated. If work hardening of the steel takes place, the yield strength will increase with a corresponding decrease in hobbability. To restore the blank to its good hobbing characteristics, annealing of the steel is necessary.

The implication is that, in many cases, several steps of hobbing may be necessary in order to accomplish the desired result. In summary, we can state that the ease of hobbing a steel can be judged by its yield strength and work-hardening tendencies.

Let us now examine each element in the operation and determine what part it plays in the overall picture.

1. The *hob* is conventionally made of an H-13-type tool steel, heat-treated to

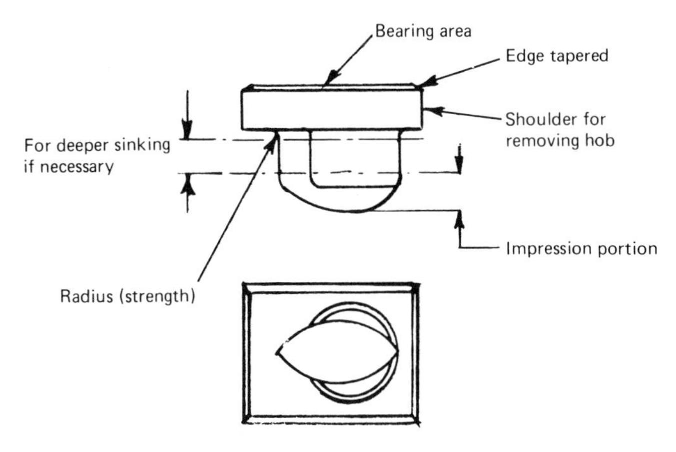

Fig. 11-2. Hob.

56 RC, highly polished, and is plated with copper to provide a lubricating surface. The bearing area of the hub (Fig. 11-2) is about 1.5 times or more of the area to be hobbed, with a 1° to 5° taper from bearing surface to outside edges. This step is particularly important for cases where the hobbing pressures reach values of 100,000 psi or more. It is well known that at the high pressure, steels (whether hardened or not) will deflect, causing a concentration of pressure at the edges. These concentrated pressures will initiate cracks, which may propagate through the hob, causing it to fail.

Deflection is a function of the modulus and since hard and soft steels have the same modulus, deflection may be overcome only by increased cross section. This, of course, is limited by part design. The problem of deflection becomes more pronounced in cases of narrow and long cavities. In these applications, if a flat straight bottom is desired, it may be necessary to have the bottom of the hob convex in order to overcome the deflection resulting from the high pressures on the hob.

In the case of slender cavities such as exist in fountain pens or similar deep cavities, the method of side hobbing or swaging is employed. In such applications, the chase holding the hobbing blank has a positive taper, hardened and polished in the same manner as a drawing die. As the pressure is exerted on the blank metal, flow takes place in the direction of the cavity until the shape of hub is fully encased by the steel of the blank (Fig. 11-3).

When necessary, lettering or emblems may be incorporated in the hobbings. These features must be of a design that will not tear the cavity. Sharp corners and shapes, around which metal flow will be difficult, are an invitation to problems in hobbing.

In all cases, the moldmaker should seek the advice of experienced hobbers to obtain a realistic hob cost as well as information on the practicality of hobbing.

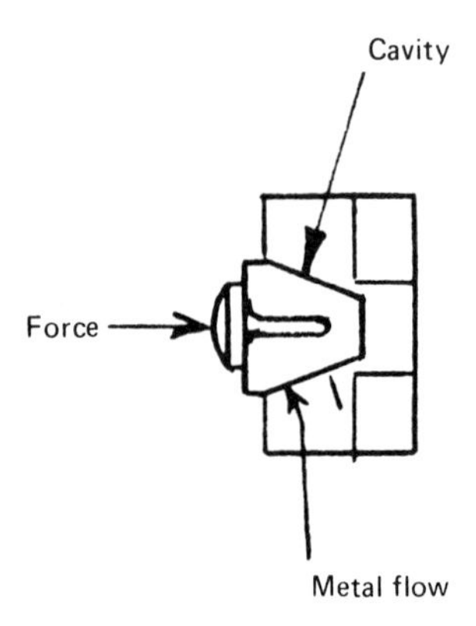

Fig. 11-3. Slender hobs.

2. *Cavity blanks* are shaped so that when the hobs are pressed into them, they will cause the metal from the blank to flow in the direction of the hob and parallel to the axis. The top of the blank has a highly polished surface, which is contacting the hob and in the long run influences the type of finish in the cavity and the flow of the steel being formed.

The types of steel generally used for the blanks are P-1 with a yield of 38,000 psi, P-4 with a yield of 280,000 psi, P-5 with a yield of 90,000 psi, and P-6 with a yield of 80,000 psi.

As stated before, the yield stress indicates the relative ease of hobbing. The thickness of the blank beyond the outline of the hob has a decided influence on the hobbing operation. Too much thickness will cause tearing, and too little thickness will necessitate too high a pressure for the impression, thereby endangering the hob.

3. The *chase* has a function of confining the movement of steel from the blank so that a faithful reproduction of the hob is obtained without tearing or initiating cracks in the blank. Cracks in the blank, however minute, when subjected to temperatures of heat-treating will enlarge and be detrimental to a cavity. In order for the chase to successfully perform its purpose, it has to be dimensioned so that, under the encountered pressures, it will have a low stress level and an insignificant elongation. The net result of this requirement is that we end up with a wall thickness equal to or greater than blank diameter. Thus, for example, with a 6-in. blank, we may end up with a chase ring of 18 in. diam or more. The ring is usually heat-treated to 56 RC and made of a tough steel. The I.D. of the ring has a taper to make possible easy release of the hobbing upon completion of the operation.

4. *Spacer blocks* of various sizes are usually needed to adapt the hob to the press or accommodate a needed blank to a standard chase, and to press platens.

5. The *press* is of the hydraulic design. The platen area in relation to the ton capacity is small when compared with presses seen in the molding and metal-working industry. This keeps the deflection of platens to a minimum. The rods connecting the press head with the hydraulic cylinder body are large in proportion to the platen area in order to minimize the stretch in the rods when the extreme loads are applied. The speed of travel of the cylinder during hobbing is in the order of 1 to 10 in./min. To those familiar with the movement of hydraulic molding presses, this is a very slow rate of travel. The tonnage ratings range from 100 to 2500 ton.

6. *Procedure.* The operation is performed by placing the chase with the blank on the bottom platen in the exact center of the press. Following this, the hub is placed concentrically on the blank and held in place while the press advances at low pressure until the hob contacts the upper platen. The necessary guards are placed in position, followed by application of the needed pressure for embedding the hob into the steel blank. The final step is removal of the blank from the chase and the hob from the blank. Some means of hob extraction is provided in its shank (tapped hole, etc.). This is arranged to accommodate the convenience of the hobber.

ELECTRIC DISCHARGE MACHINING

Electric discharge machining is the latest process being used extensively in the moldmaking field. It can be applied to soft and hard metals, and it exerts no mechanical forces that might be detrimental to frail parts. The process is constantly being improved not only in terms of new machines being capable of producing better finishes and closer tolerances at faster rates of metal removal, but also in terms of toolmakers gaining better knowledge and greater confidence in handling the equipment and producing predictable results. The operation of EDM will be described in greater detail so that it can be applied to existing molds with a view toward improving their performance. A good understanding of EDM, as well as of the controlling of parameters, by those who are involved in the production of parts and are aware of existing mold shortcomings can bring about mold refinements that will improve productivity.

Let us see how this process works. When a spark is produced across a gap between electrodes, erosion of metal from these electrodes takes place. This fact has been observed by physicists for almost 200 years. However, until the advent of the metal tap disintegrator—a tool that breaks down a fractured tap in a tapping hole for ease of removal—this observation was not used in practice.

Before explaining what is going on when metal is eroded, it is most appropriate to differentiate a spark from an arc. In too many cases, there is tendency to lump the two together. An arc takes place when the space between the electrodes is a conducting medium such as ionized gas or an electrolyte. An arc is struck by bringing the electrodes together and then separating them. The arc formation is accompanied with heat generation or burning.

On the other hand, the spark occurs when the space between electrodes is a nonconductor such as a vacuum or dielectric fluid. It occurs when the voltage or "pressure" at one electrode overcomes the resistance of the dielectric separating it from the other electrode. This takes place without noticeable heat.

In the late 1930s, an electric circuit was created that put to practical use the principle of metal erosion in the presence of an electric spark. In this circuit, capacitors (condensers) are charged by a power source, and they, in turn, at an appropriate time signal, discharge their electrons between the work electrode and tool electrode (Fig. 11–4).

The circuit in Fig. 11–4 consists of a capacitor C and is charged by direct current at terminals M and P through a resistor R. The tool electrode, also called the

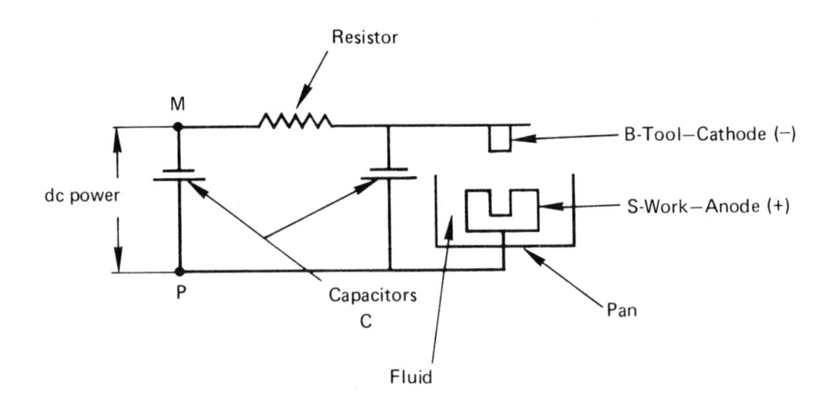

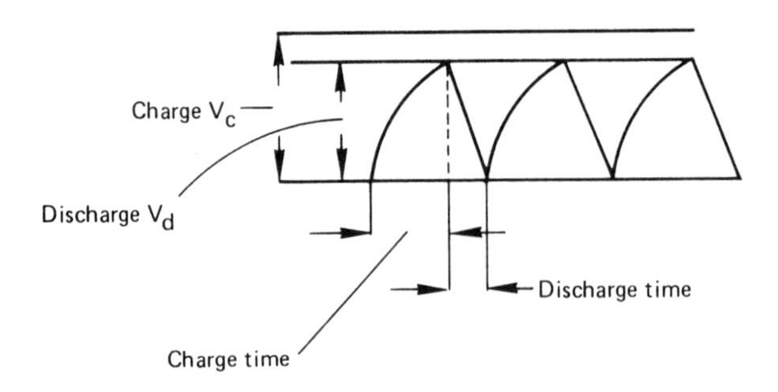

Fig. 11–4. EDM basic circuit.

cathode, and the work electrode, the *anode,* are connected in parallel to the capacitors C. From the figure, we see that the voltage imposed on the condenser brings about a charge in time T and discharges itself very rapidly (in a few microseconds) between B and S. To get some idea of the order of magnitude of the energy involved, let us consider a power source of 1 kw. With a discharge time of 0.1 to 0.001 of the charging time, each spark will hit the work with energy equivalent to 10 to 1000 times the original energy content of the 1-kw power (kilowatts divided by time in seconds equals energy).

Let us see what takes place when this amplified energy is suddenly released. During discharge, a very heavy flow of electrons onto a relatively small area takes place. This heavy flow of electrons creates an electromagnetic field, which causes the molecules of the workpiece to be put in tension. When the electromagnetic forces exceed the resistance to rupture of the work material, molecular particles are torn away from the work and pulled toward the tool. At the same time, the other electrode or tool is under compression due to the electromagnetic force; similarly, when the electromagnetic forces exceed the resistance to rupture in compression of the tool material, molecular particles are forced away, thus causing erosion of the tool. The tool erodes much slower, because the effect of the electromagnetic forces while in compression are not as productive as when in tension. This general principle is the basis of most modern EDM machines.

ELECTROFORMING

Cavities formed by electroforming are based on the principle of electroplating, which is also called *electrolytic action.* Figure 11-5 shows a schematic arrangement of a copper plating setup. It consists of a positive electrode of copper (the *anode*), the electrolyte of copper sulfate (called *blue vitriol*), and the negative electrode (the *cathode*), which is the conductive item to be plated. The imposed

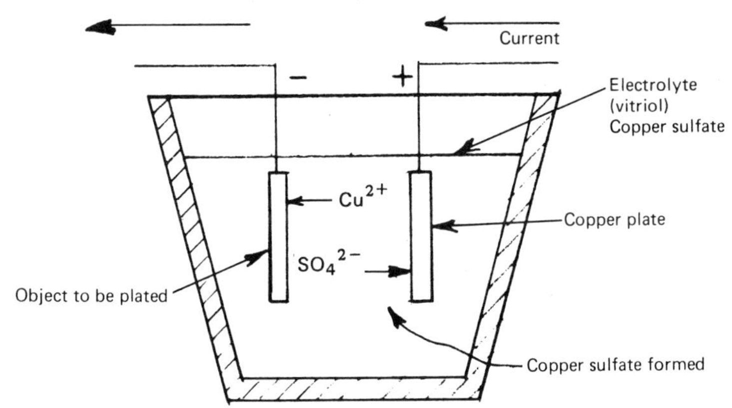

Fig. 11-8. Copperplating cell.

direct current enters through the anode, flows through the electrolyte, and leaves the setup by means of the cathode. Under the influence of the electric current, the copper sulfate is decomposed into its constituents, namely, the copper ion with a positive charge and the sulfate ion with a negative charge. As the current flows through the electrolyte, the positive copper ions move toward the negative electrode and there remain deposited and held in place by the opposite charge. At the same time, the negative sulfate ions move in the direction of the positive copper electrode, where it goes into solution with the copper to form copper sulfate and restores the metallic solution of copper sulfate to its proper strength. This process continues until the desired thickness of plate is built up on the part that is being plated.

The rate of deposition and the properties of the plated material are dependent on the metals being worked with; the current density; the solution temperature; and other factors, such as the specially shaped electrode for uniform thickness to be deposited, etc. In electroplating, we usually deal with thicknesses of 0.0001 in. or even less.

Electroformed nickel cavities require a master or pattern with the shape, finish, and dimensions that are expected to be present on the faces that will be used for molding purposes. The master for the mold configuration must have a conducting surface. If made of a nonconducting material, it must be coated with a conducting film or graphite. The conducting pattern is placed in the electrolyte, and the metal is deposited to a desired thickness. For some shapes, the plated part can be stripped from the pattern and the pattern reused. Other shapes of masters may require their removal by chemical means or, if made of a low-melting-point material, by melting.

The electroformed working shell is usually plated to a thickness of 0.15 to 0.2 in. This buildup in thickness has to be slow in order to insure a uniform wall thickness over the master. It takes from 1 to 4 weeks to deposit the working thickness of the cavity.

Advantages of electroformed nickel cavities are:

1. The nickel used in electroforming cavities can deposit a uniform thickness, be stress-free, and possess a hardness of 45 to 50 RC with good elongation. All these characteristics are important for a cavity material.

2. The deposition of nickel on the master form is done by ions, which are parts of a molecule. This means that these very small particles can fit on the surface of any configuration and thereby reproduce faithfully any irregularity of surface, including highly polished areas that appear to the eye as perfectly smooth.

3. Complexity of shape does not increase the cost of cavity-making.

4. Once a size and shape are established, there is no possibility of distortion or size change due to heat treatment or polishing.

5. Cavities that are prohibitive in cost by other conventional methods can be produced for reasonable prices by electroforming.

6. Unusual parting lines, which are very difficult to match by conventional methods, can be faithfully reproduced by embedding the master in clay up to the parting line and making an epoxy cast over the remaining portion. With the clay removed, the master with the epoxy casting is placed in the working bath for generating the one part of the cavity shell. The next step is to remove the epoxy casting and plate the other part of the cavity against the master and first portion of the cavity shell. With this system, a most irregular parting line can be accurately and economically produced.

One disadvantage of electroformed shells for cavities is that they require building up with backup-plated material, which is machined to a shape that will permit easy mounting in a steel holder. Such a holder provides the necessary reinforcement to insure the maintenance of cavity integrity when subjected to molding pressures.

The backup-plating is usually a fast-plating nickel followed by a hard copper. The thickness of electroformed sections plus backup material is on the order of 3/8 to 1/2 in.

The application of electroformed cavities should be carefully evaluated for an overall comparison with other processes in the following cases:

1. When the surface finish of a cavity is required to be 6 to 10 μ-in., profilometer roughness and accessibility for polishing are very difficult.

2. When the cost of the cavity runs three to four times higher than the average cavity of comparable size.

3. When it is imperative to have a faithful reproduction of a model in which variables from polishing could be objectionable.

4. When emblems, insignia, or surface effects call for exact duplication.

5. Like any other method, the overall cost comparison of the cavity mounted in the base, ready for use, should be analyzed and evaluated.

The present frequent application of electroformed cavities is for gears, screw threads of unusual depth and shape, worm threads, pen and pencil barrels, and products of unusual shape and surface. One of the oldest uses of the process is for molds to make musical records.

VAPOR-FORMING NICKEL SHELLS

Another process of forming nickel shells for molds is that of chemical vapor deposition of nickel on a heated form. Even though the chemical vapor deposition of nickel is over 80 years old, it is now coming into the foreground principally due to the demand for reaction injection molds (RIM), in which nickel has

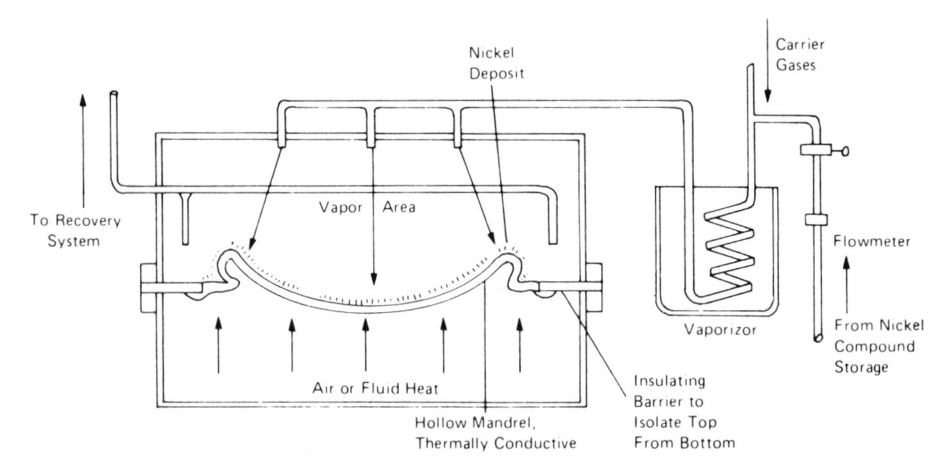

Fig. 11–6. Schematic representation of equipment for nickel chemical vapor deposition. *(Courtesy of Formative Products Co., Troy, Michigan)*

been found to be a preferred surface against which to mold, and the smooth finish of the formed shell as reflected by the finish of the mandrel is also very desirable.

Chemical vapor deposition is a gas plating of nickel accomplished by letting the vapors of nickel carbonyl contact a heated surface that is being plated. This is done in a closed chamber. The heated mandrel (the plating object) receives a uniform layer of nickel from the decomposed vapors, and after a period of time a shell is formed that accurately reproduces the surface of the mandrel. The rate of deposition of thickness can be as high as 0.010 in./hr. The process has the ability to uniformly fill internal corners without the excessive buildup on projections. The vapor-formed shells respond well to annealing and retain exactly the shape in which they were clamped during anneal. Nickel shells by this process are being made for automobile frontal sections and dashboards. They can be made in considerably larger sizes. The chemical vapor deposition of nickel is a process worth investigating and considering for application. The chemicals and products of decomposition are highly toxic, and so the process has to be carried out under closely controlled conditions. A schematic diagram of the process is shown in Figure 11–6.

CAVITIES FOR SAMPLE AND/OR LOW-ACTIVITY MOLDINGS

The section "Bases for Interchangeable Cavities" in Chapter 9 discussed inserts suitable for low-activity production needs. In that system, the object is to make the least expensive production cavities capable of producing parts that meet all specifications. For sample purposes, however, even this simplified system is too costly. The approach to making cavities for samples should be determined by such consideration as:

1. Is a model available, and is it to be used for cavity-making?
2. Is shrinkage allowance an important factor?
3. Will the product from the sample cavity be used for performance evaluation, appearance, or assembly checking?
4. What skill and facilities are available at the location where the making of the sample cavity is contemplated?

The answer to these questions will determine what method of cavity-making will be most suitable for the need.

1. *Cavities from Kirksite.* The zinc alloy known as Kirksite is used in industry to a large degree to make drill jigs, fixtures, forming dies, punching dies, and similar short-run tool applications. It is an easy casting alloy and may be cast into molded sand, special composition plaster, and into steel. It can be machined, ground, and polished to a smooth finish without difficulty.

The alloy has excellent flow properties in molten state and consequently will reproduce faithfully contours and configurations. When cavities are cast from Kirksite, there is little if any machining required for the molding surfaces of a great many sample runs. This is the case if the pattern has a good finish, and the material into which the alloy is cast will accurately reproduce the pattern surface. The shrinkage of the alloy is 0.14 in./ft, which has to be allowed for in the pattern. When cavities are fabricated from the Kirksite alloy by qualified foundry people, the end result can be a satisfactory product at a relatively low cost. The alloy can be plated for additional wear resistance and hardened by the "Iosso" method.

The Iosso treatment of zinc alloys has been claimed to produce hardness up to 72 RC, and the depth of penetration has been claimed possible up to 0.020. With this kind of hardness and penetration and proper mounting in a holder, the Kirksite cavity could be considered for intermediate production runs, e.g., 10,000 pieces per year. The process, according to the literature, is patentable and is a proprietary method of the International Processing Co. of America, Elk Grove Village, Illinois 60007.

In performance characteristics, this type of treated zinc casting could be compared with a carburized hobbed cavity with a soft core (about 10 to 15 RC hardness).

The creep resistance of the alloy is low. To overcome this weakness, the cavity should be well encased in the mold base so that it will not change in dimensions when subjected to molding pressures. In other words, the cavity has to be surrounded by steel from every direction so that there is no space into which the zinc alloy can creep. After the cavity of Kirksite has produced the necessary samples, the alloy can be remelted and used over again.

2. *Aluminum castings* with a good surface finish can also be considered for sample moldings, but they are somewhat at a disadvantage in comparison with zinc alloys for cavity application. They tend to be more porous, and therefore they could present uneven spots in mold temperature. They do not reproduce shapes as faithfully as the zinc alloy and so may involve additional machining or finishing. There are companies that specialize in casting aluminum cavities for low-activity parts and also for molds with pressures of injection on the order of 2000 psi, and where heat conductivity is of major importance.

3. *Silicone rubber* of RTV type (room-temperature vulcanizing) can be used to make cavities and cores from a model. The complete cavity shaped in RTV can become the means for casting a few epoxy samples, which can be evaluated for appearance and even functional purposes.

4. *Machined aluminum cavities* are frequently used whenever the machining time of such cavities will offset the time of making a model (pattern), casting the cavity, and finishing it ready for application. The aluminum alloy for this type of use could be Alcoa No. 7075-T73 forging, a free machining grade that will not lose sizes due to machining stresses. Again, the machinist's experience with optimizing feeds, speeds, grinds of cutting tools, etc., will determine whether the expected economies can be attained. Broadly speaking, it can be stated that the making of models, samples, and prototypes is in a state of specialty art and depends upon the skill and inclination of the craftspeople involved. What is easy and inexpensive to one can be complicated and prohibitive in cost to another.

FINISHING MOLD SURFACES

After the cavities and cores have been fabricated, they are hardly ever ready for molding. Molding materials faithfully reproduce surfaces against which they are molded. Any imperfections in the mold will show up in the product.

Before a mold is released from a tool shop, it has to impart into the cavity a surface finish that will produce an acceptable product. Each individual has a different concept of what finish is good and acceptable. To avoid misunderstandings, the industry is providing a number of comparative standards that can be specified on tool drawings and put into effect by tool shops.

These standards are:

Mold Finish Number SPI-SPE	
1	Nat. Bur. Std. Grade Diamond Compound No. ¼ to 3
1.3	Compound No. 3

1.6	Compound No. 6
1.9	Compound No. 9
2	Compound No. 15
2.3	Compound No. 30
2.45	Compound No. 45
2.5	500 grit aluminum oxide cloth plus buffing wax
2.6	1000 grit abrasive stone
2.7	500 grit aluminum oxide cloth
2.8	500 grit abrasive stone
3	320 grit abrasive cloth
4	320 grit abrasive stone
5	240 grit blast
5.5	180 to 80 grit blast
6	24 grit blast

Comparison Kit

Some mold shops have a mold in which these finishes exist and will mold a sample of the proper material and color for selection of the finish by the user of the end product.

The drawing specifications for finish fall into two categories: one for the cavity and the other for the core. The cavity surface usually determines the outside appearance of a product and may call for a No. 1 finish. On the other hand, the core is hidden from the view, and the only function of its surface finish is that it will perform well in molding and therefore, could be, for example, No. 3 finish. The difference in cost between the two is appreciable. The specifications of surface finish should be just high enough to answer the need.

Suppliers of polishing materials have reciprocating and rotary electric hand tools that will accommodate stones, polishing sticks, felts, files, brushes, small carbide cutters, mounted points, etc. Any of them will aid in speeding the operation of polishing wherever usable.

The operation of finishing of molding surfaces is not as standardized as the ones for fabricating them. Each shop has its own "best" ways of doing the job.

First of all, cleanliness is a most important phase of this work. Felts, brushes, stones, sticks, etc., once they are used with any abrasive size, must be thoroughly cleaned before changing to another one. Better yet, each one of these items should be separated and confined to one specific abrasive number. The possible mixing of abrasives can undo whatever work has been accomplished before. It can also cause pitting or indentation and thus possible damage. The steps involved in polishing cavities by different fabricating methods will vary to some degree and will be discussed separately.

POLISHING

Machined Cavities

The finishing operation during machining, if carried out with precision ground and honed cutting tools, light feed, and light depth of cut, will leave a surface that will require relatively little metal removal for finishing. In these circumstances, an allowance of 0.002 for polishing will be satisfactory. Otherwise, an allowance of 0.003 to 0.005 would be necessary. A first step will be to use moldmaker's rifflers for filing any rough spots, steps between cuts, or other ridges, and thus have the surfaces prepared for stoning. Any areas that will be inaccessible for stones or other smoothing tools should be filed to as fine a finish as possible. For cavities that are to be heat-treated, 20% to 25% of the amount left for finishing should not be removed prior to heat-treating, so that they can be brought to the desired finish after hardening. Heat treatment should be carried out in a controlled atmosphere or "bright hardened" so that the steel surface retains the imparted finish.

Stoning is carried out by stones soaked in kerosene before use. During the stoning, the pressure should be light, and the stone should be checked frequently for "loading." They are available in grits of 150, 240, 400, and 600. The first stoning operation is performed by the coarsest grit needed for the finish at hand. They are silicon carbide vitrified stones, in several shapes, and, when needed, they can be dressed to special shapes. With the stone of the first grit used, the operation should be done in one direction only until all marks of machining and filing are removed. Upon completion of this step, wash the cavity thoroughly. The washing step is performed with each change of grit number.

The next step is to use the higher grit stone, and work it at 45° to the first direction until all preceding stoning marks are removed. As the third step, use third higher or No. 400 grit size and stone at 90° to the second direction until all scratch marks from the second stoning are gone.

In the final step, use the highest grit number, and stone in the direction of the first step until all previous marks are removed.

Stones must be kept clean, separated by grit size for storage in kerosene and with all signs of loading removed prior to use. The diamond polishing should first be carried with a 30- to 60-μ paste followed by 8- to 22-μ and finished with 4- to 8-μ compound. Pressure during diamond polishing should be light so as not to produce a burnishing effect. Remember, polishing is accomplished by the cutting edges of the small abrasive or diamond particles. They must move over the metal with light pressure and be permitted to make minute cuts or else they will embed themselves into the metal and cause pitting. Should pitting inadvertently develop, the steel would have to be stress-relieved at a temperature of 100°F below the tempering heat, restoned, and repolished.

After the cavity and core are polished to desired specifications, the first acceptable parts molded should be given to the mold finisher as evidence of proper polishing performance. The reason for doing so is that after the mold is run for a certain length of time one can encounter molded items with a nonuniform surface finish. Molding conditions are usually responsible for *mold deposit,* in which a mold coating forms in some areas of the mold. Mold deposit usually occurs when a heat-degraded particle of the material combines with the air in the cavity to form a new compound that adheres to a molding surface; this alters the contour of the mold and gives it the appearance of poorly finished and polished tool. The coating can also enter the vents, clearances around knockout pins, and other irregularities of a mold that would affect the venting of gases from the moldings. Following are some of the causes of and solutions to mold deposit:

1. Excessive residence time of material in the chamber at high temperature. The cure for this is to go to smaller-size machines successively in ounces and to arrange the pyrometers so that only the last one has a higher temperature needed for the job.
2. Improper placement and spacing of vents. The vents for a mold deposit job should be placed along the circumference of the molding about 1 in. or less apart. Good venting may eliminate most of the problem.
3. Improper functioning of heaters and thermocouples for the chamber. These should be checked out for proper functioning.
4. High back pressure and screw rpm. These should be kept on the low end of the required range.
5. Overheating of polymer due to gates and length of land. The length of land should not be more than .050.

Whenever a mold deposit does occur, an automotive chrome cleaner or trisodium phosphate or similar cleaning compounds may be used for removing the coating.

Hobbed Cavities

The finish of hobbed cavities begins with the master hob itself and the finish of the blank top. Master hobs will be polished in the same manner as described for machined cavities. Extreme care must be exerted in removing the minutest particle of abrasive by thoroughly washing and visually examining under a toolmaker's microscope. After hobbing, most cavities are carburized and hardened to the specified application hardness. Any signs of surface impairment due to the heat-treating should be removed by liquid honing and followed by a light diamond polish to restore the original hobbed finish.

EDM'd Cavities

Cavities and cores, when produced by electrodischarge machining in two steps—namely, rough burning and finish burning—are in many cases satisfactory for use in the condition as finished. Several types of finishes are attainable that may be acceptable in "as is" condition. These finishes are based on standards that are supplied by companies making electrode material. Stoning of EDM surfaces may be called for to improve the finish and give direction that coincides with withdrawal of parts. Occasionally, stoning is done to remove the hardened, thin skin generated by the spark instead of stress-relieving to maintain uniform hardness throughout the steel. If a 3- to 5-μ finish is desired, the same procedure is to be used as in machined cavities except that in all likelihood the starting point will probably be the No. 400 grit stone.

Final Note

There are other polishing aids such as abrasive paper and cloth, abrasive drums, spiral-wound cones, felt loaded with abrasives, rubber loaded with abrasives, and abrasive compounds with various grits, and they are used by polishers who prefer one material over another. Each polisher feels that he has a fastest way to reach a needed surface. The method described herein will provide some concept of the polishing operation from which an estimated cost for polishing cavities can be derived, based on the differing processes of fabrication. This will enable the mold designer to choose the method that, from an overall point of view, will be most economical.

WELDING

A change in product design, excessive wear in the mold, damage of the mold in operation, and similar causes may be a valid reason for considering welding. However, welding has certain disadvantages and may even endanger the whole mold if all possible side defects are not fully compensated for. Wherever an insert will correct the deficiency, it should be considered instead of welding. As can be seen in the instructions for welding H-13 steel, certain steps have to be followed to insure a good welded job. These are:

1. When the section to be welded can be annealed, proceed in the following manner. Preheat to 1000°F if possible, or to as high a temperature up to 1000°F as is practical. Use a furnace to assure uniform, stress-free preheating. If necessary to preheat with a torch, preheat a large area around the weld area to reduce thermal shock. For critical working areas, weld with uncoated H-13 rod, or similar-analysis material, using shielded arc equipment (atomic hydrogen or heliarc). Keep the temperature of the die above 600°F by reheating until the welding is

completed. Then, retard cooling by either furnace cooling or by covering the welded areas with asbestos, lime, ashes, or some other insulating material. For noncritical areas, welding with coated stainless electrodes using electric arc equipment is satisfactory as long as the preheating and slow cooling precautions are taken to minimize thermal shock.

Follow with a full pack anneal at 1550°F. After finishing to size, the die can then be heat-treated in a conventional manner.

2. When annealing of the die is not practical, weld in the hardened condition using the following recommended procedures. Preheat the dies to as high a temperature as possible, preferably in the range of 800° to 1000°F. Again, furnace preheating is safer, but torch preheating is widely practiced. Weld with uncoated H-13 rod or similar air-hardening rod, using shielded arc equipment; or, if desired, noncritical areas can be welded with stainless rod and electric arc equipment. After welding, place in a furnace at the preheat temperature, equalize, and then cool slowly to room temperature. It is beneficial to reheat again to just below the tempering temperature and then to air-cool again. This second reheating will not only serve as a stress-relieving treatment but will also reduce the hardness of the welded area in line with that of the base metal. Local welding conditions will make variations in welding methods necessary, but again the principle of preheating and slow cooling should be respected to minimize risks.

The procedure for welding an alloy tool steel should follow the recommendations provided by the steel company. The main principle for welding any tool steel is to preheat to around tempering heat and weld in a shielded atmosphere, using the same composition rod (in small size) as the base material. Welding is to be done with the tool at 600°F, and the welded structure is permitted to cool slowly to below 200°F. The steel should be reheated to tempering temperature and held there for 0.5 hr for each inch of steel thickness. This is to be followed by slowly cooling to room temperature.

In addition to these points, some other details are of importance. In preparing the surface for welding, all the cracks should be ground away with a rounded wheel, so that no sharp *V* is formed. At least 1/8 to 3/16 in. depth for weld metal should be provided. Cleanliness of the welding rod as well as of the surface to be welded is of utmost importance. If possible, work should be positioned so that the beads are laid slightly upward. Welding procedure should be such that minimum heat is generated. The rate of travel should be slow and straight. It is preferred to use several stringer beads over heavy deposits. Slag should be removed frequently. After each bead, while metal is still hot, a light peening action is in order.

The tool steel welding is a specialty that should be done by experienced welders who respect the varying steel compositions, who accordingly select welding rods that are compatible with the steel. In addition, they should have the know-how to compensate for possible ill effects of the welding heat.

TEXTURED SURFACE

A great variety of mold surfaces are in demand by the users of plastic products. Typical examples are handle grips, leather grain, matt surfaces, etc. They can be produced by a chemical process, and the type of texture is selected from master samples. It is another specialty that is in a class of art and is carried out by companies devoted to the field. The one caution that has to be exerted by the company involved is to insure that removal of the part from the mold is not adversely affected.

HARD CHROME PLATING

The increased usage of glass-reinforced thermoplastics will find more call for hard chrome plating. There is one basic fact that is true in all plating, namely, that the chrome will faithfully reproduce the surface over which it is deposited. If a smooth chrome finish is expected, the underlying surface has to be smooth. The most common thickness deposited is 0.001 in. It may vary from 0.0002 to 2 without excessive buildup at edges. This is another field that is still in the art stage, and it is important to deal with a qualified plater experienced with mold plating. On some steels, hydrogen embrittlement encountered in the plating operation should be guarded against, and the plater should be requested to provide the treatment that will eliminate same.

CONCLUDING NOTE

Whatever operation is performed on a cavity and core, the questions should be asked: Is there a possibility that harmful stresses have been introduced? Are steps being taken for stress relieving? Remember that finishing cuts, which remove small amounts of material and consequently generate little heat, will minimize fabricating stresses.

12
Mold Temperature Design

The ability to control the mold temperature is an important factor in plastics production. In some molded materials, it can be used as an effective tool for maintaining minimum cycles, which means high production rates. However, in some other materials, particularly those that are known as engineering materials, this factor can be employed as a vital means of both controlling dimensions and properties such as tensile strength and brittleness and minimizing molding stresses. In most materials, the uniformity in appearance is largely dependent on the consistency of the mold surface temperature.

The material processing data indicate mold-temperature ranges for each material. The relatively new thermoplastic materials–namely, those that will withstand higher temperature usage (above boiling water)–demand closer attention as far as uniform heat distribution and heat transfer in the mold are concerned. Careful planning of heat transfer means that the mold will pay dividends in terms of appearnace, quality, lower stresses, better use performance, and favorable production rates. In planning the heat-transfer arrangement, it has to be recognized that each mold presents a different problem in designing the heating-cooling system.

The following sections review some basic principles of heat transfer and their application to the molding process. Chapter 17 will enumerate the side effects of improper heat transfer.

BASIC PRINCIPLES

Heat flows from a body of higher temperature to one of a lower temperature. It is the temperature difference–and not the amount of heat contained in separate bodies–that determines the flow of heat. The greater the difference in temperatures between two bodies, the greater the rate of heat flow between them. It may be compared to the difference in two water levels and the flow of water.

Heat can be transferred from one medium to another in three ways: radiation, conduction, and convection. Heat transfer by radiation is similar to the passage of light through space. Radiant heat may be deflected or shielded against in a manner similar to that achieved with light.

FIGURING A COOLING SYSTEM

The mold designer must calculate the heat introduced by the plastic material to the mold and determine the cooling method that will bring about its rapid solidification for ejecting the product from the core. The cooling method can be figured from formulas and related information found in each heat-transfer condition. As a first step, let us consider those cooling arrangements that have been found from experience to be successful in performance.

1. Whenever mold inserts are used, all effective cooling passages should be in the inserts (cavity and core). Cooling channels may also be provided in plates that hold the mold inserts as a means of preventing cooled cavities and cores from losing their own capacity to solidify the plastic.
2. Experience and calculations for deflection have indicated that the cooling channels should be 1-1/2 to 2 diameters from the molding surface and the pitch should be 3 to 5 diameters, in steel formulated for thermal fatigue. In other mold materials, the same procedure can be used as outlined in the text. See Fig. 12–1.
3. For efficient cooling, a turbulent rather than laminar flow should be used. The turbulent flow can be achieved by having a minimal flow for 1/8 pipe size (11/32 in.) of .566 gpm; for 1/4 pipe size (7/16 in.) of .72 gpm; and 3/8 pipe size (19/32 in.) of 1.0 gpm. The water in those cases is 50°F. These flows would give velocities of 1.95, 1.53, and 1.13 ft/sec for each size. The turbulent flow has a property of throwing particles of water against the circumference of the passage, thereby at least doubling the heat conductivity. See Fig. 12–2.

A layout of a specific mold showing the selection of fluid passage as well as distribution between cavity and core will provide the necessary data for a cooling system. The only missing component is the length of total passage, which can be figured by multiplying the velocity of fluid times the cycle in seconds.

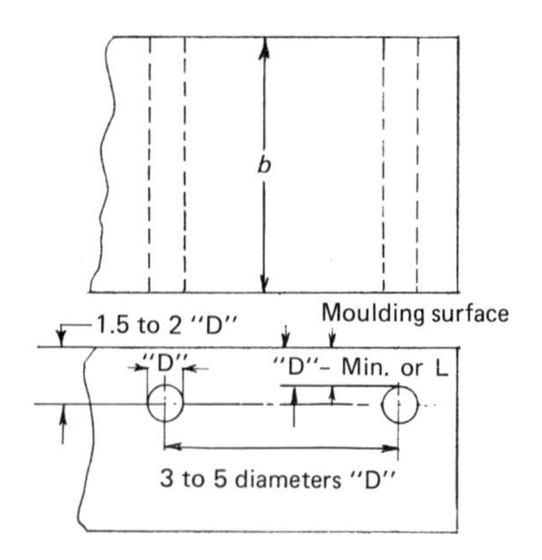

Fig. 12–1. Coolant-hole distance from molding surface.

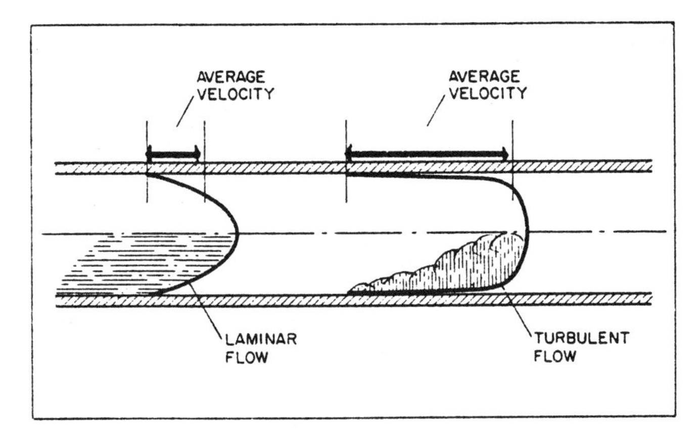

Fig. 12–2. Distribution of velocity in a tube. Note that average-velocity curve of laminar flow is parabolic, while turbulent flow has a much flatter curve, having a higher average velocity.

Preceding are the fundamental steps for determining a cooling system; these will be enlarged by additional information in the text. When a mold is designed along these lines and is ready for tests, results can be compared to a table giving a favorable cooling rate for each thickness of material. Any loss in productivity that shows up in the comparison should be investigated and corrected. The comparative table has been derived by a computer and software developed by the "Moldcool" process, in which the calculations are made for the various cooling factors. The calculations, which would be cost- time-prohibitive to figure by hand, are done in a relatively short time by computer. The table and instructions are shown in Fig. 12–3.

The following paragraphs describe the elements of heating and cooling. Some data are shown in the "Material Processing Data" chart in Chapter 18; other data sheets can be obtained from material suppliers.

Heat Content

The heat content of the plastic material entering the mold can be figured from the following formula: 1

$$H = P(t_1 - t_2)S + h \tag{1}$$

where H = content of material heat and is figured in British thermal units (BTU) per hour, per pound, and per °F.

P = weight of plastic material per hr in lb, or area of the molding in in.^{2} times its thickness in in., times its specific gravity.

Plastic Molding Minimum Cure Time

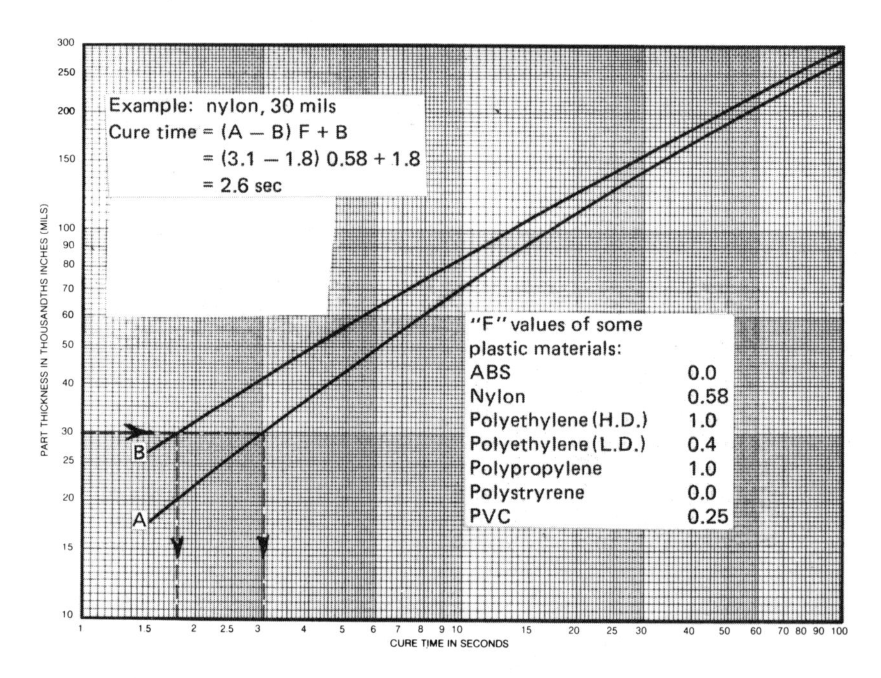

Minimum Actual Cooling Time in Seconds

Wall Thickness Maximum		ABS	Nylon	Low Density Poly-ethylene	High Density Poly-ethylene	Polypro-pylene	Poly-styrene	Polyvinyl Chloride
Mil.	Inch							
20					1.8	1.8	1.0	
30	1/32	1.8	2.5	2.3	3.0	3.0	1.8	2.1
40		2.9	3.8	3.5	4.5	4.5	2.9	3.3
50	3/64	4.1	5.3	4.9	6.2	6.2	4.1	4.6
60		5.7	7.0	6.6	8.0	8.0	5.7	6.3
70		7.4	8.9	8.4	10.0	10.0	7.4	8.1
80	5/64	9.3	11.2	10.6	12.5	12.5	9.3	10.1
90		11.5	13.4	12.8	14.7	14.7	11.5	12.3
100		13.7	15.9	15.2	17.5	17.5	13.7	14.7
125	1/8	20.5	23.4	22.5	25.5	25.5	20.5	21.7
150		28.5	32.0	30.9	34.5	34.5	28.5	30.0
175	13/64	38.0	42.0	40.8	45.0	45.0	38.0	39.8
200		49.0	53.9	52.4	57.5	57.5	49.0	51.1
225		61.0	66.8	65.0	71.0	71.0	61.0	63.5
250	1/4	75.0	80.8	79.0	85.0	85.0	75.0	77.5

Fig. 12–3. Minimum cure time. *(Courtesy of Application Engineering Corp.)*

t_1 = melt temperature in °F.

t_2 = product temperature at ejecting.

S = specific heat in Btu/lb/°F.

h = latent heat of fusion in Btu/lb.

The melt temperature t_1 on the high side would be required for thicknesses of .08 and less and a flow length of more than 3 in. The melt temperature on the low side would be applicable to thicknesses of .080 up to .250 and a relatively short flow length around 5 in. or less. Product temperature at ejecting (t_2) indicates the retained heat in the product while it is being ejected. It can be determined by using the high limit of mold temperature if the heat-deflection temperature is below it; otherwise, the average of the two–i.e., the high mold temperature and heat-deflection temperature–should prevail. S is also taken from the material processing data (Chapter 18).

The latent heat of fusion h is defined as the hidden amount of heat required to melt a unit mass of the substance without any change of temperature or pressure. Thus, it takes 144 Btu/lb to melt ice of 32°F into water of 32°F. For amorphous material, $h = 0$; for some of the frequently used crystalline materials, the values are:

Material	*Btu/lb*
Acetal	70
Nylon	56
Polyethylene, high-density	104
Polyethylene, low-density	56
Polypropylene	90
Thermoplastic polyester	54

In formula 1 the thickness of a product appears as a factor. It should be noted that the cooling requirement goes down in proportion to the square of thicknesses. During the construction of a mold, the wall thickness usually has a tolerance so that a mold designer could specify a minimal wall that would result in a lowest cycle time. In the same formula, all the generated heat finds itself into the complete mold. When the product is flat in shape, each half of the mold will dispose of half the generated heat. On the other hand, when products and therefore molds have a depth of over 1/2 in., as found in cups, containers, and similar objects, the distribution of heat demands attention.

During a complete cycle, the following occurs: When the injection pressure is on the material, the heat is equally distributed between the cavity and the core. Upon freezing of the gate, part in the mold starts shrinking, pulls away from the cavity, and at the same time grips more tightly to the core. In this portion of the cycle, at least 2/3 of the heat will go to the core augmented by an additional time segment while the product is on the core before ejection takes place. This gives a greater heat-absorbing percentage of the total Btu to the core. A core depth of about 2-1/2 in. will absorb 55% of the total. The distribution in deeper molds can be 60%; when inside ribs and other configurations are present on the inside, it can be as high as 66-2/3% in favor of core over cavity.

Cooling Design

In practice, heat transferance by conduction, convention, and radiation all occur at the same time, but in theory they are treated separately.

1. *Conduction.* Conduction takes place when two bodies are in close contact with each other and a high temperature flow is transferred to a low temperature area. In molding we have a plastic material at high melt temperature under the injection pressure that is contacting the mold surface; from there through a distance L, the heat is dissipated to the coolant that is circulating in openings provided in the mold.

The factors that play a part in the heat flow by conduction are:

1. The heat conductivity of material through the distance L or the mold material
2. The area exposed to the heat flow or the area of the cavity in the mold
3. The difference in temperature in °F between incoming plastic and the mold temperature.

The formula applied to this condition is

$$H = KA(T_1 - T_2)T/L \qquad (2)$$

where
H = quantity of heat in Btu per hr, per lb, and per °F.
K = thermal conductivity of materials of interest to molding, in Btu per hr, per ft^2, and per °F.
A = area in ft^2 of cavity
T_1 = melt temperature
T_2 = temperature of coolant
L = distance between the molding surface and the upper end of circulating hole in ft
T = time in hr, or cycles/hr

This formula applies to surfaces parallel to each other and maintained at constant uniform temperatures. In actual practice the temperature varies with time, and we have a heat flow in an "unsteady" state. The values obtained with equation 2 closely approximate the true conditions. The "unsteady" calculations are complicated, suitable only for computer figuring.

Table 12–1. Thermal Conductivity of Materials (Btu/°F/ft²/hr).

Copper (pure)	222	Water	0.39
Copper alloy*	187	Polystyrene	0.07
Aluminum 2017	95	Polyethylene	0.18
Brass	69	Nylon	0.14
Beryllium Copper	64	Polypropylene	0.07
Steel (1%C)	26	Air	0.14
Tool Steel P20	21		
Tool Steel H13	12		
Stainless 316	10		

*This alloy consists of 99.14% copper, 0.85% chromium and 0.01% silicon. It has a 70,000-psi tensile strength and is produced by Resistance Welding Assoc.

The K factor changes with the temperature at which conduction takes place. The materials listed in Table 12–1 were tested at 100°C (212°F), which is a reasonable temperature for molding.

The distance L of the circulating fluid opening is determined by strength considerations, namely, by limiting the deflection to a very low value and by thermal fatigue.

l can be figured by viewing the condition shown in Fig. 12–1 as a beam fixed at both ends and with the load in the middle. From the *Machinery's Handbook* ("Beams, Stresses, and Deflections"), we find the stress in the middle to be

$$S = Wl/8Z \quad \text{or} \quad WD/8Z$$

For this purpose, $D = 7/16$ (.4375) and W = load on 1 in.² of hole opening = 16,000 psi., i.e., the pressure inside the cavity. The normal gauge pressure exerted on the cylinder is 20,000 psi, but the pressure drop to the inside of cavity is about 4000 psi.

l = length of beam = D = 0.4375 in.

$b = 1/0.4375 = 2.29$

$d = L$

S = Safe load stress = 1/5 of tensile strength = 150,000/5 = 30,000 psi

$Z = WD/8S = 16{,}000 \times 0.4375/8 \times 30{,}000 = 0.029$

also

$Z = bd^2/6$ or $bL^2/6 = 2.29L^2/6 = 0.029$

$L^2 = 0.029 \times 6/2.29 = 0.076$ in.

Thus for a 7/16 hole, the calculated length $L = 0.276$. Now let us see what the deflection is at this value. The deflection is given by

$$a = Wl^2/192\,EI$$

where E = modulus of elasticity = 30×10^6

I = moment of inertia = $bd^3/12$

or

$$2.29L^3/12 = 2.29 \times 0.276^3/12 = 0.004$$

Substituting in these and previous values, we get

$$a = 16{,}000 \times 0.4375^2/192 \times 30 \times 10^6 \times 0.004 = 0.000125.$$

This deflection in confined areas would be excessive. For this reason, the distance from the molding surface is made to be *D,* which gives a deflection value of 0.00003.

Calculations have also been made for the 11/32 and 19/32 openings, and the deflections were very close to the 0.00003 obtained above, provided that in each case the $L = D$ substitution was made and the steels were considered to be P-20, H-13, and stainless 306. All of these have been formulated for thermal fatigue. When other materials are used, calculations similar to these lead to satisfactory results.

In half a dozen different molds, the author has observed a leaking crack developed in spite of the fact that the thickness over the hole was 5/16 in.

2. *Convection.* Convection cooling occurs where the heat exchanger acts to solidify the molten plastic by means of a circulated fluid.

There are a number of complex formulas pertinent to this method of cooling that only a computer can solve in a reasonable time. There are also formulas that will provide answers close enough to the computer solutions for practical evaluation and are suitable for doing by hand. The following equation will provide additional needed data:

$$H = MS(T_2 - T_1)T \qquad (3)$$

where H = Btu/hr

M = weight of material that circulates in the passage at T cycles/hr and produces a temperature difference of $T_2 - T_1$ in °F at the inlet and outlet of the mold

S = specific heat value of circulating medium; for water, $S = 1$

M can be expressed in terms that relate to the molding:

M = Volume × Specific gravity, or Area of exposed passage to heat pickup × Velocity × Specific gravity × Duration of flow

All measurements are in ft. Velocity, selected to provide an efficient heat exchange, is determined by a so-called dimensionless Reynolds number. The formula for that is

$$R = 7740\ Vd/n \text{ or } 3160\ Q/dn.$$

where V = fluid velocity, ft/sec

d = diameter of passage, in.

n = kinematic viscosity of water (see Table 12–2)

Q = flow rates in gpm; specific gravity of 1 ft^3 of water = 62.41

T = Duration of flow in hr or cycles/hr.

It is generally agreed that the Reynolds number for convection cooling should be around 4000 which corresponds for a 7/16 hole to 1.53 ft/sec. See Fig. 12-4. Substituting for M in equation 3, we have

$$H = 1/12 \times d^2\pi/4 \times V \times 62.41(T_2 - T_1) \times 180 \text{ (cycles/hr)}$$

The diameter of passage is now well standardized, i.e., the 7/16-diam hole is found on all molds that fit presses from 500 down to 50 tons. For 50 tons or less, the 11/32-in. holes predominate. For over 500 tons, the 19/32-in. openings are applied.

Table 12–2. Viscosity of Water.

WATER TEMPERATURE (°F)	VISCOSITY, n	WATER TEMPERATURE (°F)	VISCOSITY, n
32	1.79	100	0.69
40	1.54	120	0.56
50	1.31	140	0.47
60	1.12	160	0.40
70	0.98	180	0.35
80	0.86	200	0.31
90	0.76	212	0.28

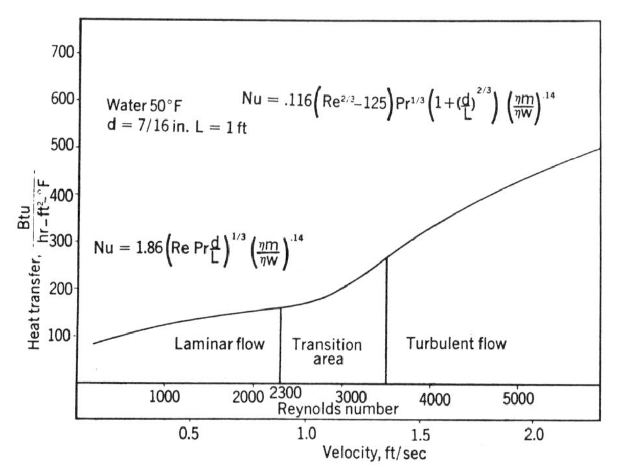

Fig. 12–4. Influence of Reynolds number on heat transfer.

The velocity in ft/hr is $V \times 3600 = 1.53 \times 3600 = 5508$ ft/hr. Reducing this to one cycle would give $V \times$ cycle = ft of passage in a mold. Solving the rest of the equation would give the difference $T_2 - T_1$ and therefore an indication of how many circuits are desirable for cavity and core. The temperature rise per circuit should be around 4 to 5°F.

To control the temperature rise of each circuit, the pressure drop should be the same.

The coolant at inlet should be modified for ambient temperature. When the coolant is 50°F, 10% more coolant should be added for ambient 70°F or 55°F. If the coolant is above ambient temperatures, then 10% should be deducted to allow for radiation from mold surfaces in figuring the actual temperature.

3. *Heat Transfer.* Heat transfer depends on internal resistance of the mold to conduction and the resistance between the mold and coolant. It is a combination of conduction and convection that can be expressed as the overall heat-transfer coeffieient times the area in ft² of the circumference and of its length of passage opening, times the difference between mold and average coolant temperature:

$$H = UA\,(t_m - t_c) \tag{4}$$

where H = the heat generated per hour Btu/hr

U = the overall heat-transfer coefficient, Btu/hr/ft²/°F of coolant passage

A = the heat-transferring surface area of cooling passage = $d\pi/12 \times$ length, ft)

t_m = mold temperature, °F

t_c = mean temperature of coolant liquid, °F

In this equation the only unknown is U, whereas the other factors have been developed in other previous equations. The U can also be expressed in the relationship

$$1/U = 1/d\pi h + 1/SK_s$$

where d = diameter of passage, ft

h = heat-transfer coefficient in cooling passage

S = shape factor, or the arrangement of cooling passage L and the pitch of same

K_s = conductivity of mold material

4. *Radiation.* In a mold, radiation takes place as a natural convection from the exposed mold surface in ft^2 to the coolant circulating medium. Where mold temperature is higher than ambient temperature, radiation to the mold surface is considered to be satisfactory if it is within the limits of human touch. In both cases when the outside limitations exceeds the normal expectations, additional coolant should be provided to overcome the added load. This load can be figured by using 1 Btu/ft^2 of mold surface exposed to the atmosphere in ft^2.

In summary, formula 1 gives the heat to be dissipated; formula 2 shows the conductivity of mold material and the strength of L; formula 3 provides the length of coolant passage per cycle and the division into circuits; and formula 4 gives the overall heat-transfer coefficient.

All the elements of a cooling system have now been presented. Their implementation will be shown in the section "Examples of Cooling," later in the chapter.

Difficult Cooling

Ordinary cooling passages are frequently not possible, and so other systems have been devised to answer that purpose.

1. The *bubbler and cascade* cooling methods are functionally the same except for connection to the mold. In both methods, coolant is admitted through an inside brass pipe, and the return or cooling part goes against the cavity or core to be cooled. See Fig. 12–5 and suppliers' catalogues. The velocity of the cooling part should be figured according to a Reynolds number of 4000 for efficient heat transmission.

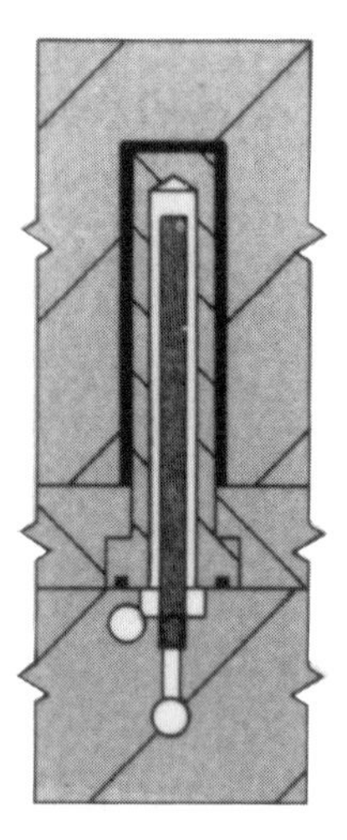

Fig. 12–5. Typical applications of bubbler tubes.

2. *Brass plug baffles* are used to split the drilled water holes into two semicircular channels. The areas of the channels minus the thickness of baffles should correspond to the calculated areas of the passage holes plus a 10% to allow for the greater resistance shapes have to a semicircular than to a circular opening. In this design also, the velocity of a 4000 Reynolds number should prevail. See Fig. 12–6. The cross section at the tip of the drilled hole and the baffle should also be the same as the calculated passage.

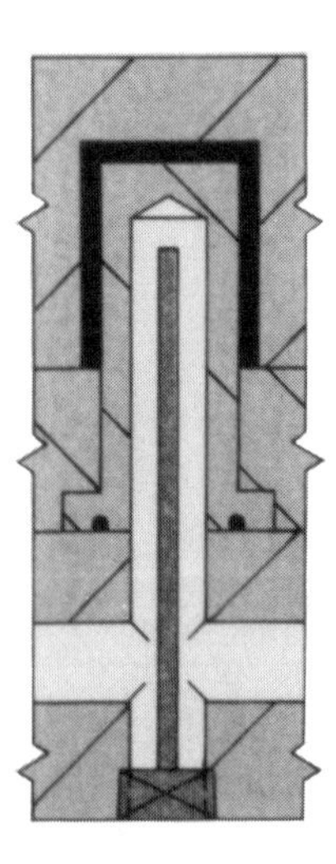

Fig. 12–6. Typical applications of brass plug baffles.

3. The *copper alloy* mentioned in Table 12–1 is being used successfully for cored holes and inserts whenever ordinary cooling is not possible. Because its high strength and heat conductivity closely approach those of pure copper, and because it is cooled in a remote manner, this alloy provides adequate cooling to keep openings from distorting. Insulating the copper alloy with an air gap, in places where heat transmission is not necessary, will improve the cooling action with this type of core and inserts.

4. The *heat-pipe principle* is now being applied to a variety of uses such as nozzles for machines, nozzles for runnerless molding, and of course inserts for difficult-to-cool molds. The heat pipe functions in this manner. A liquid under its own vapor pressure is sealed into a container, the inner surface of which contains a capillary wicking material. The fluid enters the pores of the capillary material and wets all its surfaces. Heat is then applied along the surface of the heat pipe, causing the fluid to boil and creating at that point a vapor state. When this happens, the fluid picks up the heat of vaporization, and the gas, which now has a higher pressure, moves inside the container to a cooler portion. In this place it condenses, giving up the latent heat of vaporization and subsequently moving the heat from the input to the output end. The condensed vapor, which is now a liquid, returns to the point of origin through the pumping action of the capillary wick. See Fig. 12–7.

Heat pipes have a very high heat-transfer rate in comparison with solid metals. When considering their application, the details of manner of use and cost can be obtained from companies such as Kona, Noren Products, and others active in the field.

In addition to figuring out the elements of cooling, other information needs to be stressed.

Coolant Passages

Each cooling circuit should have the same pressure drop so that differential cooling be eliminated and with it, all side effects such as warpage and stresses are minimized. The passages themselves should be uniform in dimension and finish. Thus they should be gun-drilled and reamed for a smooth inside surface and always kept clean for repeatability of cooling action. This requirement is important not only for passages in the mold but also for all the connections to and from the mold. It also means that there should be no cross-sectional restriction in the path of the flow.

The pressure drop in a circuit can be figured by the following empirical formula:

$$p = F\frac{l}{d}\ \frac{V}{2g}$$

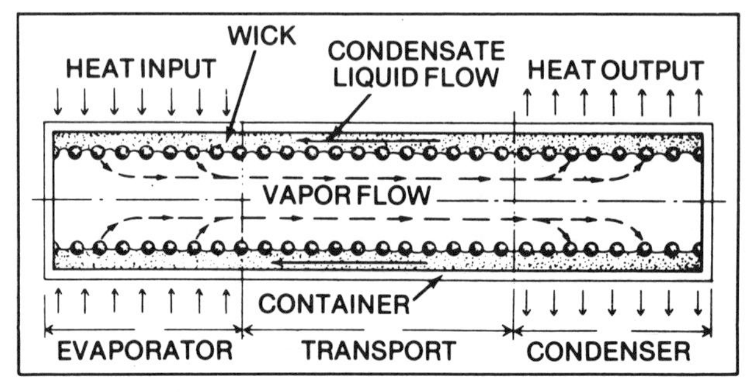

a

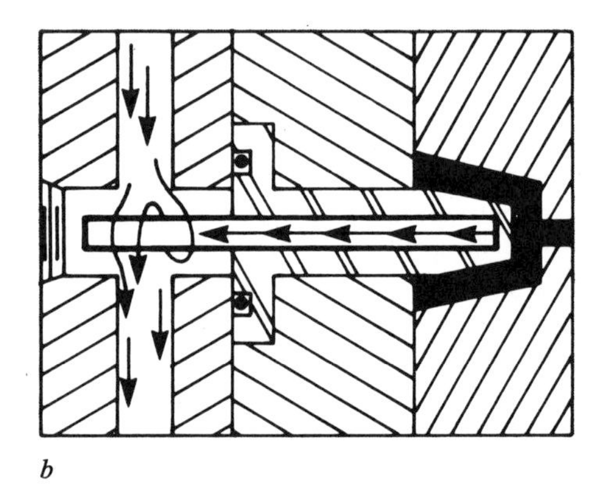

b

Fig. 12–7. *a,* Heat-pipe principle: cross section shows internal functions of a heat pipe. *b,* Application of a heat pipe: standard thermal-pin heat conductor.

where p = pressure drop, psi

F = $0.3164/(\text{Re})^{1/4}$;

l/d *ratio* = length of passage over diameter in same dimensions

V = velocity, ft/sec

g = gravity constant = 32.2 ft/sec^2

Re = the Reynolds number.

By maintaining each factor at the same level, the pressure drop in a circuit will be the same.

EXAMPLES OF COOLING

The product is a single-cavity food container with dimensions and a wall thickness of .075 in. It is to be made of high-density polyethylene or 6/6 Nylon. Following are the data for this example.

Parameter	*High-Density Polyethylene*	*6/6 Nylon*
Melt temperature	490	650
Mold temperature	80	95
Product temperature at ejecting	200	215
Weight of shot, lb	.41	.5
Weight of shot per hr	73.8	90
Specific heat	.55	.4
Duration of cure, sec	17	17
Cycles per hr	180	180
Water temperature	50	65
Average water temperature	55	67

The product has a low wall thickness also a long flow, and both dictate a high melt and a relatively low mold temperature. Therefore, the melt temperature was selected to be on the high side and the mold temperature was chosen to be slightly above the low limit because of the wall thickness and to enhance the flowability of material. The water temperature is 30°F below mold temperature but would depend somewhat on the ambient surroundings. For polyethylene the temperature of product at ejecting is 200°F since its heat deflection is below the maximum mold temperature; for 6/6 Nylon the temperature is the average of 220 and 210, or 215°F.

Applying the formula 1, for high-density polyethylene, we have

$$H = P(t_1 - t_2)S + h = 73.8(490 - 200) \times .55 + 73.8 \times 104$$

$$= 11{,}771 + 7875.2 = 19{,}446.2 \text{ Btu/hr}$$

Because of depth, we take 60% of this figure for the core and 40% for the cavity and get 11,667.72 and 7778.48 Btu/hr, respectively. For 6/6 Nylon we have

$$H = 90(650 - 215) \times .4 + 90 \times 56 = 15{,}660 + 5{,}040 = 20{,}700$$

or 12,420 Btu/hr for core (60%) and 8280 Btu/hr for cavity (40%).

For formula 2 we select a block of P-20 steel for size of product, strength, and thermal fatigue. The L distance for this product should be 7/16 in. since it will have a circulating opening of the same size.

For formula 3, the following selections are made: The circulating hole is 7/16 in. since the mold will fit in a 500-ton press. The Reynolds number is 4000 to give a turbulent velocity with a speed of 1.53 ft/sec. The coolant is water. The length of circulating hole per mold is $V \times$ cycle = 1.53 × 20 = 30.6 ft. The temperature rise of water will be

$$H = .4375/12 \times \pi/4 \times 1.53 \times 62.41(T_2 - T_1) \times 180 \text{ cycles/hr}$$

For polyethylene
$$19{,}446.2 = 492.06(T_2 - T_1)$$
$$T_2 - T_1 = 19{,}446.2/492.06 = 39.52^\circ\text{F/cycle}$$

For Nylon
$$20{,}700 = 492.06(T_2 - T_1)$$
$$T_2 - T_1 = 20{,}700/492.06 = 42.07^\circ\text{F/cycle}$$

The formula for the overall heat-transfer coefficient is

$$H = VA(t_n - t_c)$$

For polyethylene
$$19{,}446.2 = U \times .4375\pi/12 \times 30.6(80 - 55)$$
$$U = 19{,}446.2/87.60 = 221.99$$

For Nylon
$$20{,}700 = U \times .4375\pi/12 \times 30.6(95 - 67)$$
$$U = 20{,}700/98.117 = 210.97$$

Cooling of cavities is shown in Figs 12–8a through c. Since cavities are surrounded by walls and a bottom, they very seldom present a problem as far as cooling passages are concerned. The heat-transfer distances are normally relatively short, and as a general rule it can be assumed that control of cavity temperature will be relatively simple.

Cores, on the other hand, are a completely different story. First of all, they absorb most of the heat from the plastic; second, the heat has to be dissipated from the inside of the core for effective cooling. Figures 12–8d through h show some means of conducting the heat from the core. Figure 12–8d represents a beryllium copper core in sizes up to 0.25 in. diam; its cooling is done by passages practically surrounding the head and part of the core body. Figure 12–8e shows a more effective way of cooling of a beryllium copper core: having an extension from the head submerged into the fluid. The choice of either method would depend on the amount of material that envelops the core. When the thickness over the core exceeds 5/32 in the choice of Fig. 12–8e is indicated.

Figure 12–8f is a similar arrangement to 12–8e except that the core is a steel pin with a copper insert. In this case, care must be exercised to insure that there is an intimate contact between the copper and steel for the full inserted length. In case of doubt as to the attainment of a good contact by a light press fit, silver soldering or brazing may be considered provided that the hardness of the steel core is not adversely affected. In most cases, the use of a solder with a melting temperature below 800°F is required.

Figure 12–8g shows a baffle construction in which the flow takes place along the walls of segments created by placing a partition in the middle of a round opening. When inserted with a light press fit, the baffle will be a conductor of temperature from the area of the core where contact takes place. The stipulation is that the baffle is a good heat conductor like brass or copper and that cross-sectional area through which flow takes place is of dimensions that will create turbulent flow. The main opening into the core may be of almost any reasonable size and the baffle proportionately thick as long as the flow passages have the proper area for turbulent flow. It should be repeated that the baffle contact surface will be almost as effective in heat transfer as the contact of the fluid with the core.

Figure 12–8h depicts a large core such as found on wastebaskets, buckets, etc. In this type of core, because of the large area, it is essential to have a uniform heat transfer from all over the core. With this in mind, the fluid is introduced at the center and top of the cooling core in the area where the gate will be introducing the hot plastic. The cooling core is a conical block inserted into the shell of the core for enhancing the ability to dissipate the heat from the molding core.

The cooling core has at its top center a recess from which grooves radiate; these conduct the fluid along the inner surfaces of the core shell and lead it to the outlet. Here again, each groove has to be so dimensioned as to create a turbulent flow. The groove width should be about one-third of the arc into which the circumference is subdivided. The cooling core itself should be made of a material with good thermal conductivity, such as brass or beryllium copper press-fitted into the core shell, and should end up as an efficient heat dissipator. This type of core could also be made with the cooling core having a spiral groove from top to bottom with a cross section to take care of the full inlet delivery as required to absorb all the Btu transmitted to the outer core. The spacing or pitch of the spiral could be twice the width of the groove.

The copper alloy shown in Table 12–1 has proven to be a most satisfactory material for pins and inserts where high heat transfer is a problem.

TEMPERATURE VARIATION IN MOLDS

Mold temperature affects a plastic part in so many ways that its consistency during operation has to be emphasized over and over again. The elements affected by consistent mold temperature are uniform shrinkage, cycle time, warpage, flow of the plastic, weld-line strength and appearance, level of molding stresses, and general appearance. This is a formidable list of elements and should

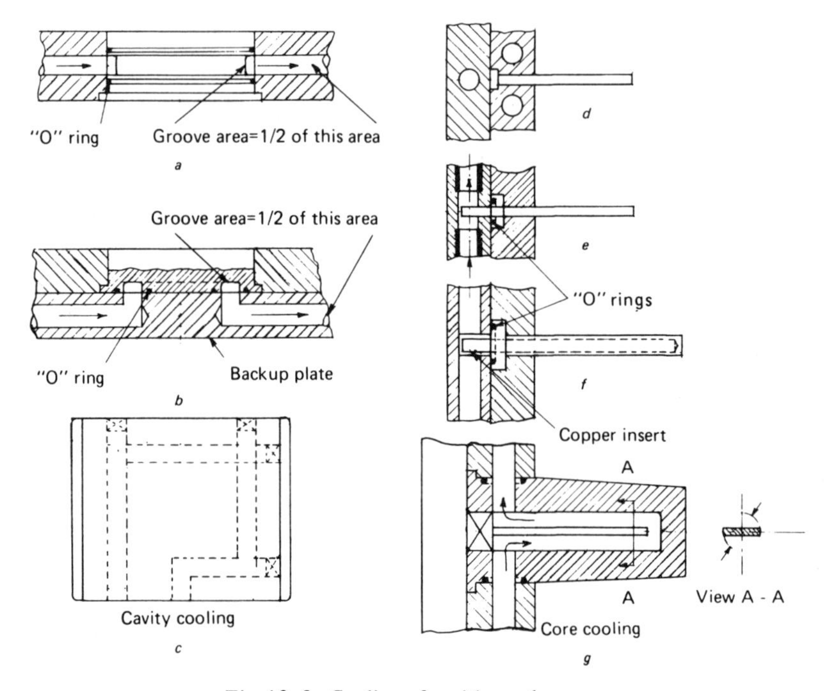

Fig. 12–8. Cooling of cavities and cores.

induce mold designers to thoroughly familiarize themselves with the principles and examples discussed in this chapter before adopting any particular cooling-heating method.

The desired location of the heating-cooling passages is in the mold inserts themselves. They should be located in the proximity where most of the heat has to be dissipated—that is, where most of the material is.

The inclusion of fluid passages in the A and B plates as well as in their supporting plates adds to the ability to control cavity temperature but not to the degree one might expect. This is because steel surfaces always have some heat-insulating film and the contact between them is never such as to induce best conductivity. This was verified in practice by interposing a sheet of soft copper or brass between the B plate and its supporting plate on the core side and checking for temperature while all other conditions remained the same. The average drop in temperature was found to be 25°F, and the core came close to the temperature of the cavity. Prior to this change, the core was running considerably hotter than the cavity. This made it possible to reduce the cycle by 30%. When a core consists of numerous thin sections that are difficult to arrange for individual control, the addition of a good heat conductor between plates

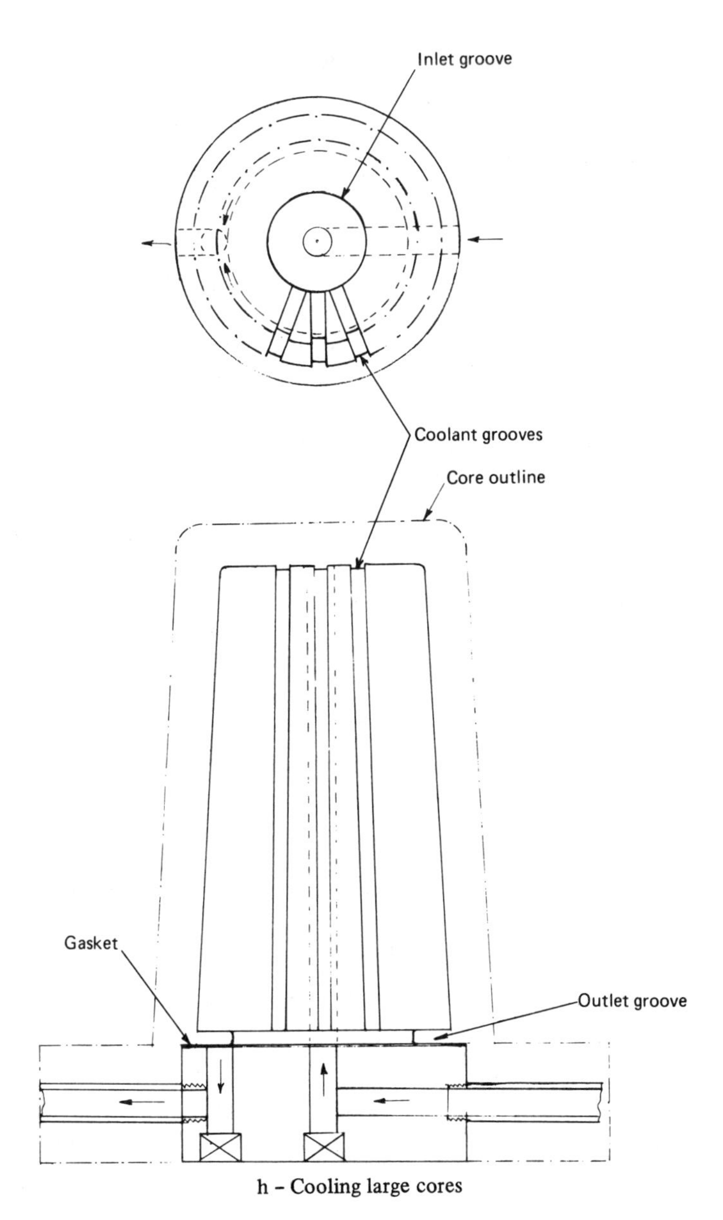

h - Cooling large cores

Fig. 12–8. (continued)

may accomplish the desired result, provided that there are enough passages in the plates to make a good heat exchange possible.

Fluid passages for effective mold and part cooling should be placed to cover most of the molding surface and to be close to the mold face. The distance between mold face and fluid passage opening has to be large enough to resist distortion or flexing of the metal under injection pressures. The inlets and outlets for each cavity should be connected in parallel to their source of supply, thereby insuring uniform heat transfer. The dimensions of the fluid passages should be such as to create a turbulent flow. A turbulent flow will dissipate about three times the Btu as compared to the laminar flow.

Whenever there is a specification for straight, smooth, and dimensionally correct openings, the cores making them will call for special attention to temperature control. The nature of cores is such that the material shrinking over them produces an intimate contact, and the bulk of the heat from the plastic is conducted into them. This condition necessitates an efficient way for dissipating the heat from cores. The following example will demonstrate what the results can be if core cooling is treated lightly. Figure 12–9 shows a schematic arrangement for an elbow-shaped object. In this type of a design, the cores are expected to maintain the mold-base temperature through the conduction of the large contact areas between mold base and cores. This assumption is proven erroneous, and, after about half an hour's running time, the side-cam-actuated core has gained 60° in temperature, an average over its entire length. At these conditions, we find that the core will expand in length in the amount of the new length,

$$1 = \text{Coefficient of linear expansion} \times \text{Length} \times {}^\circ\text{F}$$

From *Machinery's Handbook* ("Expansion, Linear"), we find a value for steel

$$= 6.33 \times 10^{-6}/\text{in./}^\circ\text{F}$$

$$\text{Expansion} = 6.33 \times 10^{-6} \times 6 \text{ in.} \times 60^\circ\text{F}$$

$$= 0.0023 \text{ in.}$$

Since there is an interference to expansion of 0.0023, a compressive stress will be created in the core. The compression formula from the handbook ("Compression Formula") is

$$S = \frac{P}{A}$$

The forced shortening caused by this stress is given as

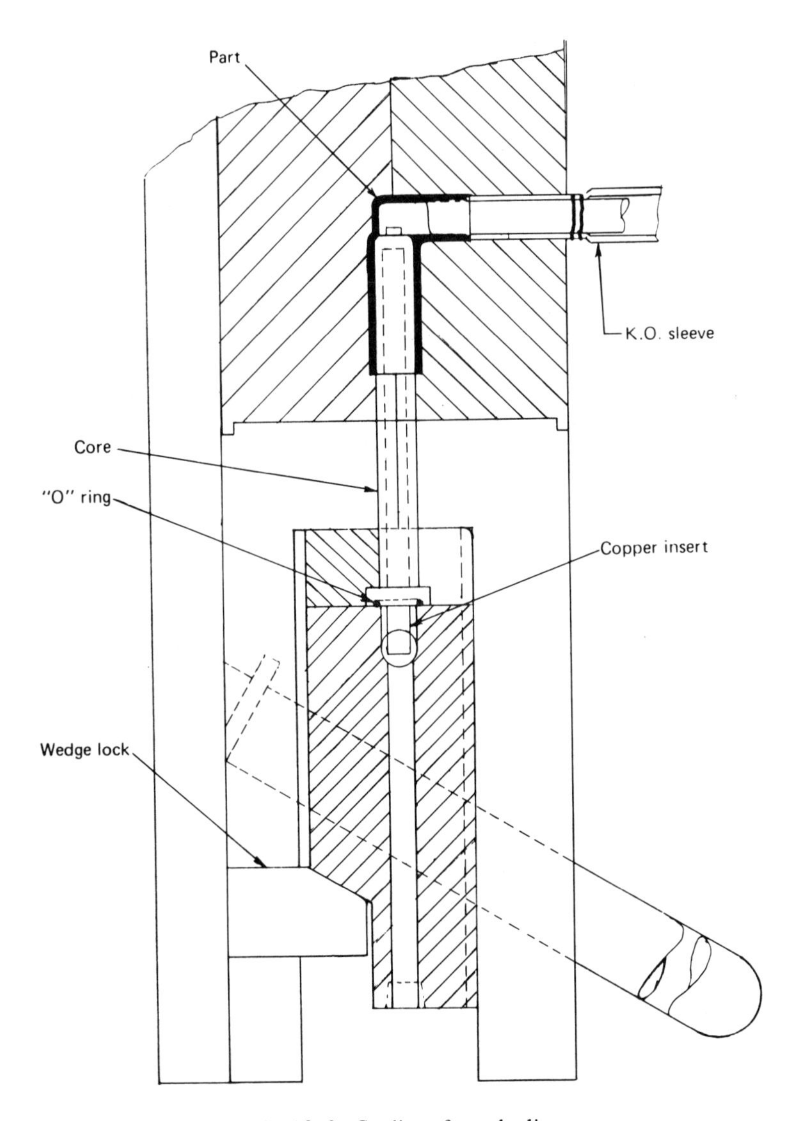

Fig. 12–9. Cooling of cam bodies.

$$e = \frac{PL}{AE} \text{ or substituting } \frac{P}{A} = S$$

$$e = \frac{SL}{E} = 0.0023$$

From this, we find

$$S = \frac{0.0023 \times E}{L}$$

where E = modulus of elasticity for steel = 30×10^6 lb/in.^{2}

L = length of core

$$S = \frac{0.0023 \times 30 \times 10^6}{6} = 11{,}500 \text{ psi}$$

A force of 11,500 lb on 1 in.^{2} is exerted as a result of the compressive action. There will be a force on the core of 5/8-in. diam.

$$P = \text{Stress} \times \text{Area}$$

$$= 11{,}500 \times (5/8)^2 \times 0.7854 =$$

$$= 11{,}500 \times 0.307 = 3530 \text{ lb}$$

This force will cause the stationary core (against which the moving core is shutting off) to bend and bring about a variation in wall thickness. The repeated bending may cause it to crystallize and fail after a relatively short usage. Temperature control of the core would totally eliminate the problem. Long cores that shut off against mold elements should be carefully analyzed for temperature control whenever close-tolerance openings and a reasonable core life are expected.

Cavity and core temperature control are also important to the proper functioning of the mold base. If cores are permitted to exceed the temperature of the cavities, the high heat of the cores will ultimately transfer onto the plate containing them. The *B* plates also hold the bushings for the leader pins. It is conceivable and frequently occurs that there is a difference of 30°F between mold halves. What would this mean to a 24 in. mold base? The expansion of the hotter side will be:

Expansion = Linear expansion × Length of mold × Temperature difference in °F

From the handbook ("Expansion, Linear"), we find a value for steel

$$= 6.33 \times 10^{-6}/\text{in./°F}$$

$$= 6.33 \times 10^{-6} \times 24 \times 30$$

$$= 0.0046 \text{ in.}$$

This difference in expansion will cause binding, misalignment, difficulty in mold opening and closing, and, in the long run, excessive wear on the components that work together.

Another mold component that is affected by the temperature of the mold halves at the parting line is the stripper plate. In most cases, the stripper plate is near to—or at—room temperature. For the majority of cases, the temperature difference in the stripper plate is compensated for by having an adequate clearance between plate, pin diameter, and pinhead to allow them to freely move to a position that conditions would dictate. There are cases, however, where this clearance provision does not apply. For example, when sleeve ejection is needed and the core over which the sleeve slides is attached to the rear clamping plate, temperature control of the clamping plate and stripper plate becomes a necessity. Another way to approximate the desired condition is to insulate the clamping plate with a material like transite board (about 0.5 in. thick), and let the steel of the base absorb enough heat so that it will permit free working of the sleeves over the core pins.

It is best to calculate the elongation of plates under the particular condition and to decide on the basis of figures what action should be taken.

By way of an example, we can get some concept of the gpm needed to dissipate a certain number of Btu. A wastebasket made of high-density polyethylene weighs 1.75 lb. How much flow will it take to control the temperature of its core? The Btu content of the basket molded at 480°F melt and with a starting temperature of 80°F will be:

$$\text{Weight} \times \text{Temperature difference} \times \text{Specific heat} + \text{Heat of fusion}$$

$$1.75 \times (480 - 80) \times 0.55 + 104$$

$$385 + 104 = 489 \text{ Btu}$$

$$\text{or with 120 shots/hr, Btu/hr} = 58{,}680$$

If the mold temperature is maintained at 80°F and the circulating water is at 50°F, it will take

$$\text{Pounds of water/hr} \times \text{Temperature difference} \times \text{Specific heat} = 58{,}680$$

$$\text{lb/hr} = \frac{58{,}680}{1 \times 30} = 1956$$

$$\text{or gpm} = \frac{1956}{8.33 \times 60 \text{ min}} = 3.91 \text{ gpm}$$

where 8.33 is the weight of a gallon of water.

From the discussion of equation 3 earlier in the chapter, we have

$$R = \frac{3160\,Q}{Dn} = 5500$$

Since the core absorbs roughly two-thirds of the Btu, the volume of water for the core will be two-thirds of 3.91 or 2.6 gal.

$$5500\,D = \frac{3160 \times 2.6}{1.3}$$

$$D = \frac{3160 \times 2.6}{1.3 \times 5500} = 1.15$$

or 1-in. pipe size.

There are some cores that are made up of thin section inserts and core inserts, and that contain a multiplicity of K.O. pins; all of them are so placed that it is impractical to introduce cooling passages into the block itself. It is even difficult to have passages in the backup plate in close proximity to the core block. In such cases, bending a copper tube to suit conditions and having it placed in a milled groove of the backup plate and flattened to a D shape will provide considerable cooling help. If the copper tube is of considerable length, interposing a flat sheet of copper will tend to diffuse the cooling action over the entire area of the core.

The following combination of sizes worked well in several applications. A groove of 5/8-in. diam X 5/16-deep was machined to the contour of the bent tubing in the backup plate. A copper tube with 1/2-in. O.D. (3/8-in. nominal water tubing) was filled with mold tryout plastic or a similar easy-melting material, bent to the desired shape, laid in the groove, and flattened in a press to insure a smooth flat surface of the D. The material was molten out of the tube, and the ends fitted with connections.

Cores and Cavities. When plastic materials flow over a core in a uniform and symmetrical manner, the hydraulic forces caused by the moving material usually neutralize each other, and the core itself retains its undisturbed position. In many products or better stated molds, the conditions for uniformity do not exist, with the result that side forces come into play that cause all kinds of difficulties. The molds are difficult to open, wall thicknesses are uneven, guide pins are wearing excessively, production is frequently interrupted, etc. The causes for this type of malfunction are numerous, but essentially the difficulty is a result of the material flowing to one side of the mold first, exerting a force on the core that will shift it in the direction of the unfilled side. The force is equal to the injection psi multiplied by the projected area that is covered by the plastic. On a wastebasket or bucket core, the force can amount to many tons of pressure.

Basically, the reason for the unbalanced flow is that the opening through which the plastic is flowing is uneven, and, where the path of least resistance exists, that is where the core is covered first. This means that, in the case of a wastebasket, for example, the core top is not parallel to the cavity bottom; or the side wall is not uniform in its thickness; or the sprue bushing is not concentric with the core; or there are hot spots on the cavity or core that may induce flow ahead of other places. To overcome any or all of these deficiencies in combination, conditions must be corrected that will bring about an umbrellalike flow that will move reasonably close to a cylindrical form.

On small cores, such as a pen barrel, three or four small gates are used on the circumference, thus locking the core in proper position. To keep cores from shifting, deflecting, or bending, each case has to be reviewed as to how the flow may influence the desired position of the core, and what corrective measures to take must be decided.

13
Runnerless Molding

Runnerless molding can be described as a system of molding in which the melt from the cylinder is brought directly to the cavities. Thus, the molten materials are delivered through auxiliary heated passages and into the parts that are being molded.

There are many advantages to this type of system, of which the most significant are:

1. There are no runners to regrind and consequently no need for use of a mixture with regrind in molded parts. Experience shows that use of regrind in a mix with virgin material increases the reject rate fivefold compared with the use of virgin material alone.

2. Quality of parts is improved since the melt is delivered to the cavity at optimum flowability. Each plastic has its own limited flow length, and, when runners are long, the flow in the cavity may be working on the tail end of polymer flow length and thus require higher pressures for filling and proper density.

3. Melt may be used at the lower range of the temperature, thus providing the potential of lower cycles. Furthermore, only solidification of the part rather than that of the runner determines the duration of the cycle.

4. The system lends itself to automatic operation, therefore providing an important positive cost factor.

5. Press-plasticating capacity is improved because no volume is required for sprue and runner and available heated volume contributed by the auxiliary manifold system is increased.

6. Molding problems associated with sprue and runners are eliminated.

7. Injection-pressure requirements are lowered because the melt fluidity is maintained right up to the gate. Good fluidity of melt at the gate reduces injection-pressure values, which in turn is reflected in lower clamp-pressure needs, which means smaller clamp sizes.

8. Greater freedom of gate location is afforded as a result of the melt being fully fluid at the entrance to the part.

9. Parts requiring long flow paths may be made with a single gate, whereas, in conventional runner molding, because of loss of heat in the runner, multiple gating may be necessary.

10. Requirements for plastic material grinders are reduced, and handling of regrind is eliminated.

There are some disadvantages:

1. Mold costs are somewhat higher.
2. Controls for manifold and even nozzle temperatures are required.
3. The initial debugging time is longer than in a conventional mold.

Runnerless molding may be accomplished by the use of (1) a *hot manifold* and (2) an *insulated manifold.*

CONSIDERATIONS FOR RUNNERLESS APPLICATION

Almost every case of runnerless molding presents problems that usually differ from previously applied conditions. For this reason, it is desirable to seek the advice of material suppliers in connection with application of at least the first three molds that are being constructed along principles with which the mold designer has had little experience.

1. The recommendations of the material supplier for polycarbonate and modified PPO (Noryl) advocate a hot runner mold similar to the one shown in Fig. 13–1. The prime reason for recommending this type is to ensure a continuous flow of the polymer without any chance of entrapment. The entrapped material would in a short time degrade, causing the polymer to lose properties and in most cases affect the appearance of the product. With a continuous and unimpeded flow, the action of decompression would favorably affect the gate by minimizing its drooling. Furthermore, when a run is completed, the content of the mold could be readily purged with, for example, general-purpose polyethylene to keep the mold available for the following production.

When runnerless molds for polycarbonate and Noryl are being designed, they should be submitted to the supplier for constructive criticism.

2. Insulated runner applications are usually intended for polyethylene, polypropylene, and polystyrene. These molds are built to enable an operator to quickly withdraw the insulated runner when solidified so that a new cycle can be initiated. The author has observed several molds in which the plates containing the cavities were bent out of shape so that contact along the circumference could not be maintained. As a result, flash at some parting lines was present.

Practically and theoretically this can be explained. Let us assume that we have a 5-in. cup to be molded out of polyethylene in a four-cavity mold. For these cups,

$$\text{Holding Pressure} = \text{Number of cavities} \times \text{Area of 5-in. diam} \times 2 \text{ tons/in.}^2 \text{ (for polyethylene)}$$

$$= 4 \times 19.635 \times 2 = 157 \text{ tons}$$

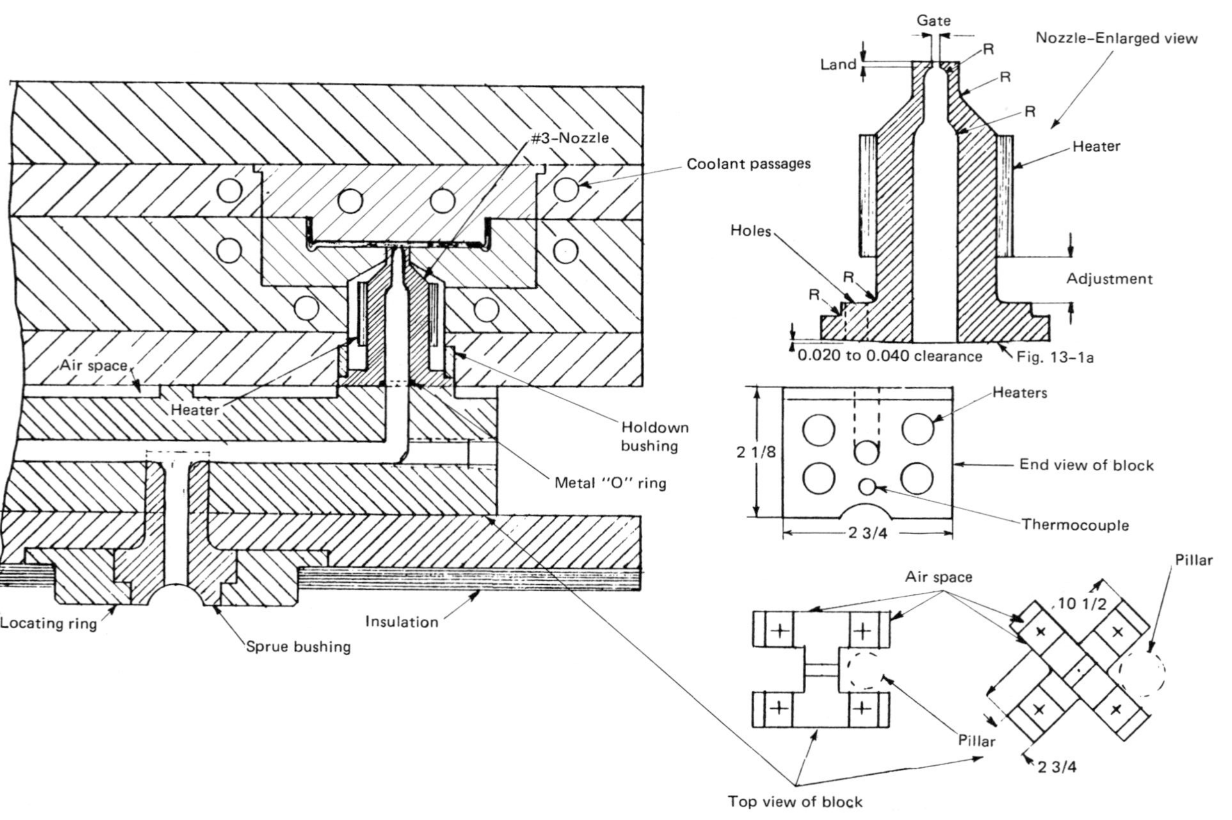

Fig. 13-1. Heated manifold.

The insulated runner is about 23-3/4 in. long with a 1-1/8-in. effective diameter during molding. The pressure at the inlet of the insulated runner is 16,000 psi. The tonnage under these conditions would be 1-1/8 X 23-3/4 X 16,000 = 213.75 tons. The projected area is 1-1/8 X 23-3/4. If the full pressure is required to fill the mold, we have an imbalance of 56.75 tons ready to open the mold and admit flashing, which would cause the mold to bend out of shape. The way to overcome this difficulty is to place the mold in a higher-capacity press that would counteract the higher tonnage.

3. This difficiency led to the developemnt of the "Cool One." In this system the passages of the plastic are in a solid plate so that separation does not take place. The major difficulties of the system are color or material changes. The DME Co. has now designed a self-contained material passage fitted into a cover plate that accelerates the problem of material and color changes. Another potential problem in the system is an approach from the probe to the gate that at present is wedge-shaped as in Fig. 13–4 instead of the one shown in Fig. 7–13. The new shape brings the hot material close to the gate and with this action is responsible for lowering the injection pressure and minimizing the tendency to drooling. This concept has been put into effect on several molds and found successful in performance.

4. A heater cast into beryllium-copper shape is produced by the Mold Masters Ltd. This arrangement includes machining of gates into the casting and has proven to give satisfactory performance in the industry. The life of the heater is prolonged beyond average expectations, and consequently the whole unit is of long duration. When considering the use of this system, it is best to work with the supplier of the units for most advantageous application.

5. The "Kona" system depends on the heat-pipe principle to conduct the heat throughout the path of material flow. The heaters that are the generating source for the heat pipe are placed in a convenient, accessible place for easy checking and if need be for easy replacement without disturbing the mold. This is another instance, in which close cooperation with the supplier is needed for successful operation of the system.

6. Some products have a requirement that the breakoff at the gate be flattened so that the surface appear smooth to the eye. Some manufacturers make valves that perform this requirement. These may be activated by hydraulics, air, springs, etc. Each one of these companies make its own hot-runner system. Some of these companies are DME Co., Husky, Incoe Corp., "Kona," Mold Master Ltd., and Spear System.

Final note. Any hot runner system is only as good as the principle, workmanship, and materials employed in them. For this reason, it is best to solicit the experience of leading material suppliers and their customers regarding cost, prolonged service time, and their experience with trouble-free operation of a particular unit.

The following sections describe in detail some hot-runner systems.

HOT MANIFOLD

In the hot manifold, as schematically outlined on Fig. 13–1, the plastic material enters a regular sprue bushing containing a seat for the machine nozzle, and from there enters into a heated manifold. The manifold is heated by four cartridge heaters that are of sufficient wattage to keep the melt fluid and controlled by a thermostat to maintain desired temperatures. From the manifold, the material moves into a feeding nozzle, which in principle may be constructed, heated, and controlled similar to a machine nozzle; from the nozzle, the plastic moves into the cavity. A set of conditions has to be attained with the hot manifold so that the feeding nozzle will not freeze during the cycle and be plugged, and yet the melt must not be too fluid or drooling will take place upon removal of the parts from the cavities.

The following details in the manifold design can assure good performance of the system without extensive experimentation during startup:

1. The machine nozzle, sprue bushing, and all the passages from the part and No. 3 nozzle are made without restricting openings so that the "melt decompress" feature may be effective and minimize the tendency of drooling.
2. As stated previously, the manifold can be considered an extension of the cylinder and its nozzle into the cavity. When the manifold is operating at high temperatures, it is necessary to insulate it from the surrounding structures. The tip of the nozzle has to be maintained at a temperature reasonably close to that of the cavity and yet warm enough to permit the initiation of the shot. This would call for minimal contact of the nozzle with the cavity and its plates.

In most cases, in insulating the hot runner system, air space is used as the heat barrier; in so doing, pressure pads and shoulders are employed to absorb the compressive forces exerted by the clamping tonnage of the press. This means that the manifold steel and the steel in the mold base, against which the pressures are exerted, should be of a quality that will resist embedding of the pads. In addition, the contact areas of the pads should be of sufficient size so that depressions will not be formed when subjected to the tonnage of the press. Therefore, it is a good idea to stamp a specific mold with the maximum tonnage of the press that can be used safely during molding.

The steel in the manifold should be of the 4100 type heat-treated to 300 Bhn. All passages should be smooth and polished to minimize pressure drops and hang-ups. All bends should be rounded, and components that are mounted on the manifold and are part of the flow system should have continuous flow passages without shoulders or other impediments to a smooth flow. The allowable pressure of the pads should be 15,000 psi.

At this stress level, there would be an indentation of 0.0005. When the load is removed, the indentation disappears because the stress is well below the yield strength. Thus, a 200-ton press should have 26.67 in.^{2} of pressure areas. The total pressure area will consist of the area provided by eight rails, one washer over the main nozzle, and the areas of pressure by the four cavity nozzles. The seal between the heated nozzle and manifold may be a stainless "O" ring or copper ring similar to those used for spark plugs. The drawback is that, each time the manifold is disassembled from the mold, a new set of rings is required, which are a potential source of hang-ups. A preferred arrangement is that shown in Fig. 13–1a in which a metal-to-metal seal is employed similar to that used on the front of the machine cylinder for prevention of leakage between joints.

The nozzle steel should be H13 heat-treated to 52–55 RC and located in the supporting plate so that the heat transfer is minimal to maintain fluidity of the polymer. For sealing and spacing purposes, a pressure has to be exerted on the shoulder of the nozzle; this pressure should be within the abovementioned limits. The tip of the nozzle, being part of the cavity, should have a finish comparable to that of the cavity and be light-press-fitted into the cavity to prevent leakage. The length of the tip should be on the order of 3/16 to 1/4 in., and its diameter should be in proportion to the gate with a relatively small wall (3/32 to 1/8 in.). The entrance to the tip should follow the recommendation of the material suppliers as outlined in their machine nozzle design.

Those dimensions are dictated by the need for quick response to temperature changes during flow and solidification (cure) of the plastic. All other areas including the surfaces over the heater should have an air space for heat-insulation purposes. To provide good control of the cycle, the nozzle tip temperature should be somewhat under the heat-deflection temperature of the material given in the supplier's literature.

The I.D. of the heated nozzle as well as the diameter of passages in the manifold should preferably be no less than 3/8 in. and should not exceed 1/2 in. The heater requirement for the nozzle is between 70 to 90 W/in. of nozzle length and should extend over its full straight length. Care should be exerted that the temperature of the nozzle does not exceed 800°F, so that the heat treatment is not affected. A control of the variac type or solid-state type is essential in order to maintain conditions that are correct for a specific material and for good processibility. In general, conditions prevailing in a machine nozzle should approximate those of the manifold nozzles. The heaters in the manifold should be arranged as shown in Fig. 13–1 so that they are uniformly spaced in respect to the material passage. Their wattage should be between 70 and 100 W/in.^{3} of material (steel) within the passage portion under consideration. A thermostatic control should be located in a critical area to accurately maintain melt temperature. Provision should be made for accessibility to the vital areas of a manifold by means of a portable pyrometer for checking of any malfunctions as well as an aid in start-up.

A practical consideration for choosing the heaters for the manifold and nozzles is that they be of a diameter and length so that the wattage may be changed by ±30% of the selected sizes without the need of changing physical dimensions in the mold structure. This allows for optimizing heating conditions with relatively little expense. Too many assumptions have to be made in calculating heat losses, conductivity, type of contacts, etc.; thus, a ±30% variation in heat requirements would not be an unreasonable expectation.

Combination Hot and Short Runner

With these broad principles in mind, hot manifolds can be arranged for a variety of gating systems including edge gating or combination of short runner and sub-gating as shown in Fig. 13-2. Once the general requirements of a hot runner system are fully appreciated, many deviations may be successfully designed to suit a specific need.

Figure 13-2 shows how a combination of the hot runner system with a standard gate approach method can provide 98% of the benefits of runnerless molding while at the same time allowing great freedom in the selection of gate type and its location. Some of the problems of hot runner molding–such as drooling, delicate temperature control of nozzle tip, and possible presence of cool material in the stream– are also eliminated. The regrind created by this method constitutes a small fraction of the total runner involved, particularly when used in conjunction with feeding several cavities from one point. This method can be truly called an extension of the machine cylinder and nozzle since it conforms in every detail to the feeding of material to cavity and includes a cold slug well as well as a most favorable condition for the effectiveness of melt decompress action.

Any runnerless system should be carefully analyzed for possible deficiencies that will be listed. Needless to say, the effect of such deficiencies will depend on the type of product that is manufactured. An item whose utilization is short-lived and which acts as a container or packaging item will have different quality requirements from a part that is used in an instrument and that has to perform a vital function. On many products, the most important consideration is low-cost production and acceptable appearance. In such cases, it is a simple matter of economics–namely, what system will produce the largest quantity for the lowest price?

The analysis of a system should include these points:

1. Is solid or cool material from the tip of the nozzle and gate getting into the stream, thus reducing the melt temperature at some area? If so, it can result in

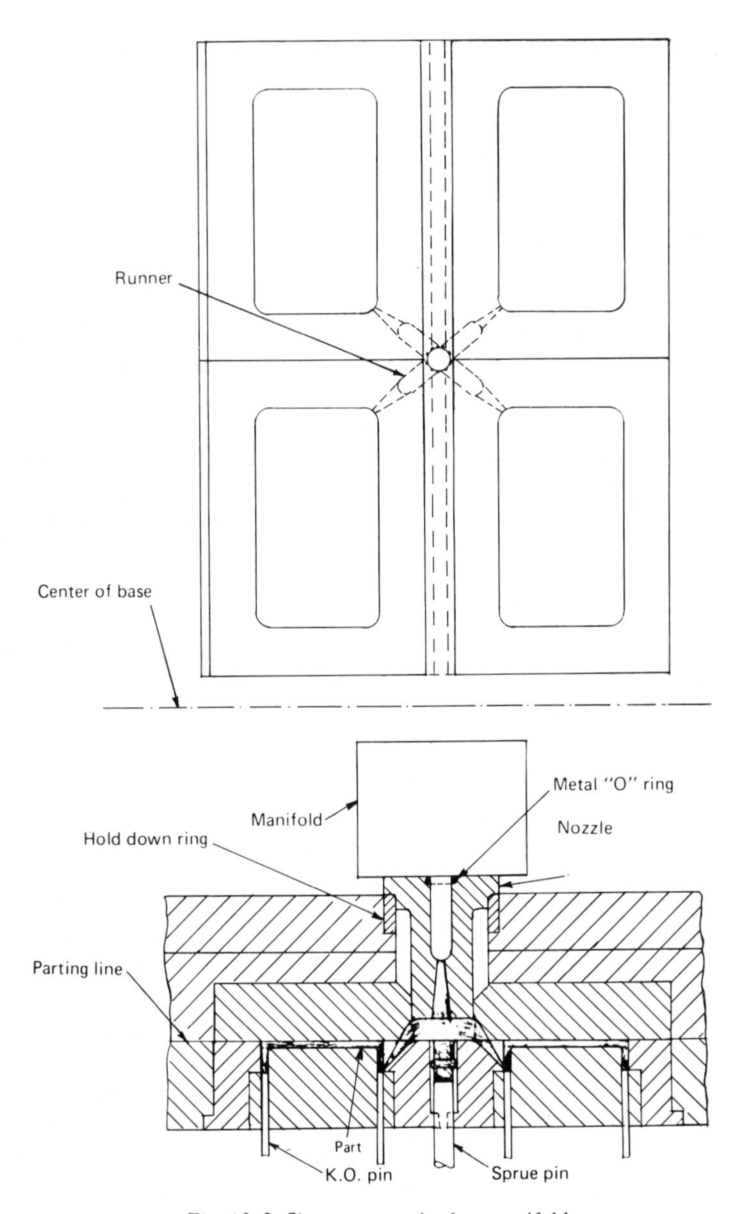

Fig. 13-2. Short runner plus hot manifold.

variation in cooling rates of the part surface and the associated generation of stresses.

2. Is there a chance for material to be hung up in the passages of the system? In due course of time, will it degrade, with the flakes from degradation getting into the stream and causing defects?

3. Is it possible for heater hot spots to develop and cause degradation of material, particularly during the time of cure?

4. How is the problem of expansion and contraction compensated for in relation to hot and cool components to avoid stressing of the involved areas?

5. Is it desirable to be locked into a specific heater size without the assurance that the actual needs will be properly served? Running the mold for a longer time is a proven way to determine the exact heater requirements.

6. Cost comparison between a commercial system and in-house construction as well as performance evaluation of each system are necessary in deciding which system to choose.

7. Last but not least, what is the opinion of the material supplier of the system under consideration? What is the supplier's own experience with the system? What is the experience of the supplier's customers?

After the answers to these points have been established, a decision can be made on the direction in which to proceed.

There are several commercial hot runner units on the market. Most of them are of the type where the material is heated from inside out, meaning that the material flows over a heated probe unit, before entering the cavity.

One commercial system (Mold Masters Ltd., Georgetown, Ontario, Canada) uses a different approach. The nozzle unit consists of a stainless-steel tubular heater, which is pressure-cast into a beryllium copper alloy. This casting method assures good contact between the metal and heater. The rough casting is machined to suit the needs of a particular mold and heat-treated for favorable use properties. A so-called hot edge gating opening is provided in the nozzle, which permits the placing of the gate in almost any point along the outside configuration of a part. One nozzle can be made to feed from one to several cavities. Figure 13-3a shows one nozzle feeding two cavities from a manifold. The enlarged view points out the details of gate arrangement and also some means of compensating for the higher rate of thermal expansion of berylium copper. It also points out the entrapped plastic material in the air-insulating space as well as the retained frozen material that is forced into the stream and molten during the following shot. The proportions of gate size in relation to part thickness can be also judged from this view.

The same company also makes a unit in which a gate valve is employed to smoothen the gate breakoff point. The valve functions as follows: the injection pressure causes the rearward movement of the valve pin until free flow of

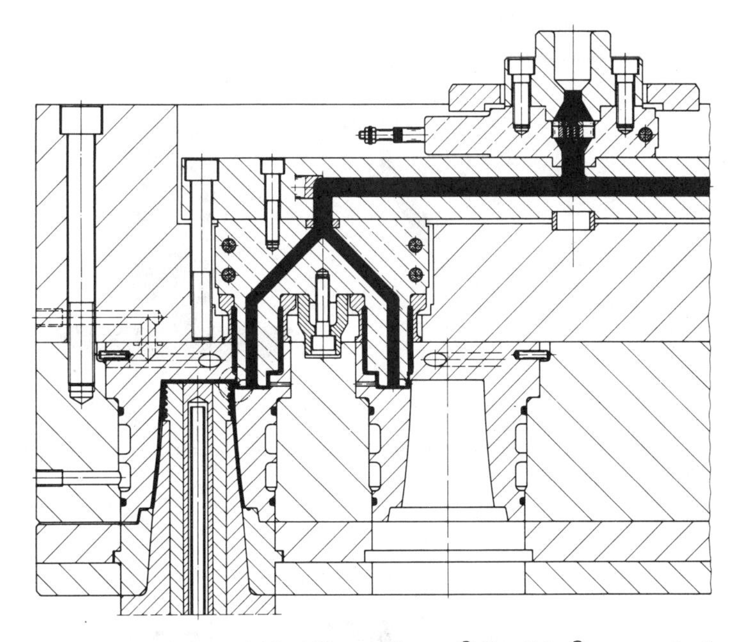

A typical Mold-Masters System® Hot Edge® gate method applied to an eight cavity mold. Note the multi-nozzle, modular design concept and the climination of a mold back plate.

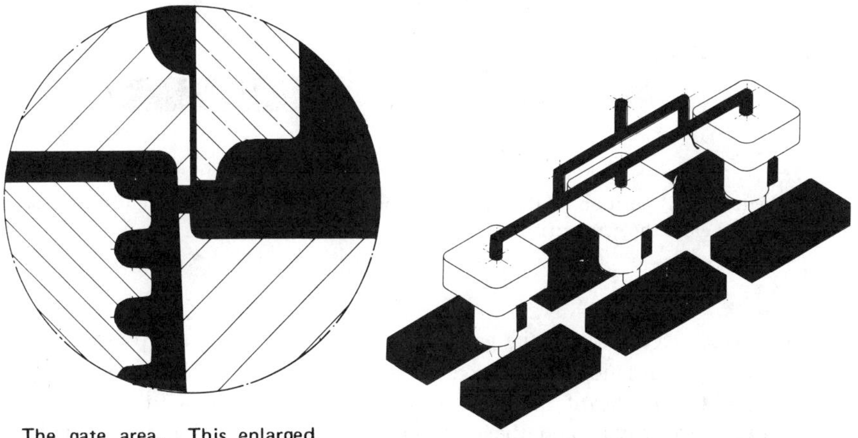

The gate area. This enlarged view illustrates the relationship of the Be-Cu nozzle end to the gate, land, and cavity walls.

(a)

Simple layout of six parts to be Hot Edge® gated with three Be-Cu heater casts.

Fig. 13–3. *a*, Edge gating with a heated manifold. *b*, Gate valve. *c*, Gate valve. *(Courtesy of Mold Masters Ltd., Georgetown, Ontario, Canada)*

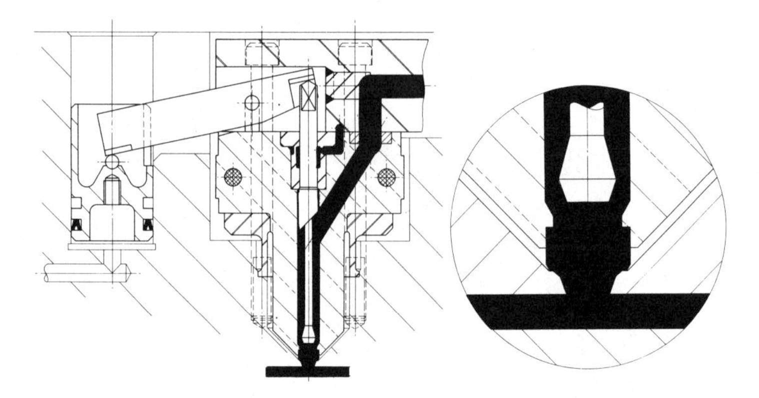

(b) The 'gate open' position. The valve pin is retracted by the injection pressure alone.

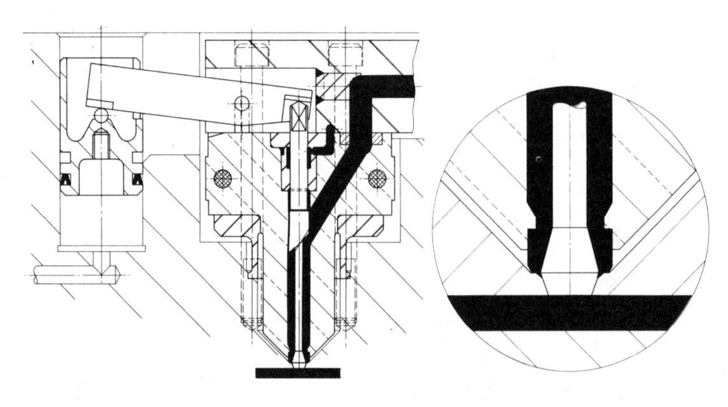

(c) The 'gate closed' position. An air activated piston and lever assembly returns the valve pin in order to close the gate orifice.

Fig. 13–3 (*cont.*).

material takes place through the gate opening. At the end of the pressure hold time, a three-way solenoid air valve receives a signal to admit air to a piston that will actuate the closing of the gate orifice by the valve pin. A titanium seal is used between the cavity and the cast beryllium heater unit because of its low heat conductivity as well as good wear properties. It is claimed that gate sizes from 0.04 to 0.5 diam can be successfully used with this type of valve gating. See Fig. 13-3b and c for the gate open and closed positions. The type of hot runner to be employed has to be decided on the basis of (1) which system will insure best properties in the finished product, (2) which one will best suit the complexity of the product and material moldability characteristics, and (3) the overall cost and consistency of operation.

INSULATED MANIFOLD

The insulated manifold also consists of a manifold that is fed by a machine nozzle except that the passages are not heated (see Fig. 13-4). The runner system in the manifold is made considerably larger in diameter than is the case with the heated manifold. After the first shot, when the system is filled with molten material, the center portion of the material remains fluid due to the passage of freshly plasticated material with each cycle. The outer portion of the

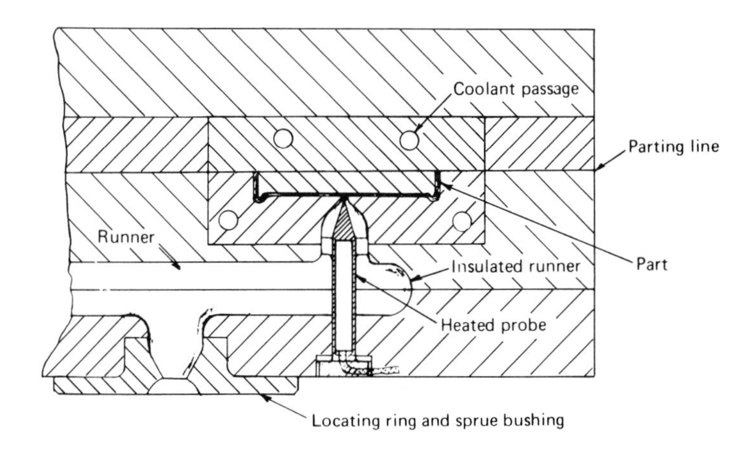

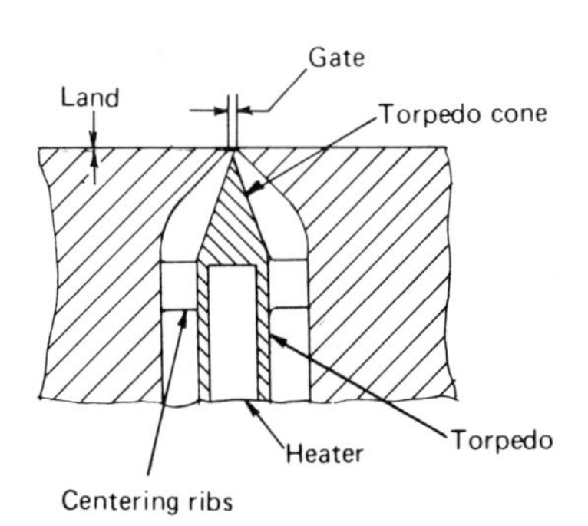

Fig. 13-4. Insulated runner.–Schematic presentation.

material in the manifold forms a solidified shell or tube through which the molten material delivered with the shot will flow into the cavities. The heat-insulating characteristics of plastic materials are used to good advantage by having the outer portion of the runner, which is frozen, insulate the hot molten material, thus making it possible for the plastic to flow freely with each shot. An interruption in the cycle of several shots' duration will cause the full runner in the manifold to freeze, thus necessitating its removal before molding can be resumed. This condition calls for a provision of easy removal of manifold runner and rapid rejoining of plates. Quick-acting latches and movement of the press are employed to accomplish the removal of the runners with little delay.

Some insulated manifolds are equipped with electrically heated probes at the entrance area to the cavity in order to reheat the material for best flow into the parts.

To be considered suitable for insulated manifold, the materials must have a broad range of melt temperature, must not degrade under prolonged heat exposure as is the case with the insulating "tube," must have "long flow" properties, and in general must not discolor under these conditions of operation. The material used in this system must have flexibility in setting molding conditions and allow delays in cycling without thermal degradation. It should have a low specific heat and a high thermal conductivity so that it can be melted quickly and attain temperature uniformity. The material should also have a high heat-deflection temperature, so it may set up (cure) in the relatively warm cavity in a short time for economical cycles. The insulated runner, although limited to certain materials in application, involves lower mold costs and a minimum need for temperature controls. On the other hand, the hot runner manifold affords greater ability to controlling melt temperature, which is a prerequisite for precision and quality of parts.

In conclusion, it can be stated that runnerless molding, when properly executed, will result in saving material and improving production. At present, polyethylenes, polypropylene, and polystyrene are the best candidates for the insulated runnerless molding. This conclusion is based on two assumptions: (1) The area of the insulated runner is somewhat smaller than the molding areas of the cavities. (2) If the mold occasionally flashes at the parting line and the plate holding the cavities is not rigidly fastened to its backplate, then this cavity plate could move forward during the slight opening of the mold and admit some plastic between the mentioned plates; this would cause bending and distortion of the cavity-holding plate. All this would be brought about by the injection pressure of about 20,000 psi. When multiplied by the area of insulated runners, this could produce large tonnages to distort and bend the cavity plate.

Out of a dozen molds equipped with insulating runners, more than half were found by the author to have the cavity plate distorted so that during the running of parts in the same areas were these parts flashing and in others the parting line

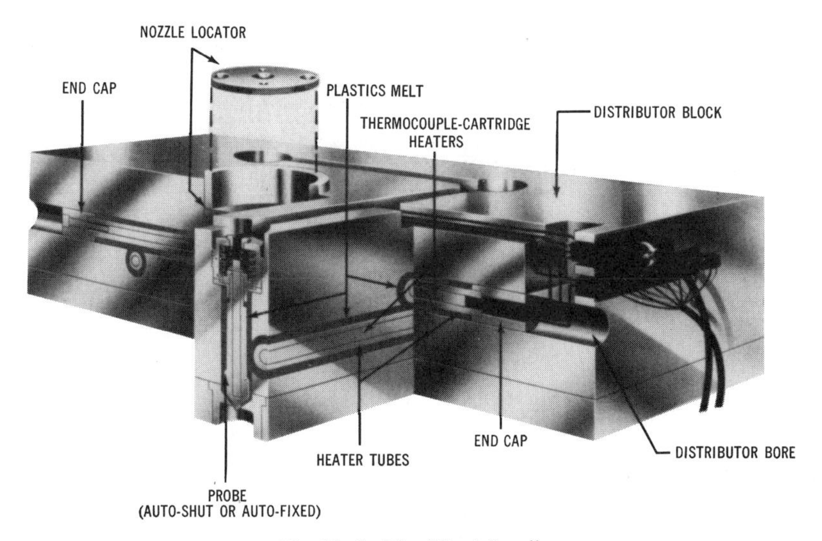

Fig. 13–5. The "Cool One."

was metal-to-metal-fitted. In addition to these problems, there is another drawback to the insulated runner, namely, that when the operation is stopped for a few minutes, half of the mold has to be dismantled to remove the solidified runner and start the operation all over again.

All these disadvantages gave impetus to a new development supplied by the DME Corp. known as the "Cool One" (Fig. 13–5). This development not only eliminated the weak points of the older system but also took a step forward in standardizing the components for the system so that tool cost can be accurately anticipated and a mold life with reasonable expectations in production can be expected. The system includes 1-1/4-in. and 2-in. bored holes in a plate containing the hot runner. We shall confine the description to the 1-1/4-in. bored hole since that is the most popular size. Fig 13–6a and b show a schematic arrangement of the principle involved. The 1-1/4-in. holes are bored usually in a plate 3-3/8-in. or even larger if circumstances demand it. One bore usually runs the width of the plate and connects to the seat of the nozzle. The other two or more are running perpendicular to the first one and are connected to each other by milled portions of each tube that would bring their centers to 1-1/16 in. This means that bores overlap each other by 3/16 in. In Fig. 13–6b four cavities can be fed; if two more horizontal bores were provided and the crossbores were correspondingly larger, eight cavities could be accommodated. The system provides considerably flexibility, and combination of cavities can be arranged. Only one probe is shown in the illustration for clarity purposes. The Fig. 13–6b shows the schematic arrangement of components in the bore. The centers of the bores with respect to the probes are .906 to .953.

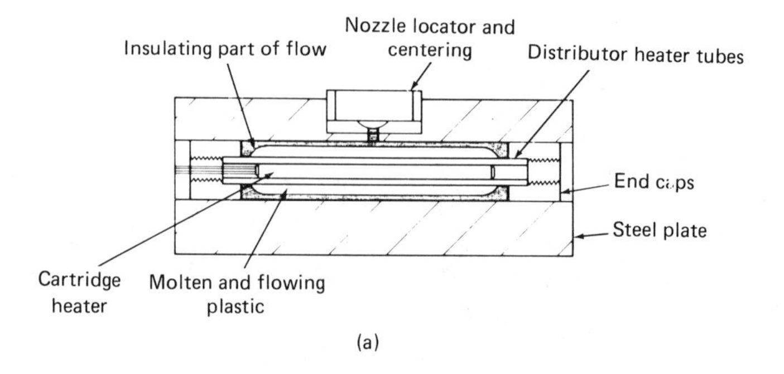

(a)

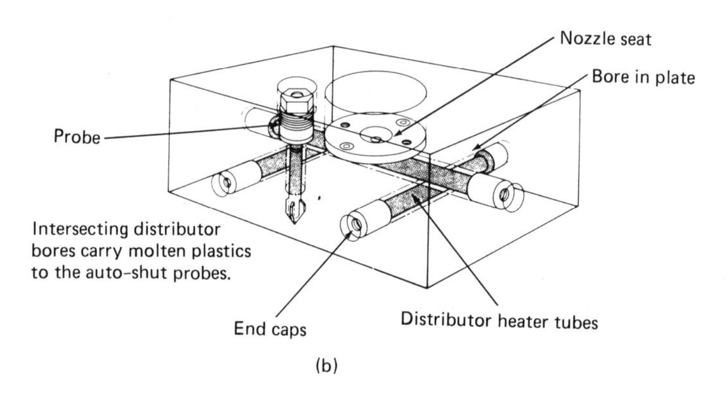

(b)

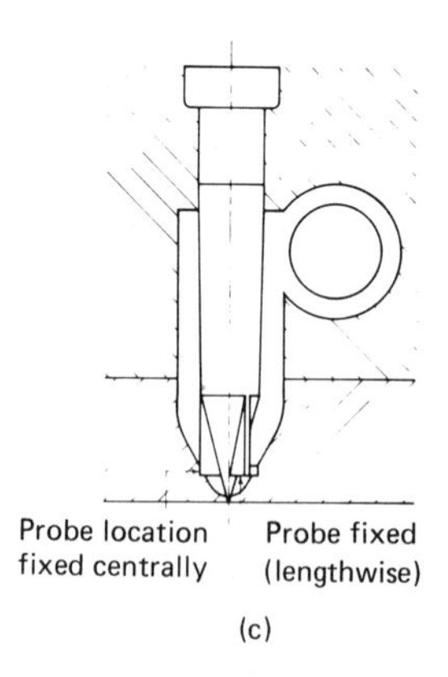

(c)

Fig. 13–6. The "Cool One" using 1-1/4-in. bored holes. *a,* Principle of operation. *b,* Schematic distribution system. *c,* View showing auto-fixed probe and distributor conduit.

All the components that make up the complete assembly are standard and available out of stock.

The approach of the gate to the cavity as dimensioned in the supplier specifications has a small cold slug that under normal conditions is not visible, nor in many cases would it have any ill effects on performance of the product. On the other hand, when clear parts are molded or close-tolerance parts are being manufactured, the presence of a cold slug such as prevailing with the probe of standard design is objectionable. For this reason, the author designed the approach to cavity as shown on Fig. 7–13. The outcome of this is a negligible cold slug and full response of the plastic from the front of the tube to the suction of melt decompress. Another drawback in the standard designed probe is that tramp metal or other clogging orifice material when caught in the gate will call for disassembly of one-half of the mold in order to remove the obstructing material from the gate orifice. To overcome this problem, the author has designed the dropoff from the main conduit by making a nozzle similar to an injection nozzle and placing it on top of the cross channels (Fig. 13–7). In addition to placing the nozzle on top of the conduit, contact with the heated steel tube has been made so that easy flow takes place along the inside of the cavity. To open the gate, all that is required is placing the right drill in the gate and making a few turns to open it. This modification has proven to be quite successful in plant

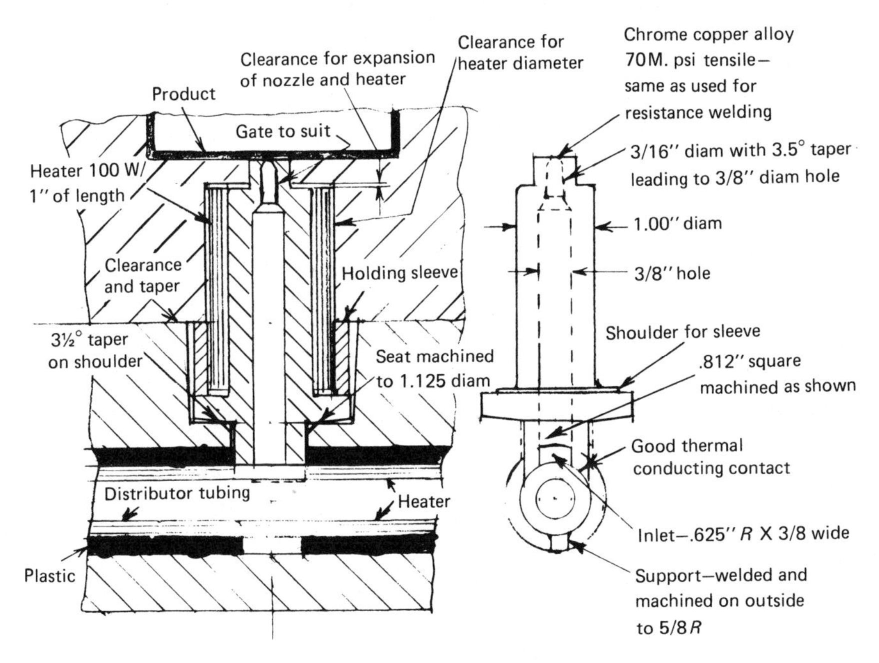

Fig. 13–7. View of nozzle and tubing.

operation, especially when regrind is used in combination with virgin material. The figure also shows some details pertaining to the altered design.

The cartridge heaters for the system should have a thermocouple built into them so that the temperature in each circuit can be accurately controlled. If for some reason the thermocouple heaters are not available, it is best to test each heater with its steel shell prior to placing it into the mold. In this test the control setting should be so set as to provide molding temperature at the steel shell for a particular polymer and should be kept at that temperature for about 1/2 hr and the control setting recorded for each assembly of heater and shell. With that type of information, small changes in the control setting while the heaters are in molding position would insure that the temperature of each assembly is within the limits of the polymer and that no degradation would take place.

Let us now review some of the design details shown in Figs. 13–1 and 13–7. Most of the commercial hot runner systems heat the material from inside out. The plastic is in close contact with the heater or a thin layer of metal over the heater, which for all practical purposes is almost the same as if there is direct heating of the plastic by the element.

Some material suppliers do not recommend this type of heating for their heat-sensitive polymer. They reason that most cartridge heaters tend to have hot spots on the surface. Such hot spots can cause hang-ups or the degradation of some materials. When the heaters are new, this characteristic may not be as noticeable as when usage progresses. A thermocouple does not pick up such spots, but material in contact with them can be affected.

It takes at least 0.25-in. wall per side over the heater to average out the heat over the surface. The design in Fig. 13–1 literally brings the type of a nozzle used on the machine directly to the part as is the case with sprue-gated single-cavity molds. If the length of the nozzle is less than 2.5 in., the chances are that a beryllium copper nozzle could be used without a heater band and would absorb the necessary heat from the manifold itself. Should a heater band be needed, it could be obtained of a length that would make possible a slight variation of nozzle tip temperature by moving it toward the bottom flange. In order that the flange does not diffuse too much heat from the body of the nozzle, eight or ten 3/16-in. holes can be drilled in it to reduce heat conductivity.

Considering the coefficient of expansion, the arrangement shown in Fig. 13–1 permits the nozzle to expand into the cavity by having a clearance between the shoulder of the nozzle and the corresponding counterbore of the cavity. If a recess of a few thousands on the part is objectionable, then the nozzle tip during room temperature would have to be below the cavity to allow for expansion. The lengthening of the manifold would find the bottom of the nozzle stationary and sliding on the smooth contact face.

To facilitate the adjustment of the nozzle heater band without disturbing its seat, the A plate should be attached to its backup plate by screws from the parting line side. At the same time, the backup plate should be independently attached to the clamp plate, so that the assembly behind the A plate forms a solid unit. When metal "O" rings are used to seal the nozzle against the manifold, then, anytime the assembly is disturbed, it invariably means replacing the ring or else leakage takes place. The dimensions for the space of the "O" ring should be obtained from the manufacturer of the ring so that the correct compression is obtained.

Cartridge heaters should be placed a distance from the plastic passage that compares to the distance of fluid passages from the face of a cavity (about 7/16 in. diam). This is important because the pressure from the plastic could decrease the opening for the cartridge heater to make its removal for replacement impossible. Each arm of the manifold should be provided with an opening to receive a thermocouple. While only one is used for control purposes, the others are needed for troubleshooting needs.

If the pressure pads provided on the manifold are insufficient to absorb the full clamping force, support pillars of proper cross section and number can be added to prevent indentation in the plates and the possible disturbance of seals and fits.

The figure shows two shapes for manifold for the four-cavity mold. The H-shape has an advantage over the cross-shape of using half the number of heaters, but it may not heat the passage connecting the two arms well enough to provide good fluidity of the plastic. This potential weakness could be overcome by placing copper rods above the heaters, and through and across the whole manifold to improve the heat conductivity of this section.

The cross-shaped layout has the advantage of having four fewer 90° bends for the plastic flow than the H-shape, and of having passages for a more effective melt-decompress action. The melt-decompress action is important in a heated manifold because it keeps the hot material from drooling. The nozzle itself in a hot runner system is subjected to considerable pressure, thus every inside corner should have a radius of 1/32-in. to avoid excessive stress concentration in heat-treating.

The gate size will vary with the material, but in general it should be somewhat smaller than the standard recommendations because of the high fluidity of the resin. The land should be about half of the gate diameter. For some materials, it is advantageous to provide a chamfer on the tip of the nozzle in the amount of 10 to 20 mil for an easy break-off. The inside of the nozzle should not only have a very fine finish, but should also offer smooth transition from one area to another.

Figure 13–1a (upper right-hand side) shows the bottom of the nozzle tapering toward the outside, so that the pressure can be concentrated around the sealing opening. When the mating surfaces of this arrangement are properly machined and finished, the seal is good and long lasting without the need of "O" rings.

Let us now check the design for pressure-pad needs. On this assembly, the pressure will be absorbed by the four holding rings of 1-7/8-in. O.D. × 1-9/16-in. I.D., and the center pad of 1/2 in. × 2-3/4 in.

The areas of these are (2.761 - 1.917) 4 + 1.375

$$\begin{array}{r} = 3.376 \\ 1.375 \\ \hline 4.751 \text{ in.}^2 \end{array}$$

This mold is to run in a 200-ton press. The permissible stress on the plates is 15,000 psi.

$$S = \frac{F}{A} \text{ or } 15{,}000 = \frac{400{,}000}{A}$$

$$A = \frac{400{,}000}{15{,}000} = 26.7 \text{ in.}^2$$

The area subjected to the pressure in the mold is 4.751 in.^{2}, which leaves 21.949 in.^{2} to be absorbed by pillars. Per cavity, the area will be 5.487 in.^{2} or a pillar of 2-5/8 in. diam for each cavity will be needed.

In Fig. 13–2, which is a combination of a short runner between a group of four cavities and a relatively long, hot runner between the two groups of four cavities. It is an excellent compromise between normal long runners and full runnerless system. It is an arrangement that is much easier to control and debug than the system in Fig. 13–1. In addition, it affords about 90% of the significant benefits from full runner molding. It has the flexibility for application of several gate types and minimal need for heaters and their controls.

The method shown in Fig. 13–2 applies an unheated beryllium copper nozzle and depends on the heat derived by conduction from the manifold. The amount of reground generated by this arrangement is less than 25% of that of a normal cold runner method. The manifold is a simple long bar equipped with four long cartridge heaters. It will be easier to insulate, control, and shield against radiation. This system suggests the possibility of applying a heated sprue bushing for two-, three-, or four-cavity molds, similar to the method used for the four of the eight cavities shown in the figure.

Note that, for the insulated runner with a probe, the probe or torpedo should have four ribs that will center the tip with respect to the gate. It should be arranged to have the torpedo adjustable vertically, so that the area of the gate may be varied for most favorable fill and break-off. Several stainless shims (0.005 in. thick) under the head of the torpedo could serve such a purpose. The land can be about half or the gate diameter. The finish of the torpedo surface should be of same quality as a good cavity finish.

HEATERS

There are some problems connected with the hot runner system, mostly due to either incorrect selection of heaters or their mounting. Therefore, it is desirable that we look deeper into the subject and examine how to eliminate some of the problems.

First, let us establish the size of heaters needed. We are using a wattage per cubic inch of manifold material as a basis for arriving at the heating needs of the manifold. Here, we will calculate the Btu and see how it compares with the empirical formula. We will use the manifold dimensions shown in Fig. 13–1 as the example for calculations, and the material being processed will be polycarbonate. The dimension of the manifold in the cross shape is 10-1/2 in. long X 2-3/4 in. wide X 2-1/8 in. deep. We will allow the manifold to come up to temperature in 30 min. The processing temperature will be 570°F, and the starting point will be 70°F.

The amount of heat for the job is determined by the need for bringing the manifold to operating temperature in the required time and by the amount necessary to maintain the temperature under operating conditions. Whichever of the two is larger will be the one selected for the job.

The steps for the calculations are:

1. *Required capacity for bringing the job up to operating temperature in the desired time.*

$$\text{Kilowatts} = \frac{\text{Btu}}{3412 \times \text{Hours allowed for heatup}}$$

$$\text{Btu} = \text{Weight of manifold} \times \text{Specific heat} \times \text{Temperature rise, °F}$$

$$= \text{in.}^3 \times \text{lb/in.}^3 \times \text{Specific heat} \times \text{Temperature rise, °F}$$

$$= \text{in.}^3 \times 0.284 \times 0.12 \times 500 =$$

$$\text{Pound/in.}^3 \text{ for steel} = 0.284$$

$$\text{Specific heat} = 0.12$$

$$\text{Temperature rise} = 500$$

$$\text{Cubic inches} = 4 \times 3.875 \times 2.75 \times 2.124 + 2.75 \times 2.75 \times 2.125 =$$

$$90.578 + 16.07 = 106.648$$

$$\text{Btu} = 106.648 \times 0.284 \times 0.12 \times 500 = 1817.28 + 21.6$$

Since the manifold holds a certain amount of polycarbonate, it will require additional Btu to raise its temperature to 570°F from 70°F.

$$\text{Btu}_p = \text{in.}^3 \times \text{lb/in.}^3 \times \text{Specific heat} \times \text{Rise in temperature}$$

$$= 3.2 \times 0.045 \times 0.3 \times 500 = 21.6$$

$$\text{Kilowatts} = \frac{1838.34}{3412 \times 0.5} = 1.078 = \underline{1078\ \text{W}}$$

2. *Heat losses at 570°F.*

a. Losses by convection and radiation.

Exposed surface area = 4 × 3.875 × 2.125 + 2.75 × 2.75 + 8 × 3.875 × 2.125 + 4 × 2.75 × 2.125 = 32.94 + 7.56 + 65.88 + 23.38

$$= 129.76 = 0.9\ \text{ft}^2$$

Losses by convection and radiation at 570°F for oxidized steel (curve A, Fig. 13–8) are 670 W/ft². Thus,

Kilowatts lost through convection and radiation (Square feet × 670)
= 0.9 × 670 = 603 W

b. Losses by conduction through insulation.

Thermal conductivity of transite board is 0.087 found in tables and expressed in Btu/hr/ft²/°F/ft

$$\text{Area of conduction} = 4 \times 2.75 \times 3.875 + 2.75 \times 2.75 = 42.63 + 7.56 = 50.19\ \text{in.}^2$$

or

$$\frac{50.19}{144} = 0.349\ \text{ft}^2$$

Thickness of insulation is 0.375/12 = 0.031 ft

$$\text{Kilowatts lost through conduction} = \frac{0.087 \times 0.349\ \text{ft}^2 \times 500^\circ\text{F}}{0.031\ \text{ft} \times 3412}$$

$$= 0.1435 \text{ or } \underline{143\ \text{W}}$$

The heat loss calculated = 603 + 143 = 746 W

c. The average loss during heatup is 746/2 = 373 W.

Total watts required during heatup is 1078 + 373 = 1451 W.

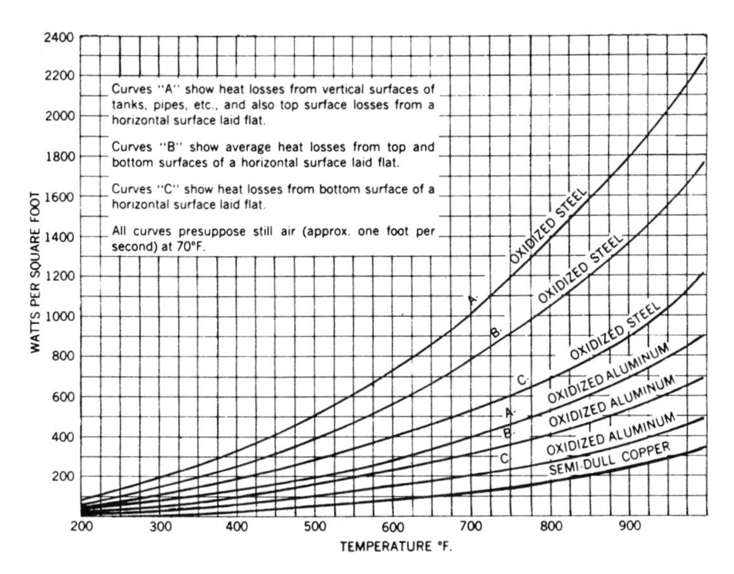

Fig. 13-8. Curve G-125S–Heat losses from uninsulated metal surfaces.

3. *Heat needed for material during processing.*

Weight of material processed per hour is 16.5 lb, the specific heat of polycarbonate is 0.3, and the temperature rise is 500. Since the material is being delivered at 570°F to the manifold, the heat requirement may only be 10% of the amount that would be called for if the plastic was heated in the manifold from room temperature to 570°F. The reason for the 10% heat is that some passage like the sprue bushing may not be at full temperature since there is no external heat introduced for it.

$$\text{The kW will be} = \frac{16.5 \text{ lb} \times 0.3 \times 500}{3412} = 0.725 \text{ kW or } 725 \text{ W},$$

and 10% of this will be 72.5 W.

The total operating heat requirement is: heat losses during operation shown in (a) and (b) = 746 W + 72.5 W from (3) calling for some heat for material during processing. The total is 818.5 W.

As indicated in (c), the heat needed for heat-up is greater than the above heat loss during processing; therefore, the 1451 W will be used. As a safety factor, the heater manufacturer recommends an addition of 15% to 20%, which will amount to about 1760 W. This figure is only about 25% of the amount obtained by using the empirical formula, which the author checked on at least a dozen working molds.

The reasons for the large discrepancy between figures derived from average industrial applications and those of the hot runner usage are the losses from radiation, conduction, and convection. It is believed that the major loss is from conduction due to the nature of mold operations. There exist water-cooled plates in contact with heated mold blocks and other features that are not conducive to efficient heat utilization.

In the manifold shown in Fig. 13-1, the heaters selected had dimensions of 0.5 in. diam × 4 in. long. In these particular dimensions, six capacities ranging from 180 to 1000 W are available. They are relatively inexpensive, and, if correction in wattage is needed, it can be readily made.

For long life and best performance, heaters with lower watt densities are definitely more desirable. Watt density is the heater wattage divided by its surface in square inches. Heater manufacturers list the watt density as part of the specifications of a heater. Watt densities over 200 W/in.^{2} are usually a potential source of problems. The way these can be avoided is by using more or larger diameter heaters, or by having a closer fit between heater and hole, and, if possible, by reducing heat requirements by shielding and insulation and using lower watt heaters.

Manifold constructions, with the principle of heating the plastic from inside out, frequently have to resort to high-watt density heaters. As a result, their performance is not always up to expectations.

Installing heaters into the manifold and connecting them into the circuit require a certain amount of attention to insure long and uninterrupted performance. Therefore,

1. The holes for the heaters should be reamed to a smooth finish and size so that the clearance per side should be 0.005 or less.
2. The heater should be spaced so that the lead end is flush with the opening. The heater should be retained in place by a clip attached to the manifold so that it will not move out of position from either direction.
3. The leads should be protected by armored covering and held in position to keep from vibrating.
4. The wires extending to the leads should be the heat-resisting type attached firmly with crimped tubing or similar method so that there is no chance of poor contact. The extension wires should also be protected by armored covering to eliminate the danger of damage to insulation and wire breakage. Stranded wire of ample size should be used. All cables should be clamped in place. Whenever insulating tape has to be used, it should be of the silicon type.
5. The wiring of heaters and their controls should have all the earmarks of good workmanship and the appearance of a permanent arrangement.
6. Capacities of wires, terminal blocks, connector plugs, etc., should be carefully checked out for current-carrying ability and correct voltage.
7. In conclusion, every phase of heater life should be carefully planned and carried out by qualified and experienced workers.

When heaters of proper wattage are selected and their mounting is carried out along suggested lines, the runnerless manifold will perform in a trouble-free manner.

14 Molding Thermoplastics

Those concerned with conversion of plastic material into a finished product have a real interest in the properties of a material and the specifics of processing so that the finished product will reflect performance requirements. All information that leads to a product with stated test bar properties is of utmost importance to the molder. The control over the chemical composition, etc., is in the hands of the raw-material maker. The molder can only use the material as furnished, apply the parameters of molding as recommended, and see that good moldability features are incorporated in the mold. Each material has its own range of molding requirements, and they are listed on "The Material Processing Data" sheets under "Thermoplastics" (pp. 320–321). These sheets also contain information on molding problems that are peculiar to a specific type of material.

The features of molds for favorable productivity have been discussed previously. The part that the molding machine plays in thermoplastics molding will be described in this and following chapters. For the nature of plastics, see Chapter 2.

INJECTION MACHINES FOR THERMOPLASTICS

The function of an injection molding machine is to convert a plastic raw material in granular form into a homogeneous fluid state for delivery to cavities of a mold. This is accomplished in a heated cylinder and under pressure. A plunger or screw forces the material into the mold that is held shut by the clamp end of the machine. The plastic is held in the mold until it cools and becomes rigid enough to be ejected upon mold opening.

PLUNGER MACHINE

The injection process of thermoplastic materials is carried out in two basic types of machines. Their difference is in the injection end only. The plunger type uses a measured shot of the required amount of material, which is moved into the heating chamber by means of a plunger or piston (Fig. 14-1). The measured shot is being controlled by the injection plunger, which actuates an adjustable piston or scooping device arranged to measure the volume needed per shot. The accuracy of these devices is not satisfactory for most jobs and has been substituted by weigh-feeding scales that would provide an accurate weight of material for

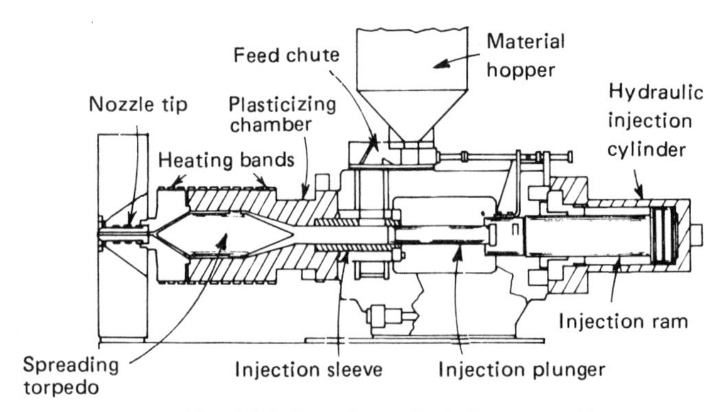

Fig. 14-1. Injection end of plunger machine.

each shot. The problem of supplying a duplicate amount of material for each shot is one of the major weaknesses of the plunger machine.

The chamber or cylinder is heated by means of electrical heating elements from the outside walls and through conduction reaches the plastic material. In order to facilitate plastication, the inside of the cylinder is equipped with a spreader-type torpedo. This creates a plastic tubular section, about 1/4 in. thick, or multiple sections of about 3/8 in. diam depending on torpedo design. With these dimensions, it is possible to induce reasonably uniform heating, which will plasticize the material ready for the shot. Each shot displaces the same amount of material from the chamber into the mold as the one received by the moving plunger and is driven into the chamber. The remaining part of the cycle is the same as for screw machines.

SCREW MACHINE AND ITS PROCESSING—ALL-HYDRAULIC

The basic requirements that each type of a machine has to meet are time, temperature, and pressure. If one type of machine is thoroughly understood, it is relatively simple to transpose the knowledge to the variation in detail of another type. Since screw machines are continuously increasing in number, they are selected for analysis of all the elements and their favorable utilization.

The operation of the screw machine will now be examined from a conceptual point of view. With the mold properly mounted in the press, the closing of the safety gate initiates a cycle. The clamp closes at a fast rate, begins to slow down as the protruding ejection return pins contact the stationary half of the mold, and stays at the slow rate until the mold is fully closed. At this time, the full high pressure is applied. Now the injection end comes into play.

We start with the hopper where material is introduced and may be dried to a prescribed temperature if needed. The tapered portion of the hopper connects to the throat, where the rotating screw is exposed to the material for pickup. The rotation causes the material to move forward, be compressed and "sheared" until fully plasticized, and prepared to enter into the metering front section of the chamber. As the plasticized material advances into the metering portion, it displaces the screw itself, forcing it to retract to a predetermined position equal to the volume needed for the shot. At this time, the screw stops rotating, and, on proper signal, it will move forward to inject the measured-out material into the mold. Backflow around the screw flights is prevented by a floating ring, which, when in the forward position, allows the material to move into the metering section, but which, when in the rear position, seals off the screw from the metered material that is being injected into the mold. The forward movement of the screw causes the ring to retract into the sealing position. The forward or injection movement of the screw is actuated by single or twin hydraulic cylinders. When retracting for filling of the measuring chamber, the plasticating screw has to overcome the frictional resistance as well as the resistance of removing the oil from injection side of the hydraulic cylinders. The resistance of the returning oil is normally adjusted upward by a flow control valve. Such increased resistance to the retracting screw results in better mixing of the plasticized material and, in a small way, in an increase in its temperature. The transition from chamber to the mold is by a nozzle that makes a tight contact against the sprue bushing of the mold. The injection itself is of short duration and under high pressure and is followed by injection "hold pressure" to permit the plastic in the mold to solidify and freeze the gate. The next time interval covers the "curing" in preparation for ejection from the mold. During the cure or complete solidification time, the screw plasticizes the material for the subsequent shot. When cure is at the end as determined by a timer, the clamp opens, slowly at first, followed by a speedy rate, and finally is slowed again beginning with the position of ejection. The movement of the ram actuates the stripping arrangement and brings about removal of the parts from the mold. Now the cycle is at the end and ready for a new start.

MACHINE COMPONENTS AND THEIR PART IN THE PROCESS

Machine components each play an important part in the process, so proper attention to them is necessary for overall good performance. Figure 14–2 on page 244 describes the operation of an injection screw machine.

Hopper and/or Dryer

Removing the moisture from the surface of such materials as polyethylene, polypropylene, and polystyrene can be accomplished by using heated ambient air, set to the desired temperature and circulating through the hopper. As the air

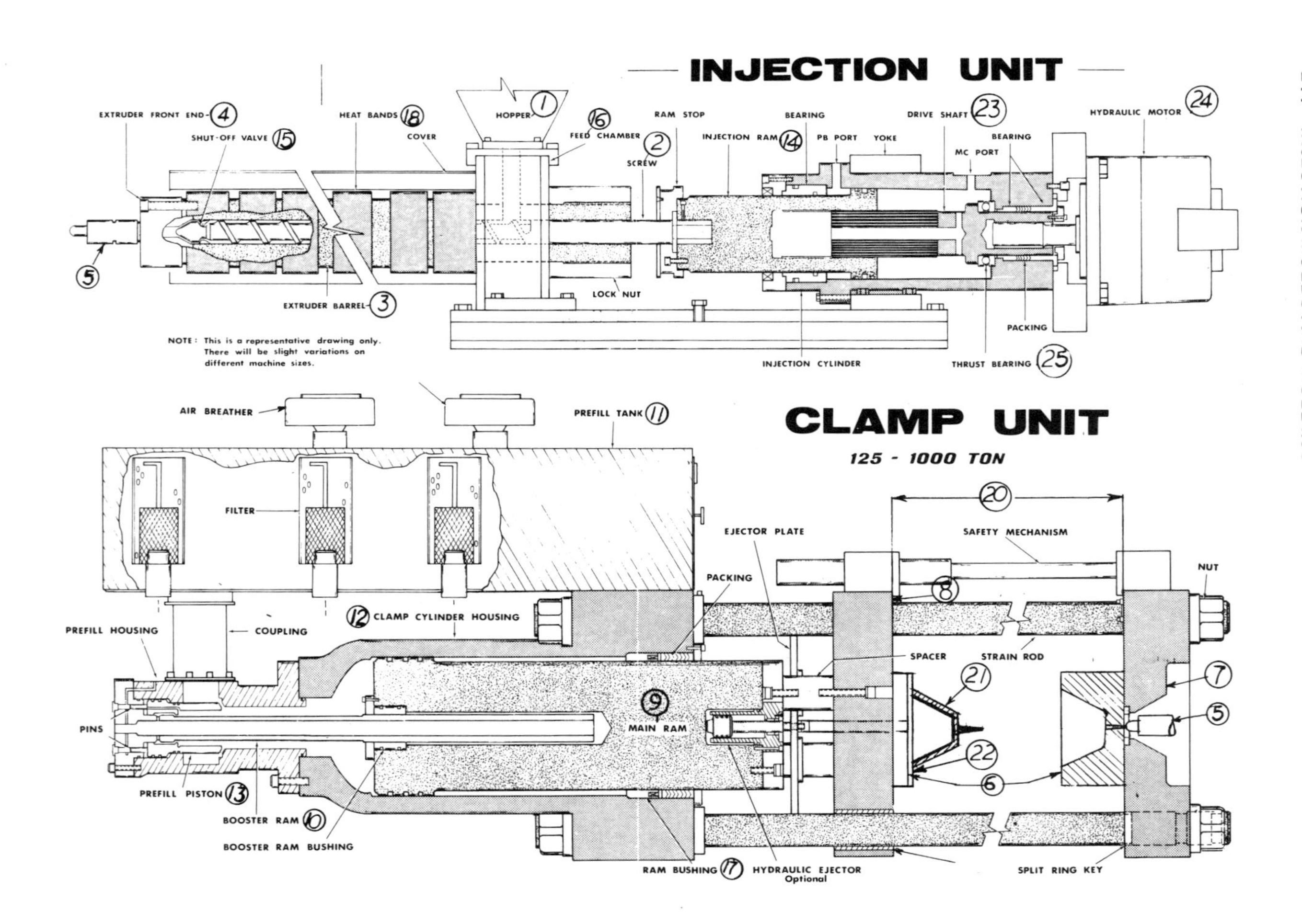
INJECTION UNIT
EXTRUDER FRONT END 4
SHUT-OFF VALVE 15
HEAT BANDS 18
COVER
HOPPER 1
FEED CHAMBER 16
SCREW 2
RAM STOP
INJECTION RAM 14
BEARING
PB PORT
YOKE
DRIVE SHAFT 23
MC PORT
BEARING
HYDRAULIC MOTOR 24
5
EXTRUDER BARREL 3
LOCK NUT
PACKING
INJECTION CYLINDER
THRUST BEARING 25
NOTE: This is a representative drawing only. There will be slight variations on different machine sizes.
AIR BREATHER
PREFILL TANK 11
CLAMP UNIT
125 - 1000 TON
FILTER
20
EJECTOR PLATE
SAFETY MECHANISM
PACKING
8
NUT
12 CLAMP CYLINDER HOUSING
PREFILL HOUSING
COUPLING
SPACER
STRAIN ROD
21
7
9 MAIN RAM
PINS
5
22
6
PREFILL PISTON 13
BOOSTER RAM 10
BOOSTER RAM BUSHING
RAM BUSHING 17
HYDRAULIC EJECTOR Optional
SPLIT RING KEY

CLAMP UNIT OPERATION

1. *Clamp open.* There is no hydraulic pressure acting on any part of the ram. The press is shown with its full daylight #20. Closing the press gate brings about the actuation of hydraulic and electrical circuits.

2. *Fast close.* A high volume pump acts through the prefill piston #13, on the small area of the booster ram #10, that causes the main ram #9 to move at high speed until a limit switch "closing slowdown" is tripped. Prefill arrangement fills the vacated space in the clamp cylinder housing #12 by gravity.

3. *Clamping pressure.* A low pressure pump, controlled by a low pressure pilot head that is actuated by "closing slowdown" limit switch, takes over on the advancing main ram #9, to close the ram the rest of the way. This low pressure protects the mold against any foreign objects that might get caught in it. When the mold is almost closed a "full pressure" limit switch is activated and brings about full clamp tonnage that builds up rapidly by a high pressure pump acting on both the booster ram and the main ram. Prefill arrangement is shut to prevent oil from backing into prefill tank #11.

4. *Mold breakaway.* When a mold is ready to open at a signal from opening timer, a low volume high pressure pump delivers oil into the pullback area that is formed in back of the ram bushing #17, to slowly separate the mold and prevent damage to the molded part.

5. *Fast return.* When a limit switch "fast open" is tripped, a high volume high pressure pump takes over to provide rapid opening. Another limit switch "open slowdown" decreases the speed. During this slowdown, the removal of product #21 takes place and is activated by the stripping arrangement of #22. The limit switch "reverse stop" brings the main ram to the starting position. The limit switch arrangement is shown on Fig. 15–3, page 287.

INJECTION UNIT OPERATION

The injection unit, also called plasticizing unit, consists of a reciprocating screw #2, injection ram #14, drive shaft #23 and hydraulic motor #24. The screw is inserted into a hard barrel that becomes a part of the stationary feed chamber or the throat #16. A material hopper #1 is mounted on top of feed chamber #16. The screw is inserted and keyed into the injection ram #14. The splined drive shaft #23 engages the splines in the injection ram bore to rotate the ram and screw assembly. A thrust bearing #25 absorbs the force placed on the drive shaft during injection and also while the plastic is being moved forward. The hydraulic motor #24 has a constant speed and thus providing the screw of uniform rotation.

The operating sequence is as follows starting with the elapsed time of the injection timer.

1. The screw remains in the forward position after injection.
2. The rotation of the screw causes the plasticized material to move forward while absorbing the heat from the heat band #18. As the plasticized material enters the extruder front end #4, it displaces the screw itself by moving the whole assembly rearward against an adjustable pressure until the volume needed for the molded product is reached. This movement consists of the screw and injection assembly that slides back on the splines until a limit switch is actuated. At this time, the screw stops rotating and is ready for injection.
3. The mold and press are locked tight to a predetermined high pressure and calls for injection to take place. Back flow around the screw is prevented by a floating shut-off valve #15, which when in forward position allows the material to move into the extruded front end #4, but when in the rear position seals the screw from backflow around the screw flights.
4. The hydraulic fluid is supplied to the injection ram #14 and brings about the forward movement of the screw to inject the plasticized material into the mold.
5. The contact between injection unit and mold is made by a nozzle assembly #5.
6. The screw moves forward injecting the plasticized material into the mold.
7. Injection pressure is held on the material until the elapse of time set on the injection timer after which a hold timer determines the duration to which backflow from cavity to the extruder barrel #3 will not take place.

The preparation of material for the following shot normally starts again from the forward position of the screw.

Fig. 14–2. Injection machine.

temperature increases, it absorbs and carries away the surface moisture of the plastic material.

With hygroscopic materials such as nylon and polycarbonate, where the moisture penetrates the surface to form a molecular bond with the plastic, a dehumidifying system is recommended. This system employs a molecular sieve desiccant with a high affinity for water vapor in the drying air. The air passes over the material to be dried and absorbs the moisture. Eventually, the desiccant becomes saturated with the removed moisture. To overcome this problem, the system is equipped with the desiccants such that while one is "on stream," the other is being regenerated for switching to future use in the drying cycle.

At present there are injection machines on the market that have incorporated in the heating barrel a devolatilizing section to remove moisture and other gases from the polymer during processing. This type of a machine involves longer barrels and screws than those normally found on similar standard machines. The screw is specially designed to have the screw shaft smaller in diameter and flights deeper at the devolatilizing station so as to bring about a lower pressure in that area in comparison with the first metering section. Fig. 14–3 shows the general arrangement of the self-venting barrel.

In considering a design of this type, one must be careful to evaluate the devolatilizing design against standard types available. Some of the consideration could be the strength of considerably longer screws and barrels, their maintenance and replacement, prolonged residence time of the polymer, operational difference this and standard designs, and other details pertinent to the new design. Only a thorough and detailed analysis of all the factors determine which model to use.

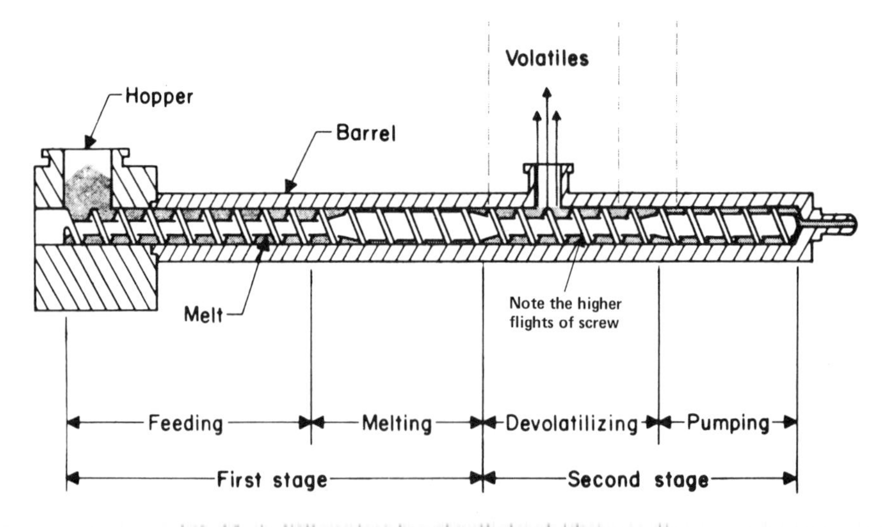

Fig. 14–3. Self-venting barrel with devolatilizing section.

The hopper is the reservoir of material that will be used for a number of cycles. Whenever a material requires drying prior to molding, the "hopper dryer" can be utilized for drying in position without additional handling. Each material has its own specifications for drying temperature and time, as shown in material processing data sheets (Chapter 18). When a machine is ready for operation, the hopper should contain an amount of dried material that will be used during the required drying time. At the same time, an additional supply of material is to be added to the hopper so that it can be dried while the original batch is being used up. If the hopper is not large enough to accommodate the necessary volume, an auxiliary removable hopper may be provided. It should be remembered that if handling of material can be saved, the possibilities of contamination are eliminated.

The material in the hopper should be maintained at a level at which refilling starts, thereby insuring that the plastic is always exposed to the required drying time before use. The temperature in the hopper should be checked so that the possibility of granules adhering to each other, which would interfere with free flow, is eliminated.

The temperature of the hopper should be the drying temperature of the material, and, in the case of ABS, in the example used below, it is 170° to 190°F. A portable pyrometer for checking the hopper temperature provides a good check on the thermostat setting of the hopper dryer.

Knowing the poundage used per hour makes it easy to calculate the pounds required for the drying cycle and thereby the space it would occupy in the hopper.

Example: Material is ABS, the weight of the shot is 10 oz, and the cycle is 60 shot/hr. The material consumption per hour will be 600 oz or 37.5 lb, with a drying time of 2.5 hr; the hopper should at no time contain less than 93.75 lb or about 100 lb. When this level is reached, it is time to add a new batch of 100 lb to keep the supply on schedule. Converting the pounds into cubic inches, the level should not fall below 2800 in.^{3} (1 lb of ABS has a specific volume of 28 in.^{3}).

Throat

The next passage point is the throat. It has a circulating water jacket around it so that its temperature can be maintained at the same level as that of the hopper, i.e., the drying temperature of the material. Too high a temperature will cause the material to "bridge" and thus bring about an intermittent supply to the screw. When the throat is too cold, it may cause condensation and introduce moisture into the material as it enters the heating chamber. This would be a most undesirable feature and could be a source of many molding problems.

Screw

The general outline of a screw design is shown in Fig. 14–4. The portion in the throat area has deep flights, providing greater volume for the bulky material re-

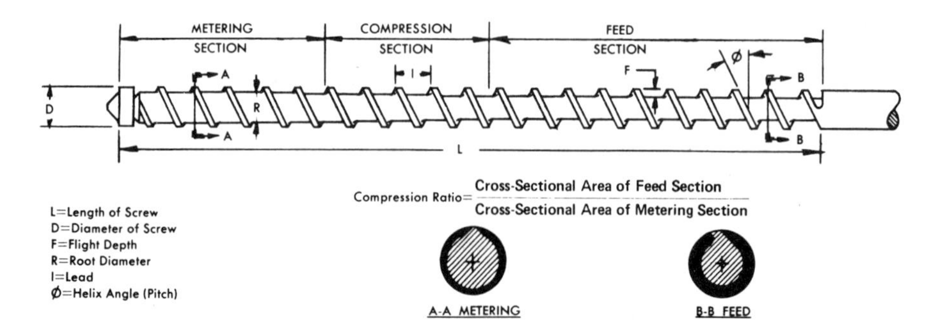

Fig. 14–4. Plasticating screw. *(Courtesy of HPM Corp.)*

ceived from the hopper. This portion is known as the *feeding section,* and, while the screw is rotating, it moves the plastic into the compression area where it is densified and homogenized, thus gaining in capacity to absorb heat. While being compressed, the plastic is also sheared, thus obtaining added energy for plastication. The material now enters the metering portion of the screw, where it is thoroughly mixed, attains the ultimate desired temperature, and is fed uniformly into the front of chamber in preparation for the shot. The desired speed of screw rotation and back pressure for the screw are indicated on the processing data sheet for each material. The rotation of the screw plays an important part in keeping the plasticized material in motion and averaging out the melt heat. It should be kept rotating up to the time of mold opening, thereby minimizing the time of heat soaking by the melt within the measuring chamber. Since plastic materials have poor heat conductivity, their exposure to heat soaking will result in temperature gradients within the material. This condition may lead to stresses, warpage, and other undesirable features in the molded product.

Once the shot is prepared as outlined, the mold closes and the material is injected into the cavity by the front of the screw, which acts as a plunger. In normal cases the screw does not come home but leaves in front 1/16 in. to 1/8 in. of material known as a cushion. Any void created in the product while cooling is being replenished by the cushion under the hold pressure during the time that the gate to the part remains open.

Some products with a uniform wall thickness do not require replenishing of sinks; for this application it was felt that a cushion could be eliminated and feeding an exact amount into the cavity would give satisfactory parts. This method is known as starve feeding. To insure that it provides quality parts, the product should present visually as being without defects, and should be subjected to any test that would determine its suitability in application. Starve feeding was originally started with plunger machines when accurate weigh-feeding machines were installed. Each cycle then dumped the same weight in front of the machine plunger. The operation was generally successful and the net result was to save from 3 to 6% of the weight of parts without sacrificing any of their properties. With screw machines and the process control available for them to starve feed, one could produce more accurate results and a decided decrease in rejects.

Fundamentally, we have a pressure sensor in the cavity that controls the uniformity of pressure on the material and shot-size control with position of screw sensor, both of which determine the exact amount of material needed in the cavity. With such an arrangement a saving of material and reduction of rejects should recommend the process control for application even when other benefits of the system are not immediately apparent. See page 306.

Back pressure is induced when the discharging material from the rotating feeding screw forces its way to the front of the metering chamber. The force of the incoming material exerts pressure on the front face of the screw and causes it to retract. The pressure from the screw is transmitted to the hydraulic cylinders that are used for injection purposes. The hydraulic oil from the injection side is being forced out through a pressure-regulating valve into the tank. The setting on the pressure-regulating valve determines the amount of back pressure to be used for thorough mixing of the plastic in the front and also the added amount of heat to be introduced into the plastic as a result of the work exerted by the screw. The pressure of the material on the screw in most cases amounts to 500 to 1000 psi, which corresponds on most machines to a gauge reading of 50 to 100 psi.

Let us calculate the psi on the material in our example. We know the diameter of the screw is 2 in. and the injection psi is 20,000. The pump that supplies the injection cylinder is 2050 psi (see machine specifications under "Machine Capacity," p. 252). The force of injection is:

Area of screw X Injection psi = Pump psi X Area of injecting cylinder.

$$0{,}7854 \times 2^2 \times 20{,}000 = 2050 \times \text{Area of cylinder}$$

$$\text{Area of cylinder} = \frac{68{,}200}{2050} = 30.634 \text{ in.}^2$$

If the back pressure gauge reads 50 psi, we have:

$$\text{Gauge reading} \times \text{Area of cylinder} = \text{psi on material} \times \text{Area of screw}$$

or

$$50 \times 30.634 = \text{psi on material} \times 3.14$$

or

$$\text{psi on material} = \frac{50 \times 30.634}{3.14}$$

$$= 487.80 \text{ psi}$$

With the basic machine specification on hand, similar calculation will give the necessary answer. In some rare cases, the back pressure on a material may go as high as 5000 psi, but care must be exerted that such a pressure does not have a deteriorating effect on the material. The back pressure element is frequently used for instant small upward adjustment in material temperature by a slight increase in the pressure reading.

The present arrangement of gauge readings is not conducive to accurate settings because the gauge used for the purpose has a scale of 0 to 3000 psi and, in the 0 to 75 range, it is difficult to make an accurate interpretation. For legible repeatability and for some heat-sensitive materials, it is important to have reasonably accurate setting information. The problem may be overcome by installing a 0 to 500 range scale pressure gauge connected directly to the back pressure-regulating valve. The point of this detailed description is that the back pressure setting and adjustment are of greater significance to some materials in the molding process than is recognized by the equipment manufacturers.

The proper function of the screw is very important in the performance of a machine. When we consider the variety of materials being processed, we find that it is frequently necessary to inspect the condition of the screw as well as the check rings that act as flow valves for the plastic (Fig. 14–5). Repair and replacement are indicated in many instances. Very seldom are the necessary precautions taken to insure that the correct clearance between screw and chamber are maintained. These clearances are the important dimensions that determine the effective working of a machine.

Injection screws are known by their length-to-diameter ratio. Some machines have a general-purpose screw with a ratio of 16:1 or 20:1. When ratios differ, the change is made in the compression section, and the other sections are the same.

The general-purpose screw shown in Fig. 14–4 was designed to plasticize a great variety of plastic materials; at the same time, its configuration and dimensioning are such as to have no optimum performance for any one specific grade of plastic. A great deal of development is occurring for screws with special-purpose requirements, and when so applied, they are usually quite successful. One such screw made by Union Carbide Corp. is detachable to the end of a general-purpose screw (see Fig. 14–6). On a 300-ton press, the last three turns of the screw were replaced with 7-1/2-in. section of fluted design (Fig. 14–7). This design consists of parallel grooves machined into the added extension. Half of the grooves are opened towards upstream end (inlet) of the screw and the other half towards the downstream end (outlet). As material comes to a dead end in those open, upstream grooves, melt pressure forces it over a barrier into the groove that is open downstream. Localized shear assists in dispersing colors or other similar ingredients. For shear rate-sensitive polymers, a design that reduces mixing action is being recommended. The benefits of such as fluted section are (1) that lower coloring per pound produces the same effect as

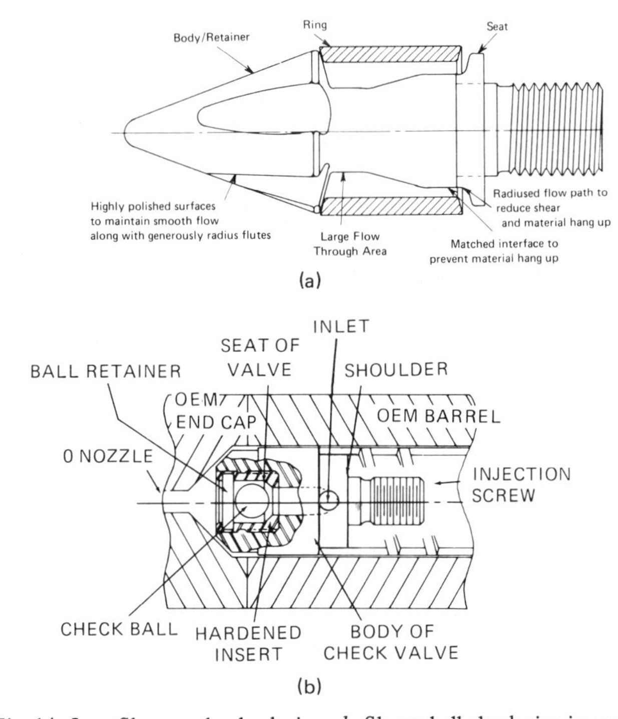

Fig. 14–5. *a,* Shows only check ring. *b,* Shows ball check ring in assembly.

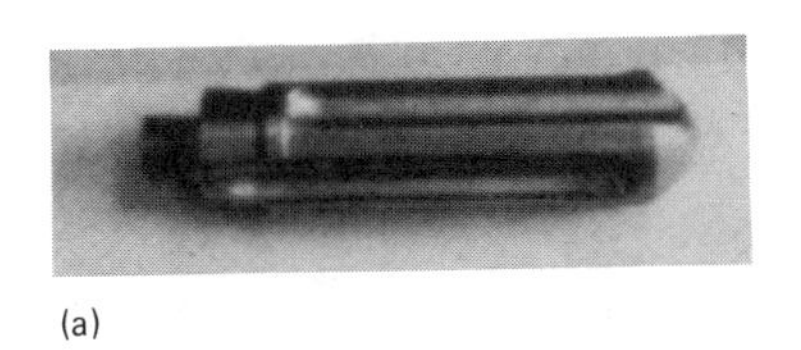

(a)

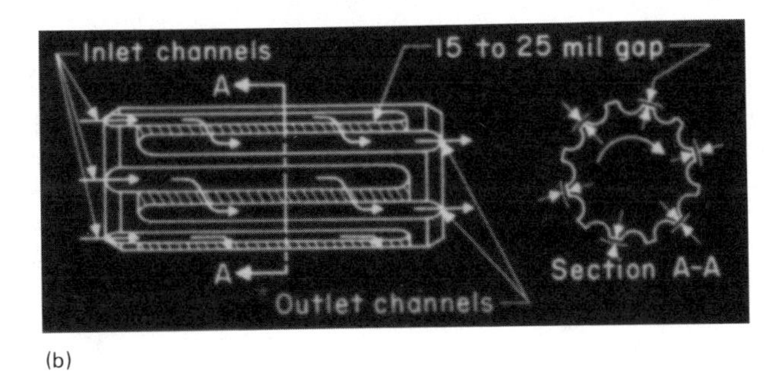

(b)

Fig. 14–6. Details of Union Carbide fluted extruder-screw barrier mixing section. *a,* Photograph of actual mixing section. *b,* Schematic diagram. *(Courtesy Union Carbide Corp.)*

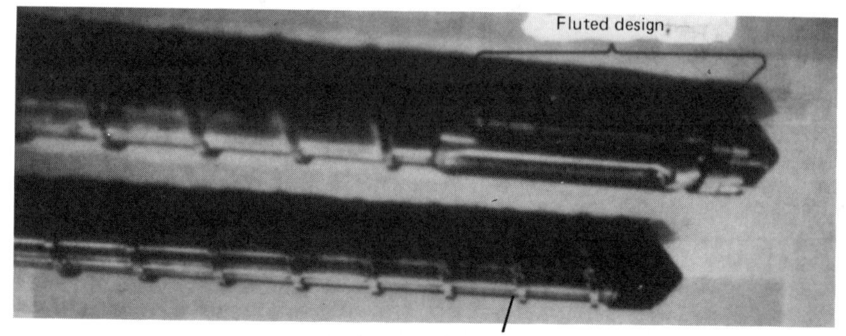

Fig. 14–7. Comparison of fluted and standard design screw ends.

higher concentrates in a general-purpose screw; (2) this new section produces parts with good appearance and optimum properties. The improvements are attributable to operating at more uniform melt temperature, less tendency to mold in stresses, and more efficient color dispersion. (3) All these benefits are also obtained at lower cycles. It is an addition to a screw worthy of consideration.

As a guide to the tolerances of a 2- and 3-in. screw system, the clearances are listed:

	Screw O.D.	*Check Ring O.D.*	*Chamber I.D.*
Nominal 2 in. diam	1.993–1.994	1.996–1.995	2.000–2.001
Clearance/side	0.003–0.004	0.002–0.003	NA
Nominal 3 in. diam	2.992–2.999	2.996–2.995	3.000–3.001
Clearance/side	0.0035–0.0045	0.002–0.003	NA

NA=Not applicable.

Machine Capacity

Machine capacity relates to the cubic inches of space for maximum shot, rate of injection, plasticizing capacity, etc. In all explanations of the process, examples are used in connection with a specific size of machine. With this method in mind, it is believed best to list the specifications of interest of the particular machine and the appropriate formula of converting the data into variables encountered in practice. The information and formulas are especially useful when a mold is transferred from one machine size to another. The specifications for a 200-ton, 14-oz machine are:

Injection capacity (in.^{3})	25.1
Injection capacity (oz of general-purpose polystyrene)	14.0

Plasticating capacity (estimated lb/hr)	240
Recovery rate (estimated oz/sec)	1.07
Injection pressure, psi max. (adjustable)	20,000
Injection stroke, max (in.)	8.00
Screw diameter (in.)	2.00
Barrel length-to-diameter ratio (L/D)	20/1
Injection rate (in.^{3}/sec)	22.0
Screw rpm (Range No. 1/Range No. 2) (adjustable)	215/155
Clamping force (tons) (adjustable)	200
Clamp opening force (tons)	13.6
Ejector force (mechanical tons)	13.6
Ejector force (hydraulic tons)	5.4
Ejector stroke, (in.)	4.0
Clamp speed (in./sec)	
Closing fast	31
Closing slow (adjustable)	3.4
Opening slow	1.6
Opening fast	29
Pumps	
Clamp	42 + 5 gpm at 2425 psi
Injection	42 + 14 gpm at 2050 psi
Injection hold	5 gpm at 2050 psi
Heating capacity of cylinder	14.4 kW
Nozzle heating capacity	0.4 to 1.4 kW

With the aid of a pocket calculator, many calculations can be made speedily and information can be obtained that will reduce to a minimum the cut-and-try method, which is wasteful of material and time-consuming.

Machine capacity as it relates to shot size means how many ounces of polystyrene can be plasticized per second. The number of ounces of polystyrene plasticized per second divided into the number of ounces of machine capacity will provide an estimated time that a full shot of polystyrene will take to be prepared for injection into the mold. The machine specifications tell us that 240 lb of styrene can be plasticized per hour. The number of pounds multiplied by 16 will give ounces; divided by 3600, it will equal ounces per second or

$$\frac{240 \times 16}{3600} = 1.07 \text{ oz/sec}$$

Fourteen ounces will require

$$\frac{14}{1.07} = 13.1 \text{ sec}$$

to plasticize the full shot of polystyrene.

In order to establish how this capacity will relate to other materials, we will compare the heat requirements in Btu for polycarbonate and nylon 6/6 with those of polystyrene.

$$\text{Heat input} = (T_2 - T_1) \times \text{Specific heat}$$

where T_2 = average melt temperature (for polystyrene, 420°F; for polycarbonate, 580°F; for nylon 6/6, 600°F)

T_1 = material temperature in hopper, 80°F

Specific heat in Btu/lb/°F for polystyrene is 0.32; for polycarbonate, 0.3; and for nylon, 0.4.

Specific heat, from the same source as melt temperature.

Melt temperatures from material processing data sheets (pages 320–321).

Thus, the Btu requirement for polystyrene:

$$(420 - 80) \times 0.32 = 109 \text{ Btu/lb or } 95.4 \text{ for } 14 \text{ oz}$$

for polycarbonate:

$$(580 - 80) \times 0.3 = 150 \text{ Btu/lb or } 131 \text{ for } 14 \text{ oz}$$

for nylon 6/6:

$$(600 - 80) \times 0.4 = 208 \text{ Btu/lb}$$

In addition to the 208 Btu, there is a "latent heat of fusion" for nylon in the amount of 56 Btu/lb, making a total of 264 Btu/lb or 231 for 14 oz.

The latent heat of fusion is comparable to the heat that has to be added to a pound of ice at 32°F in order to convert it to water of 32°F. Just comparing the Btu values, the downgrading of capacity would be very large; in practice, however, it has been found that a reduction to 75% of rated capacity will cover most of the frequently used materials. Such factors as heat conductivity at melt temperature have an important influence on the result.

One way to minimize the reduction of capacity is to have the hopper at drying temperature of each material, and the total Btu requirement will thus be reduced.

For polycarbonate, the drying temperature is 250°F, and for nylon 6/6, it is 160°F. Substituting these values for T_1, we obtain:

$$\text{Polycarbonate: } (580–250) \times 0.3 = 99 \text{ Btu/lb or } 86.6 \text{ for } 14 \text{ oz}$$

$$\text{Nylon 6/6: } (600–160) \times 0.4 = 176 + 56 = 232 \text{ Btu/lb or } 203 \text{ for } 14 \text{ oz}$$

Note: The latent heat of fusion applies to crystalline materials. These materials usually have a high and dual shrinkage rate, one in the direction of flow and another perpendicular to it. This latent heat of fusion is different for each material of that type.

The examples cited indicate that there is a valid reason for the downgrading of machine capacity. With the variables of plasticating time and weight of material, there is considerable room for manipulation. Another factor not to be overlooked is that heating the material in the hopper to drying temperature will raise the plasticating capacity of the machine.

Time of plastication is equivalent to curing time, and there are not many polycarbonate or nylon 14-oz jobs that will solidify in 13 sec as calculated for the 14-oz polystyrene shot. From a practical point of view in most cases, the limiting element as far as plastication is concerned is the cure of the parts. When machine is used on the low end of capacity (about 25% of rating), the L/D ratio of the screw becomes an important factor because of the total volume of material in the chamber and the residence time. Some materials are not affected by prolonged exposure to melt temperature. Those that are heat-sensitive have to be watched to see whether the residence time is within safe limits of the material. The suppliers of the material have information as to how long their product will withstand melt temperature without deterioration.

Injection Rate

The rate at which a material is injected into a mold frequently determines the appearance and properties of a product. For this reason, it is important to know how to interpret the injection speed so that it can be interpolated to another machine if necessary or faithfully reproduced in the same machine.

The injection volume of oil is controlled by a flow valve of rated capacity that divides the flow into equal increments and bypasses the amount indicated on the setting. Thus, for example, if the valve has 10 divisions, each one will represent about one-tenth of the valve rated flow to be bypassed. The 14-oz machine, for example, is equipped with a flow control valve of 28 gpm and has 10 adjustment settings; each will bypass 2.8 gal/min. If the setting is made at 4, the amount of oil bypassed will be $4 \times 2.8 = 11.2$ gpm, or the flow into the cylinder will be 56 (from machine specifications) - 11.2 = 44.8 gpm. This will mean a reduction in delivery rate of $44.8/56 \times 22$ (22-injection rate from machine specifications) = 17.6 in.^{3}/sec. The valve preferably should be of the type that its flow is independent of temperature and pressure variations. At maximum injection rate, the flow control valve is set at zero, and the full flow of the pumps is utilized.

When the volume of oil is decreased, the time of injection will increase in the same proportion. In the machine specifications of the machine example, the maximum rate is 22 in.^{3}/sec, and the full shot capacity of 25.1 in.^{3} will take 25.1/22

= 1.142 sec. With the valve setting at 2, it will take about 20% more time or 1.37 sec. If the in.^{3}/sec are not given, they can be calculated in this way:

The pumps that move the injection cylinder are shown on the hydraulic circuit and specify their volume in gallons per minute. With the injection cylinder bore known, we have a relationship of

$$\text{Piston movement (in./sec)} = \frac{231\ (\text{in.}^3/\text{gal}) \times \text{gpm}}{60 \times \text{Cylinder area}}$$

In our example,

$$\text{Piston movement (in./sec)} = \frac{231 \times 56}{60 \times 30.634} = 7.037$$

In the section on back pressure (p. 249), the area of the injection cylinder in the above equation was figured to be 30.634 in.^{2}. The piston movement per second (7.037), which is also the screw movement, when multiplied by area of screw (3.14) will give the cubic inches delivered per second. Thus,

$$7.037 \times 3.14 = 22\ \text{in.}^3/\text{sec}$$

If, for any reason, the injection cylinder diameter cannot be identified, it can be established as detailed in the following paragraphs.

The force of the injection cylinder is equal (with minor friction loss) to the force of injecting the material into the mold. Force = Area X Pressure. Or

$$\text{Area of injecting screw} \times \text{Injecting psi} = \text{Area of cylinder} \times \text{Pump pressure}$$

In the foregoing example,

$$0.7854 \times 2^2 \times 20{,}000 = 2050 \times \text{Area of cylinder}$$

$$\text{Area of cylinder} = \frac{3.14 \times 20{,}000}{2050} = 30.634\ \text{in.}^2$$

We have established the time needed for the screw to travel the 8-in. stroke as 1.141 sec at its highest speed. Let us now convert into seconds per inch and seconds per ounce so that time can be figured for various screw positions and the "fast injection" or "injection high" timer can be set to correspond.

We have

$$\frac{1.141}{8} = 0.143 \text{ sec/in.}$$

or

$$\frac{1.141}{14} = 0.08 \text{ sec/oz}$$

or

$$\frac{1.141}{28.35 \times 14} = 0.0029 \text{ sec/gram}$$

Any of these results multiplied by the ratio of

$$\frac{\text{Specific gravity of polystyrene}}{\text{Specific gravity of material in use}}$$

will translate into the needs for the material being run. For example, how much time will it take in the 200-ton press for the screw to travel for a 6-oz shot of polycarbonate?

$$0.08 \text{ sec/oz} \times 6 \times \frac{1.08}{1.20} = 0.432 \text{ sec}$$

It is very important to set the "injection fast" timer to the proper value so that upon timing out, the large volume of oil can be discharged into the tank at very low pressure. This prevents waste of power and eliminates the danger of oil heating, which may cause its degradation.

To get an idea of power waste, we can figure the horsepower required to drive the injection pumps:

$$\text{Horsepower} = \frac{\text{gpm} \times \text{Pressure}}{1714}$$

$$= \frac{56 \times 2050}{1714} = 66.98$$

At current power rates, this can be very costly.

Another useful bit of information is to convert the known weight of a material into inches of screw travel:

$$\frac{\text{Material (oz)} \times \text{Stroke of screw}}{\text{Ounce capacity}}$$

This result multiplied by ratio of

$$\frac{\text{Specific gravity of polystyrene}}{\text{Specific gravity of material}} = \text{Inches of travel for material to be used.}$$

For weight in grams,

$$\frac{\text{Grams} \times \text{Stroke}}{28.35 \times \text{Ounce capacity}} \times \frac{\text{Specific gravity of polystyrene}}{\text{Specific gravity of material}}$$

Example: A screw travel of 7-oz polycarbonate in 14-oz machine is wanted:

$$\frac{7 \times 8}{14} \times \frac{1.08}{1.20} = 3.6 \text{ in.}$$

or, converting ounces to grams,

$$= 198.45 \text{ gram}$$

$$\frac{198.45 \times 8}{28.35 \times 14} \times \frac{1.08}{1.20} = 3.6 \text{ in.}$$

If each machine had its own data sheet with information described and frequently used conversion factors listed, many steps could be saved in setup and interpolating data from machine to machine.

As the "injection fast" timer runs out, it starts the "inject hold" timer. The hold timer brings into action a low-volume pump that can be set at a pressure somewhat lower then the initial injection pressure. The purpose of the "hold" pump is to replenish any voids created in the part with material and to maintain a pressure in the cavity that will prevent the flowing of material from the cavity back into the injection system. The duration of "hold" pressure should be just long enough to cause freezing of the gate. Too short a time may show up as excessive sinks on the part resulting from backflow of material. Too long a holding pressure may increase the cycle for no valid reason.

Recharging for Shot

Upon expiration of the "hold" timer, the screw rotation and plastication are instigated. There are high and low torque positions on the hydraulic motor for driving the screw. Experience in a large custom molding operation indicated that the vast majority of jobs can be run at the high torque setting with speeds that correspond to it. The importance of this operation is not only to plasticize the material, but also to repeat the distance of screw travel with accuracy so that there is no variation in the volume of material being injected. The distance of travel is governed by the setting of a limit switch. If the limit switch is of such a type that it will be actuated within a distance of ±0.002, its performance will control the distance of travel in a satisfactory manner from shot to shot. With the screw in the completed full-stroke position for the job, there is a pressure on the material anywhere from 500 psi and upward, depending on the setting of the back-pressure valve. This pressure may cause drooling during mold opening. Such drooling may cause an accumulation of material at the parting line that would prevent the mold from fully closing. In order to stop the side effects of pressure on the material within the metering chamber, a melt-decompress timer is provided. This melt-decompress timer brings about an added motion of the screw beyond the position established by the limit switch for shot size. This motion may range from 1/8 to 1 in. in length. By providing an added space, the melt-decompress movement eliminates the pressure and to some extent creates a suction that causes the material to move away from the front of the nozzle.

Shutoff Valve

When machines are used for fast cycling parts and the possibility of material hangup is not a factor, a shutoff valve is provided between nozzle and chamber. The hangup problem is connected with materials such as polycarbonate that are highly fluid at melt temperature, will entrap easily, and will degrade when exposed to the high heat for a prolonged time. Once the material is degraded, it tends to flake off and contaminate the fresh material being used.

Polyethylene, polystyrene, and many other materials do not behave this way, and a shutoff valve for those materials provides the ability to plasticize not only during curing time but also during mold opening and closing time. The shutoff valve opens to supply material to the closed mold, and it closes when the gates freeze. At this point, screw plastication starts and continues until the following mold closing. This is a valuable feature for thin parts that can be cured in very few seconds and that need additional time to plasticate the shot.

Nozzle

From the metered section of the cylinder, the material is injected into a mold through a nozzle. The nozzle contacts a relatively cool mold on the front end and a heated transition adapter on the cylinder end. The nozzle is normally provided with a heater band that is controlled by a variac for the purpose of regulating the temperature of the nozzle so that no drooling (flow of material) will take place while the mold is open. The temperature of the nozzle tips should be checked with a portable pyrometer, and the reading made when the nozzle is away from the mold for about 3 to 5 min. The temperature should not exceed the top range of material as indicated on the material processing data sheet.

There are three basic types of nozzles: (1) nylon, (2) ABS, and (3) general-purpose. Figure 14–8 shows the inside configuration of each type of nozzle. Their function is to provide a smooth transition of material from cylinder to mold, and, after flowing stops, to have the condition inside the nozzle maintained so that flow can resume on the subsequent shot.

Nozzles should have the type stamped on the outside in a position that can be readily identified when it is mounted in the machine. The types of nozzles

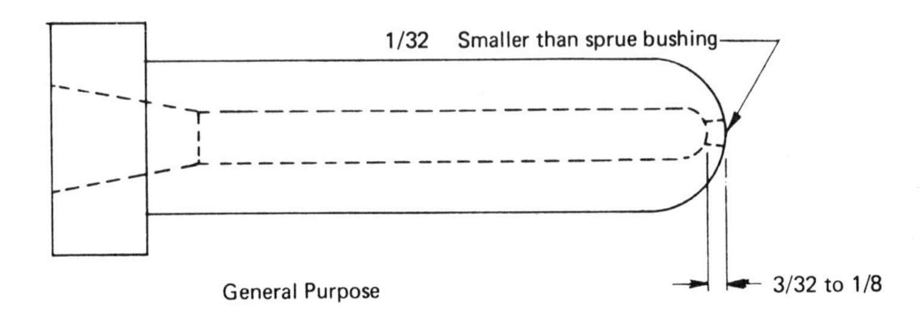

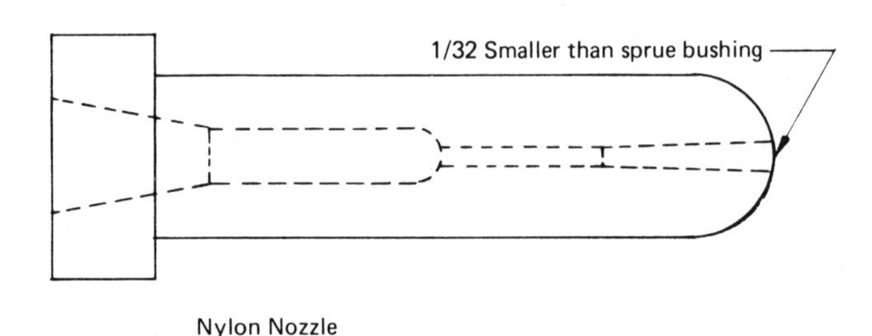

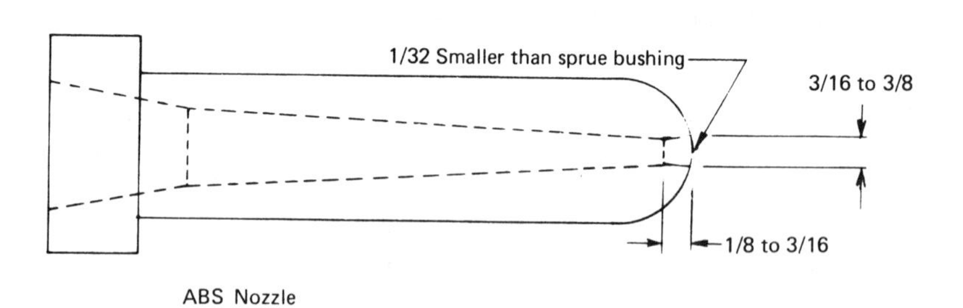

Fig. 14–8. Types of nozzles.

shown date back to the plunger machine. There is a reasonable suspicion that with the screw machines, the significance of the internal configuration of the nozzles is minimal. This is being proven on small machines (1 oz) where the same general-purpose nozzle is used with all materials without any ill effects.

Chamber

The chamber contributes most of the heat needed to plasticize the material. It is equipped with band strip heaters, which are divided into two to four zones depending on the length of the chamber. Each zone is thermostatically controlled by suitable pyrometers to insure a narrow range of temperature variation. Modern pyrometers have an accuracy of 1% of scale range.

When viewing the potential performance of the injection cylinder, we must keep in mind the purpose for which it was designed, namely, to plasticize a certain weight of polystyrene within a specified time. The interpolation of the capacity from one material to another has to take into account the Btu requirements of any material in comparison with polystyrene. This phase, which was previously described, also has to be taken in consideration while making pyrometer settings.

The setting of the pyrometers is determined by the recommended range of temperatures as indicated on the material processing data. When the capacity of the machine is used close to the nominal rating and the parts being molded have a flow length of over 3 in. and a thickness of 3/32 or less, the setting of the upper range would be indicated. All zones could be of the same magnitude. On the other hand, a part with a flow length of less than 3 in. and a thickness of more than 3/32 will have a setting of pyrometers at the low end of the temperature range and all the zones will be about equal.

When less than machine capacity is being shot, then only the metered zone should have the setting according to part dimensions; the remainder of the zone should decrease gradually in the direction of the hopper to the minimum or even less, depending on the proportion of machine being used. A machine utilization of less than 20% of volumetric capacity should be avoided since it is an invitation to problems related to product quality. Capacity utilization of less than 20% means that the material has a long residence time in the cylinder exposed to the high melt temperatures. Most of the materials will withstand the high temperatures for short duration (less than 3 min.), but, when exposed for longer periods, they will degrade and lose properties. Each material has a different degradation curve in terms of time-temperature relations. This factor has to be kept in mind in making temperature settings for conditions of long residence time in the cylinder caused by either long cycles, low machine-capacity utilization, or other elements that may contribute to a long residence time.

Consistency and correct range in cylinder temperature are extremely important, and for that reason any external influences on its variation, such as drafts, should be guarded against. It is very desirable to provide the chamber with heat-reflective, nonradiating shields that are easily removable and yet will provide adequate protection against loss of energy and outside influences.

When making pyrometer settings for crystalline materials, it is to be remembered that they have a latent heat of fusion requirement, which raises the Btu needs over and above those that may be calculated by using the specific heat and weight of material. The materials falling in this category and their latent heats of fusion are:

Material	*Btu/lb*
Nylon 6/6	56
Acetal	70
Polyethylene, low-density	56
Polyethylene, high-density	104
Polypropylene	90
Polyester, thermoplastic	54

Clamp Closing

The closing of the safety gate releases the electromechanical interlocks and initiates the movement of the ram by delivering the large-volume pump to the booster ram. The booster ram is located concentrically with the main ram and is usually about one-fourth of the main ram diameter. The large volume of oil, acting on a small area, brings about high-speed movement of the ram. During the high-speed movement, the prefill valve is open and lets oil from the tank flow by gravity into the space vacated by the moving ram.

As the mold halves approach closing, a limit switch is actuated, which brings about the bypassing of the large volume of oil to the tank while the small pump continues the movement of the ram at a slow pace until another limit switch is actuated. At this point, the small-volume pump closes the prefill valve, and the unloading valve for the large-volume pump closes and directs the flow to the back of the main ram. This builds up the pressure in the booster as well as the main ram. While the large-volume pump is assisting in building up the pressure, the small-volume pump will be directed to the pull-in cylinder so that a tight fit between nozzle and mold is maintained. When the mold is closed and full pressure is reached, the unloading valve is signaled to discharge the high-volume pump to the tank, while the small-volume pump maintains pressure on the ram and the pull-in cylinder. The high-volume pump is discharged after it performs a specified function to conserve horsepower and minimize heat generation in the oil.

Injection and timing of its operations start with mold closing as described in "Injection Rate," p. 255.

Clamp Opening

When the overall timer signals completion of the cycle, the prefill valve is opened, decompressing the main and booster rams. The low-volume clamp pump is now directed by the main valve to enter the return side of the ram and starts a slow return or opening of the ram. When the ram reaches the "fast" limit switch, the high-volume pump comes into action, and the ram moves fast until it approaches the stripping area. In this position, the large-volume pump is directed to the tank, and the small-volume pump moves the ram slowly, ejecting the parts, and brings the ram to the back position where a limit switch de-energizes the solenoids. The press is then ready for a new cycle.

Mold Protection

An optional addition to the hydraulic circuit is available that will protect the mold from having to take the clamping force if there is an obstruction to closing between the mold halves. The highest force that can be exerted on the obstruction is determined by a setting of the pressure control valve, usually about 200 psi, acting on the booster ram, and supplied by the low-volume clamp pump.

In principle, the system works this way: as the ram moves at high speed, it contacts a limit switch to change to a slow speed by having the low-volume pump take over the movement of the clamp; the same limit switch actuates the pressure-relief valve setting to about 200 psi, and the clamp keeps moving until the high-pressure limit switch is actuated. If there is an obstruction, the press will stop before the high-pressure limit switch is reached and, after a small time delay, will give a danger signal. The signal can be a visible red light or a bell to call attention to the malfunction of the press.

In the case of the 14-oz, 200-ton press, the force of the mold protection arrangement is about 1300 lb. On different size presses, the force will vary because the size of the booster ram changes and the 200 psi remains about constant. It takes close to the 200 psi on the booster ram to bring about the movement of the platen, ram, and mold half.

MOLDING PROPERTIES

Time, Temperature, and Pressure

Variable factors in molding have an important influence on product properties. Most of the variables relate to the parameters of time, temperature, and pressure. The effects of variables as well as the means of overcoming them will be discussed in the following paragraphs.

1. When a press is operator-attended, the time of mold opening, removal of parts, and mold closing may vary as much as 100%, with the result that the time-influenced segment may cause a change in the heat of the melt, the mold temperature may vary, etc. These types of variations are not conducive to consistent products. In this event, it is best to use a mold open timer that would keep the variable time to a smallest value possible. In the fully automatic operation, this is not a problem.

2. Guarding against sudden changes in ambient temperature (drafts, etc.) is important for cylinder heat and mold temperature to stay within their limits of thermostat variation. Reaction to heat change is slow, and it takes minutes to readjust to set requirements.

3. Pump pressure output usually has a tolerance of 5% of rated capacity. On a 2000-psi pump, the tolerance would be a 100-psi variation. On the surface, this does not appear to be a sizable amount, but, when translated to pressure on the material, it normally amounts to 1000 psi. This is due to the ratio of injection cylinder area to the area of the injecting screw. A total of 1000 psi in the cavity on the high side may mean flashing at the parting line; on the low side, it may cause voids, porosity, and weak welds in the product.

4. Line voltage fluctuations that are outside of permissible limits may cause malfunctioning of heaters and motors on the machine.

Voltage fluctuations and their side effects are some of the factors that have to be checked before blame is assigned to machine, mold, or material. When product properties do not come up to expectations and the causes for deficiency are difficult to locate, the investigation has to extend to other likely causes. Although some have already been mentioned, they are restated here for emphasis; they are: voltage fluctuation, air circulation, proper working of major press components, and proper functioning of mold features. The attainment of specified part properties is a goal that requires a systematic approach to the testing and evaluating of the mold, molding parameters, and external influences that will affect the end result.

Stresses and Mold Proving

Whether specifications call for close dimensional tolerances, tensile strength, impact strength, chemical resistance, electrical and thermal properties, or any combination of these, they are all affected to some degree by stresses that are molded in a part. In addition to stresses, weld line quality (whenever present) may also act as detractor on some properties.

These property degraders can be traced back to the operation of mold proving. When a mold is received from a tool shop, it is looked upon as a sizable investment, which must not be idle. There is an urgency for expeditious proving out of the mold so that it can be placed in production and start paying for itself. In most cases, the mold is placed in a proper size press, tested for proper mechanical functioning, and followed by making shots for examination of prescribed specifications. The molding parameters for pressure and temperature are interpolated to suit the application, and the cycle time is set to estimated cost values. This procedure may provide satisfactory results for short-term tests. Thermoplastics are time-dependent and thus should include tests that indicate long-term behavior.

The test on a new mold with the associated prescription for production should be conducted in a manner very closely simulating operational conditions. Regardless of the pressure for getting the job ready for production, the important item is to provide the necessary information that will insure desired performance of the product. When proving out a mold, it should be done in a press of the same shot and clamp tonnage for which the mold was designed. The press should be in first-class working condition; the operator should make sure that pyrometers and pressure gauges have been checked and oil filters cleaned within a regular maintenance period–say every 6 months. If it is suspected that line voltage may fluctuate or ambient conditions are different from normal, a check should be made to see whether test results will be influenced by them.

With the mold firmly attached to the press, the circulating lines hooked to the "in" and "out" connections, and the circuits interconnected according to the "judgment" diagram, the molding surfaces should be checked with a portable pyrometer to determine that there is heat transfer from mold to fluid and that the mold temperature is reasonably uniform. What really matters is that the mold temperature is approximately the same during molding–i.e., that no "hot" or "cold" spots are found on the mold surface while the job is running. All the injection parameters are set in line with the material processing data interpolated to the needs of the job. The melt temperature should be checked with a needle pyrometer to double-check the pyrometer settings. With all parameters in correct range, the machine and mold can start operating. Some adjustments will most likely be necessary to obtain a product of satisfactory appearance. After this, the process should be run until all conditions have stabilized and a mold temperature reading for uniformity is made. At this point, a change in the circuit of mold circulating media may be necessary for consistent mold temperature. When all the settings have reached the proper values, the press is operated for at least half an hour so that checking samples (about six) are produced and evaluated. After another half-hour or so, parts can be observed for straightness and dimensional accuracy, etc., and suitably recorded. A practical indication of stress can be obtained by exposing the parts in an oven to a heat deflection temperature as recorded for the 264-psi stress. This is for all practical purposes a stress-relieving temperature. The parts are exposed to this heat for 15 min for each 1/8-in. thickness. The temperature should be taken from the literature of the material supplier for the polymer in use. There is a variation in the heat-deflection temperature of the same generic material obtained from different sources. This is most likely due to the variation of the polymerization process.

After the annealing operation, the parts are measured and observed for other comparative evaluations. If the changes are appreciable—say the dimensions are out of tolerance, the parts are warped, etc.—then it is an indication that the molded-in stresses are high and steps have to be taken to reduce them.

What are some causes for stresses in molded parts that can be corrected or minimized? An inconsistent cycle will expose the material to variation of heat absorption and introduce a stress. Anytime there is a possibility of introducing into the melt some cold material or finding within the melt a temperature difference, then we have a variation in rate of cooling of different portions of the part and associated with it, a stress. Causes of temperature differences include absence of a cold-well space or an improperly sized one; melt lying in the measuring chamber causing it to soak up more heat on the circumference than in the center; some granules not fully molten; the material not thoroughly mixed and homogenized; or a temperature gradient on the surface of the mold.

Variations in ambient temperature, line voltage, feed of material, shrinkage due to cores, or pressure applied to the cavity can all contribute to a stress.

Weld Line

Weld-line strength has an overall influence on the quality of a part. Conditions for self-welding of the plastic material should be such that the strength of a joint is not impaired. If there is moisture on the mold surface, the flowing plastic will tend to wipe the moisture between the joining halves and adversely affect the weld. A slow injection speed will cause the material to be joined to cool and thus reduce its capacity for self-welding. Excessive lubrication of the mold will also bring about the wiping of the lubricant into the joint and thus cause too much foreign material at the joint. Vents, when not maintained in open and operating shape, will force air and gases at the joint, making welding very difficult.

Parameter Control for Properties

Generally speaking, the invisible property detractors such as stresses and weld-line quality can be kept to a low value with these steps:

1. The pressure in the cavity during injection should be maintained within 1% variation. Too much pressure will mean overstuffing and too little pressure will not produce a dense part. Upon completion of the shot, the cushion or material left in the chamber in front of the screw should be uniform; otherwise, the effective pressure on the cavity will vary. Vents should be adequate and maintained open to prevent entrapment of gases in the cavity, which will also be responsible for pressure variation.

2. The temperature of the melt entering the cavity should be uniform. During plastication, the material should be in constant motion up to the time just prior to injection so that the heat content is well averaged out. Any causes for colder material getting into the stream must be guarded against, and external variables in temperature should be eliminated. In short, the uniformity of melt temperature throughout the full volume of the shot must be insured. The heat conduction from plastic to mold has to be such that one portion of the plastic part is not cooled ahead of another. An uneven rate of cooling would create a stress. This suggests that uneven thicknesses of product and their transitions from each other call for special attention to cooling needs where thicker sections exist.

3. Time variables during any portion of the cycle must be held to zero.

4. Foreign substances in the cavity such as moisture, gases, and lubricant cannot be tolerated.

With close attention to these four points, a large part of property problems will be solved.

MACHINE MAINTENANCE

The operator's manual supplied with each machine outlines a series of maintenance requirements that should be followed. A few such requirements that determine the quality of the product itself are singled out here for special attention:

1. *Pressure gauges and pyrometers.* The pressure gauges and pyrometers are the means by which machine parameters are set for the molding operation. These instruments should be checked once at the beginning of each week or at least once at the start of each month with a master pressure gauge and a master pyrometer to insure that accuracy is maintained.

2. *Hydraulic oil.* The hydraulic system performs all the important functions of the machine. The hydraulic oil is the means by which all the hydraulic components do their work. The hydraulic oil must be in top-notch condition not only to transmit power but also to lubricate and prolong the life of pumps and valves. Cleanliness and oil fortified with necessary additives and of proper viscosity are prerequisites to proper functioning of the press.

If foreign substances are kept out of the circulating fluid, will the cleanliness problems be solved? No. Certain circumstances cause formation of a sludge, although the oil container is sealed against incoming dust.

One of the most important properties of a hydraulic oil is the chemical or oxidation stability. Oxidation is the greatest factor that determines the useful life of an oil. Oxidation is the reaction of oxygen with oil to form a multitude of compounds. It is a reaction in degrees. At first, hydroperoxides are formed, which in turn react to form alcohols, aldehydes, ketones, acids, and oxy-acids.

While soluble in oil, all of these will eventually, through polymerization and condensation reactions, form an insoluble gum, sludge, and varnish. These insoluble products will cause a hydraulic system to become sluggish, will increase wear, reduce clearances, plug orifices, and finally bring about failure of the system. The rate of oxidation of a hydraulic oil depends on operating temperature, pressure, contaminants, presence of water, agitation, and exposed metal surfaces. Temperature is the biggest offender. Below 135°F, the rate is quite slow; above that, oxidation increases rapidly. It is estimated that the useful life of an oil will decrease by 50% for every 15°F above 140°F. In analyzing the temperature of a hydraulic system, the point to be checked is the outlet of the pump. Increase in pressure brings about an increase in oil viscosity, which causes increase in friction and a rise in temperature. Increase in pressure is also responsible for an increase of air that can be absorbed by the oil. This will provide the added oxygen to promote oxidation. It therefore becomes imperative to minimize the oxidizing agents and plan periodic inspection and replacement of oil as recommended by the equipment manufacturers.

A good-quality hydraulic oil should have a broad useful viscosity range, be fortified with antioxidants, be free from tarlike material, be noncorrosive, and have good lubricating characteristics.

The condition and cleanliness of hydraulic oil cannot be overemphasized. When we are aware that sliding valve spools have to seal pressures of 3000 psi, we can readily recognize that the clearance between the valve body and the spool has to be very small. These clearances are on the order of 0.0001 to 0.00015. It does not take much to cause the malfunction of a valve with these kinds of clearances.

3. *Filters.* When we recognize that sliding spools have to seal against pressures of 2000 or 3000 psi and that rotating vanes also have to seal under these conditions, we can readily see that the hydraulic fluid has to provide good lubricating properties. Consequently, it must be free from any foreign particle size that would interfere with the free movement of spools, vanes, etc. or cause abrasive effects. The fluid must not only be contained in clean and dust-free receptacles, but it also must be continuously filtered. The filtering agents must have openings not larger than 200 mesh screen.

4. *Coolers (heat exchangers).* Since the control of oil temperature is so vital to the proper functioning of a hydraulic system, it is necessary to examine the heat exchange between the oil and the cooling medium. A typical heat exchanger is shown in Fig. 14–9. Its function is to cool the incoming oil by means of circulating water through the multitude of tubes. The oil flows over the water-cooled tubes, thus losing some of the heat to the moving water. When in good functioning condition, these exchangers do an outstanding job in maintaining oil temperatures at prescribed levels. In addition, the oil tank plays a major part in heat dissipation. It is imperative that tank walls as well as the

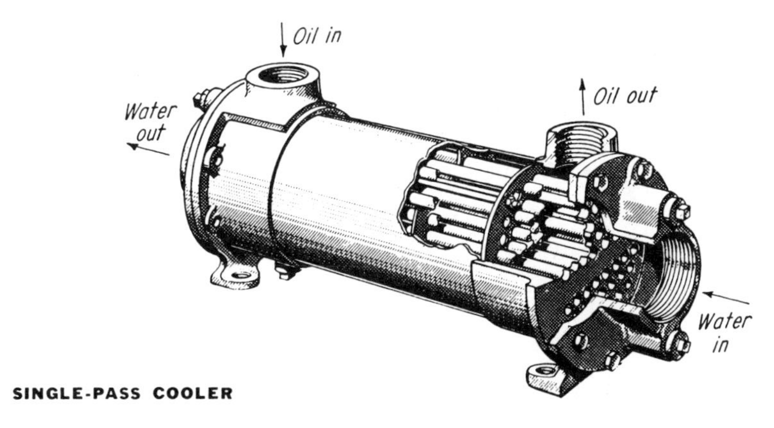

Fig. 14–9. Heat exchanger.

inside of the heat exchanger be free from sludge or other substances that may form on the conducting surface. Any substance on a heat-conducting area will act as heat insulation and thus minimize the effectiveness as a cooler. Therefore, periodic thorough cleaning of heat exchangers and tank surfaces should follow oil inspection.

In summary, the best cure for the problem is to have the least heat for dissipation. Major causes of the heating of the oil are high-pressure and high-volume requirements set for longer periods than necessary, leakages around pressure control seats, partially clogged orifices, clogged filters, or ineffective cooling. In other words, recognition of these problems should make those workers who are responsible for mold setup conscious of the importance of pressure settings, and make the maintenance supervisors aware of possible leakages within valves, of the need for clean filters and clean heat exchangers, and, last but not least, of the need for oil in first-class condition.

5. *Oil leaks.* Wherever an oil leak develops, it should be stopped immediately. Leaks cause pumps to work harder and thereby generate additional heat in the system. In addition, they are a potential hazard to machine attendants, and they contribute to poor appearance of the shop.

Leaks may be a result of worn "O" rings, loose connections, or improperly adjusted packings. A U or V-shaped packing should be free enough in its space so that the arms of the U or V will flex under pressure, thus bringing about good sealing action. It is best to obtain information on a specific type of packing from the manufacturer and to follow recommendations for adjustments to seal.

6. *Troubleshooting of a hydraulic circuit.* While analyzing a malfunction of a hydraulically actuated machine, we should keep in mind some general principles, namely, that a pump makes oil flow but that there must be resistance to flow to generate pressure. Determine where fluid is going. If actuators fail to move or

move slowly, the fluid must be bypassing them or going somewhere else. Trace it by disconnecting lines if necessary. No flow (or less than normal flow) in a system will indicate that pump or pump drive is at fault.

PROBLEMS AND POSSIBLE CAUSES

Noisy Pump

- Cavitation (pump starving)
  1. Clogged inlet strainer-filter
  2. Obstructed inlet piping
  3. Fluid viscosity too high
  4. Operating temperature too low
- Pump picking up air
  1. Low oil level
  2. Loose or damaged intake pipe
  3. Worn or damaged shaft seal
  4. Aeration of fluid in reservoir (return lines above fluid level)
- Other reasons
  1. Worn or sticking vanes
  2. Worn ring
  3. Worn or damaged gears and housings
  4. Shaft misalignment
  5. Worn or faulty bearings

Low or Erratic Pressure

- Contaminants in fluid
- Worn or sticking relief valve
- Dirt or chip holding valve partially open
- Pressure control setting too low

No Pressure

- Low oil level
- Pump drive reversed or not running
- Pump shaft broken
- Relief valve stuck open
- Full pump volume bypassing through faulty valve or actuator

Actuator Fails to Move

- Faulty pump operation (see *Noisy Pump*)
- Directional control not shifting
  1. Electrical failure, solenoid, limit switches, etc.
  2. Insufficient pilot pressure
  3. Interlock device not actuated

- Mechanical bind
- Operating pressure too low
- Worn or damaged cylinder or hydraulic motor

Slow or Erratic Operation
- Air in fluid
- Low fluid level
- Viscosity of fluid too high
- Internal leakage through actuators or valving
- Worn pump
- Pump drive speed too slow

Erratic Feed Rates
- Sticking, warped or binding ways
- Air in fluid
- Faulty or dirty flow control valve

Overheating of System
- Water shutoff or heat exchanger clogged
- Continuous operation at relief setting
  1. Stalling under load, etc.
  2. Fluid viscosity too high
- Excessive slippage or internal leakage
  1. Check stall leakage past motors and cylinders
  2. Fluid viscosity too low

7. *Plasticating screw and check rings.* The chamber, check rings, and screw are the heart of the machine and must be maintained in good working order to obtain favorable properties in the molded part. Every 6 months, the chamber, screw, and check rings should be examined carefully to see that they are in top-notch working order. Regardless of how many safeguards there may be against metal particles getting into the stream of the plastic, somehow one can always find some pieces of nonplastic material to be responsible for damage to the vital parts of the injection process. A great deal of attention should also be paid to the correct clearance between check ring and chamber as well as screw and chamber. The correct clearances are those that provide sealing against backflow and shearing for plastication.

TOGGLE PRESS

The toggle press differs from the hydraulic clamp only in the clamp end. The injection end performs its function in the same manner and is normally designed identically in both cases.

In the toggle press, a tension is created in the tie rods of the press, which provides the necessary force to keep the clamp and mold closed. This is accomplished through a series of linkages actuated by a small hydraulic cylinder. In this type of press, it is very important that the adjustment of daylight of the press be made in a manner that insures parallelism of platens and thereby exerts a uniform tension on all four rods.

The advantage of the toggle press is that the mold opening and closing time is less than the all-hydraulic ram, is somewhat less expensive, and will have a lower horsepower requirement. They are usually made in sizes of 700 tons and lower.

Their disadvantage is that they are not as flexible and may require more attention for maintenance needs.

PROCESSING OF REINFORCED THERMOPLASTICS

Plastic resins may be modified in their properties and use characteristics by a number of additives, reinforcements, or fillers. From the very beginning, thermosetting resins have used such fillers and reinforcements as wood-flour, mica dust, asbestos, cotton, linen, graphite, and, in the last 20 years, glass fibers. Each one of these when compounded or impregnated with the resin was designed for a specific range of application.

The introduction of reinforcements and other additives to thermoplastic resins was merely an extension of a practice that has prevailed from the early days of thermosets. At present, the most popular reinforcing agents in thermoplastics are glass fibers, minerals, carbon fibers, and metal fibers; the additives are Teflon and molybdenum disulfide. When combined with the resin, each one of these offers a new list of properties appreciably different from the plain resin.

The size of most glass fibers is 0.0005 in. diam and 0.020 to 0.032 in. long. For good properties in the composite material, the dispersion and bonding of the glass fibers have to be carried out under controlled conditions with the aid of appropriate bonding agents. Despite all this effort, there are possible moldability features that can disturb the structure of the compound and bring out areas in the molded product that are either resin-rich or fiber-rich. In most cases, the cause of this condition is the cross section that does not permit easy flow of the composite material. When the separation of fibers from polymer takes place, even to a partial degree, it will cause weak spots and nonuniform properties of a product.

The general direction in which the properties of reinforced thermoplastics change are as follows:

1. The modulus of elasticity or rigidity is higher.
2. The resistance to creep or deformation under load is higher.
3. The tensile strength is higher.

4. The heat conductivity is higher.
5. The resistance to elevated temperature is higher.
6. The deflection temperature under load is higher.
7. The coefficient of thermal expansion is lower.
8. The specific gravity is higher.
9. Cost per cubic inch is higher.
10. Mold shrinkage is lower, which indicates need for greater draft.
11. The impact strength at room or higher temperature is usually lower. The reinforced materials act in a brittle manner. On the other hand, at -40°F they have a higher impact strength than the plain resin.
12. The weld line whenever present drops in strength to a considerable degree in comparison with parts without weld lines.

When fibers are involved, the molding parameters as well as moldability features of the mold will necessitate detail modifications so that fiber lengths are not altered. It is the length of the fiber that gives the composite material the needed properties. In addition to guarding the fiber length, the conditions of molding must be such that no resin segregation will take place. This requirement necessitates certain modifications in the mold and molding parameters.

Mold

1. *Steel.* The steel in the mold should be a prehardened or heat-treated steel that has good abrasion resistance and polishing characteristics. For additional resistance to wear, chrome plating or nitriding is suggested. The P-21 and H-13 steels are good candidates for this application.

2. *Gates.* The gates should be increased by 20% to 40% for glass-fiber content of 20% to 40%, respectively. For rectangular gates, the increase should be in both directions in the percentage indicated. The starting dimensions for gate are those of the base polymer. The gate should be so placed that there will be no impingement on a pin or wall of the cavity. This is opposite to the requirement for the unfilled resin. The reason for the increase in gate size is to minimize the tendency of fiber breakage.

3. *Land of gate.* The land of gate should be on the small end of the base polymer.

4. *Runners.* A special effort should be made to have the shortest runners possible. Bends in runners should be avoided, but, if that is not possible, there should be a cold slug well at the turn. For precision parts in a multicavity mold, a balanced runner system is a must. Runners should be smooth and polish lines in the direction of flow. The range of runner diameter is 1/4 to 3/8 in. The reason for these modifications is the sizable reduction of the filled thermoplastics in ability to flow, especially those filled with fiberglass.

5. *Sprue bushing.* The minimum "O" dimension should be 9/32 in. diam, and the taper per side should be 3°. This is due to the low shrinkage of the filled materials.

6. *Vents.* The filled materials do not flow as well as the unfilled counterparts, and therefore the vents can be 30% to 50% deeper. There are usually more gases present in the filled materials because of chemicals used in compounding the glass fibers or other fillers with the polymer, and additional venting depth will aid in producing quality parts.

7. *Mold temperature.* Glossy appearance and better flow are results of higher mold temperatures. At the upper range of mold temperature, a skin is formed on the outer surface of the molding, and its thickness is 0.003 to 0.005 in., which gives a shiny appearance and completely hides the fiberglass.

8. *Weld lines.* For highest part strength, weld lines should be avoided. When a fiberglass-filled plastic flows around a core, a large number of the fibers arrange themselves in the direction of flow; when meeting for self-welding, they are parallel to each other. This condition reduces the "felting" action of the material and weakens it at the weld line. The random arrangement of fibers (felting) is the main feature, which adds to most of the properties of the composite material.

The Machine

The preferred machine is the screw type because of its smooth plasticating action and resulting lower tendency to break the fibers.

1. *Injection pressure.* The machine should be capable of generating 20,000 psi of injection pressure. The clamp should be large enough to exert 5 to 10 tons of pressure on the projected molding area. The injection capacity should be downgraded to 70% of rated capacity.

2. *Nozzle.* The inside diameter of the nozzle should be 0.5 in. with a rounded front portion. The outlet opening should be 0.25 in. diam with a taper of 2° per side toward the 0.5 in. diam. The nozzle length should be the shortest that the machine will accommodate. All of these requirements are in the interest of minimizing the tendency toward orientation of fibers.

3. *Back pressure.* The gauge reading of back pressure should not exceed 50 psi. All that is expected from the back pressure is to get good mixing without breaking the length of fibers.

4. *Speed of screw.* The screw rpm should be between 30 to 50 rpm. This is also for the purpose of maintaining the original length of fibers.

5. *Drying.* The addition of fiberglass has a tendency to make the composite hygroscopic even though the base polymer does not require drying prior to molding. The glass-filled materials should be dried to the high end of the range.

6. *Cycles* can be shorter in comparison with plain materials for two reasons: (1) the reinforcing material does not have heat of fusion to be released, and (2) since the composite is a better heat conductor, it will bring about faster solidification. The mold can run at higher temperatures because parts are more rigid at ejection.

General

This section has stressed the differences as far as mold and machine are concerned for the composite material. Those details not mentioned are identical with those applicable to the base polymer. When designing a mold or prescribing processing, the combined requirements should be reviewed and the pertinent information used.

15
Setup Data

After the molding parameters have been established for required properties in a part, it becomes necessary to record them so that subsequent runs can be duplicated. A sample setup sheet is shown in Fig. 15–1. Local conditions may dictate additions or even subtractions to the list. Regardless of the number of items, the setup sheet should be considered an important source of information for running a job. It should be followed, always kept up to date, and if necessary, revised.

The information that follows will explain each heading and in addition will list formulas that will aid in interpolation of the data when the need for transferring the job to another press arises. Each heading represents a record of information needed for setup.

PRESS MANUFACTURER AND SIZE

The first run or specification run should show press number, size, and manufacturer. When the job is rerun in the same press number, it should be possible to reset conditions and start running. On the other hand, even though a press is of the same size and manufacture, it may require adjustments mainly due to press manufacturing tolerances.

$$\text{Press tons } t = \text{Projected area of cavities and runner} = \text{in.}^2 \times t/in.^2$$

$$\text{For most materials, } t/\text{in.}^2 = 2$$

$$\text{For polycarbonate, } t/\text{in.}^2 = 5$$

$$\text{For nylon, } t/\text{in.}^2 = 7$$

The numbers of $t/\text{in.}^2$ are based on 20,000-psi injection pressure; when pressures are lower, $t/\text{in.}^2$ can be reduced in proportion. The shot size in ounces is meant for polystyrene, but for most materials the nominal capacity is reduced to 3/4 or 75% of polystyrene.

MATERIAL SPECIFICATIONS

The supplier's name and material number should be indicated; also, any additional information of interest to the molder should be recorded. The specific gravity of the material should be shown. With it, the travel of the screw can be figured.

Press size and make ____________________

Material specifications ____________________

Drying specifications ____________________

Cushion ____________________

Injection high-timer ____________________

Mold closed timer ____________________

Back pressure on material ____________________

Melt temperature ____________________

Screw torque ____________________

Mold daylight ____________________

Safety check ____________________

Limit switches ____________________

Shot weight ____________________

Rate of injection ____________________

Injection hold-timer ____________________

Injection psi on material ____________________

Melt decompress ____________________

Nozzle type ____________________

Mold temperature ____________________

Stripper rods, size and no. ____________________

Notes and coolant circuit diagram:

Fig. 15–1. Setup data sheet.

$$\text{Distance of screw travel} = \frac{\text{oz material} \times \text{Stroke}}{\text{Rated oz}} \times \frac{S_1}{S_2}$$

S_1 = Specific gravity of polystyrene = 1.08

S_2 = Specific gravity of material being run, e.g., Polycarbonate 1.2

Example: A 28-oz machine has a stroke of 10 in. How far will a 15-oz shot of polycarbonate go?

$$\text{Distance} = \frac{15 \times 10}{28} \times \frac{1.08}{1.2} = 4.82 \text{ in.}$$

The distances will be approximate due to the fact that the sizes we use in calculations are nominal, whereas actual dimensions are somewhat different.

Useful conversion factors:

$$\text{in.}^3 = \frac{\text{oz} \times 27.7}{16 \times \text{Specific gravity}}$$

$$\text{in.}^3 = \frac{\text{Grams} \times 27.7}{453.4 \times \text{Specific gravity}}$$

DRYING SPECIFICATIONS

Whenever applicable, temperature range and duration of drying should be indicated; this information can be found in the processing data. For a hopper dryer, the method of figuring out the level of material in the hopper is shown in Chapter 14 ("Hopper and/or Dryer," p.243).

SHOT WEIGHT

The weight of a complete shot should be recorded in ounces or grams so that the total material needed for the job can be figured. Also, if the job is run in a machine other than the one in which the first run was made, it should be recorded so that the distance of screw travel can be figured and heat settings made to suit the new conditions.

CUSHION

Cushion is the amount of material left in the front of the chamber after the shot is completed. This amount can be used for replenishing material in the cavity during pressure hold time as long as the gate remains open. The size of the cushion, indicated on the scale of screw travel, has an effect on the pressure drop as measured between chamber and cavity gate; thus, it influences the pressure available in the cavity. The pressure in the cavity is one of the most important elements controlling dimensions, properties, and warpage; therefore, it must be constant for uniform results. The average cushion is approximately 1/16 to 1/8 in.

RATE OF INJECTION

If we analyze the basic law of rheology as it relates to flow of materials, we find that the number of cubic inches of material per second delivered to the mold is important for consistent results. During mold proving, the delivery of material per second is established and should be repeated thereafter. It is especially important to be able to translate the information from machine to machine should the occasion arise for transferring the mold to a machine with different specifications for delivery. We will extract the pertinent specification and use an example for carrying out the calculations.

The shot will be a 10-oz polycarbonate:

	Machine No. 1 (200 ton–14 oz)	*Machine No. 2 (200 ton–28 oz)*
Screw diameter	2	2.5
Injection rate, in.^{3}/sec	22	25.2
Stroke, in.	8	10
Injection pump, psi	2,050	2,050
Injection volume, gpm	56	67
Flow control valve, gpm	28–10 divisions	40–10 divisions
Injection psi on material	20,000	20,000
Injection capacity, in^3	25.1	50.2

On machine No. 1, for example, the setting of the control valve was No. 3, meaning that three parts of rated capacity of 28 gpm were bypassed. Or, $3 \times 2.8 = 8.4$ gallons were bypassed. The volume of $56 - 8.4 = 47.6$ gpm will act on the cylinder. The cylinder area of the No. 1 machine is:

$$\text{Force on screw} = \text{Force on cylinder piston}$$

$$\text{Screw area} \times \text{Injection psi} = \text{Cylinder area} \times \text{pump psi}$$

$$\text{Cylinder area} = \frac{\text{Screw area} \times \text{Injection psi}}{\text{Pump psi}}$$

$$= \frac{0.7854 \times 2^2 \times 20{,}000}{2050}$$

$$= \frac{3.14 \times 20{,}000}{2050} = 30.634$$

The speed of injection with 47.6 gpm will be;

$$\text{Velocity} = \frac{231 \times \text{gpm}}{60 \times \text{Area}} = \frac{231 \times 47.6}{60 \times 30.634} = 5.98 \text{ in./sec}$$

The polycarbonate shot of 10 oz with a specific gravity of 1.2 will occupy a measuring chamber volume of

$$\text{in.}^3 = \frac{\text{oz} \times 27.7}{16 \text{ Specific gravity}} = \frac{10 \times 27.7}{16 \times 1.2} = 14.43 \text{ in.}^3$$

This volume divided by the area of the screw will result in screw travel of:

$$\frac{14.43}{3.14} = 4.59 \text{ in.}$$

With a volume of 47.6 gpm, the screw travels 5.98 in/sec. Therefore, for 4.59 in., the time of travel will be 4.59/5.98 = 0.77 sec, the time in which 14.43 in.^{3} of polycarbonate will be injected into the mold in machine No. 1.

The question is what the setting of the No. 2 control valve should be in order to reproduce the 0.77 sec for 14.43 in.^{3} In machine No. 2, the screw will travel a distance equal to the volume divided by the area of the 2.5-in. screw, or

$$\frac{14.43}{4.91} = 2.94 \text{ in.}$$

The time for travel of this distance has to be 0.77 as in the No. 1 machine in order to deliver 14.43 in.^{3} to the mold in the same time. To find the corresponding rate of travel in machine No. 2, the proportion

$$\frac{2.94}{0.77} = \frac{x}{1}$$

or

$$x = \frac{2.94}{0.77} = 3.83 \text{ in./sec}$$ is the velocity of the 2.5-in. screw while moving 2.94 in. in 0.77 sec.

$$\text{Thus, velocity of } 3.83 = \frac{231 \times \text{gpm}}{60 \times \text{Cylinder area}}$$

$$\text{Cylinder area, as before} = \frac{\text{Screw area} \times \text{psi}}{\text{Pump psi}} = \frac{4.91 \times 20{,}000}{2050}$$

$$= 47.30$$

Substituting this above, we have

$$3.83 = \frac{231 \times \text{gpm}}{60 \times 47.30}$$

$$\text{gpm} = \frac{3.83 \times 60 \times 47.30}{231} = 47.05 \text{ gpm required for velocity of } 3.83$$

Injection pumps of the No. 2 machine deliver 67 gpm. The difference between 67 and 47.05 = 19.95 gpm; that is, approximately 20 gpm have to be bypassed.

Since the flow control valve has a 40-gpm capacity divided into 10 portions, the setting at 5 will bypass 20 gpm and will create the correct rate of injection of the 14.43 in.^{3} in 0.77 sec.

Under this heading, the setting and capacity of flow control valves should be recorded. For machine No. 2, the control valve with a 40-gpm capacity with a setting at 5 will deliver 14.43 in.^{3} of polycarbonate in 0.77 sec. In the example of machine No. 1, the control valve with a 28-gpm capacity with a setting at 3 will also deliver 14.43 in.^{3} of polycarbonate in 0.77 sec.

INJECTION HIGH TIMER

The injection "high timer" controls the high rate of injecting the material into the mold. For most of the materials and part shapes, it is desirable to move the screw at its maximum speed. In the case of the 14-oz, 200-ton machine, the attainable speed with the flow control closed is according to machine specifications 22 in.^{3}/sec. The full volume of the machine is 25.1 in.^{3}; therefore, it can be injected in 25.1/22 or 1.14 sec. The adjustment on the control valve, in combination with a suitable pump volume in the circuit, will provide as low a rate as 1.9 in.^{3}/sec. The timer setting for this operation should be only long enough to fill the cavities. Otherwise, when the high-volume high-pressure pump is operating longer than seconds indicated, it will generate excessive heat in the hydraulic system, thus affecting the overall performance of the machine and the life of its components.

The accurate setting of this timer can be figured out and verified while running. The steps are:

1. Cubic inches of material per shot $= \dfrac{\text{oz} \times 27.7}{16 \times \text{Specific gravity}}$

2. Distance of screw travel = in.^3 of material/Area of screw
3. Volume of oil acting on the injecting cylinder = Total injecting volume minus gpm bypassed

4. Velocity of screw travel (in./sec) $= \dfrac{231 \times \text{gpm}}{60 \times \text{Area of cylinder}}$

5. Area of cylinder (in.^2) $= \dfrac{\text{Area of screw} \times \text{Maximum Injection psi}}{\text{Maximum pump psi}}$

6. Time (sec) $= \dfrac{\text{Distance of screw travel}}{\text{Velocity}}$

When the injection high timer times out, the pump volume is delivered to the tank at little or no pressure. The timing out of this setting actuates the injection hold timer.

INJECTION HOLD TIMER

The effect of the injection hold timer prevails until the gate is fully frozen. The injection hold pressure is provided by a small-volume high-pressure pump that is capable of replenishing sinks and voids while the gate is still open. The usual setting for the timer is approximately 5 sec or more. The small volume of the pump, even though the pump operates for relatively long times, will not adversely affect the temperature of the hydraulic system because of the low horsepower involved.

At present, the actual time setting for the injection hold is established on the basis of judgment and is adjusted after the results of the test run indicate that satisfactory parts have been obtained. However, an accurate way of predetermining the setting exists if the means for obtaining a recording of "pressure versus time" profile in the cavity are available. Such a profile (see Fig. 15-2) shows the time required to reach the peak pressure and also the time for the cavity pressure to decay. In this case, the injection high timer would be set at 2.75 (Filling) sec and the injection hold timer at 8 sec.

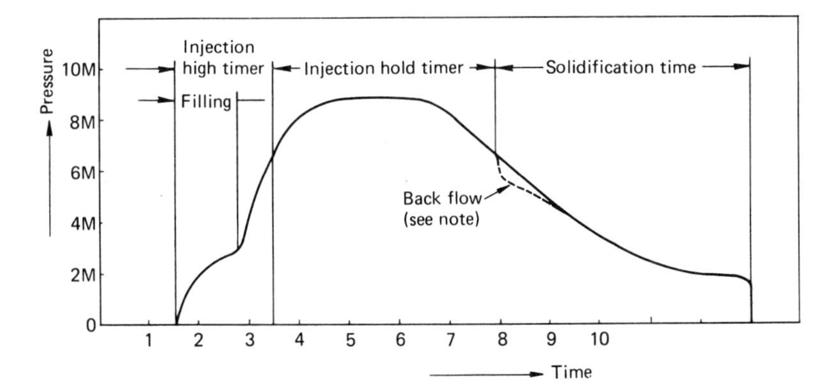

Note: Transducer placed close to gate will show a drop in pressure in contrast to smooth decay if gate is not frozen.

Fig. 15-2. Cavity pressure profile.

MOLD CLOSED TIMER

In the interest of economy, the mold closed timer should be set at the lowest possible value. This timer should call for mold opening just at the time when the change from liquid to solid has taken place in the cavity, i.e., when the cavity pressure has stopped decaying. If mold opening is delayed, the mold will act as a cooling fixture that will bring about increased dimensions in the molded parts since the dimensions in the steel are larger to allow for shrinkage. This fact is sometimes used for dimensional adjustments on the plastic parts, but when so applied, it may cause stresses in the parts that may adversely affect their useful life.

INJECTION PSI ON MATERIAL

Information regarding injection psi on material is indicated on the material processing data sheet as a range in which a material will produce satisfactory properties. This is different than the pressure gauge indication, which shows the pump pressure. On many machines, the pressure gauge dial shows the psi of the pump and the corresponding psi on the material. Where this is not the case, the psi on the material can be calculated from the following formula: the force of the screw driving the material into the cavity is equal to the force of the hydraulic cylinder that pushes the screw, that is:

$$\text{Force of screw} = \text{Area of screw} \times \text{psi on material}$$

$$\text{Force of cylinder} = \text{Area of cylinder} \times \text{Pump psi}$$

If the injection psi is 20,000, pump psi is 2050. This equation is

$$\text{Injection psi} \times \text{Area of Screw} = \text{Pump psi} \times \text{Area of cylinder}$$

or

$$\text{Injection psi} = \frac{\text{Pump psi} \times \text{Area of cylinder}}{\text{Area of screw}}$$

$$= \frac{2050 \times 30.634}{3.14} = 20{,}000$$

In reality, part of the cylinder force is used to overcome friction in the system, and the equation is modified by a factor to account for it. For this reason, the two pressure markings on the gauge do not coincide exactly according to the ratio of areas.

Example: An 8-in.-diam cylinder is used to inject a 2.5-in.-diam screw. The area of the 8-in. cylinder is 50 in.^{2}, and, if the pressure gauge reads 2000 psi, there is a force of F = Area $\times$ psi gauge reading or $50 \times 2000 = 100{,}000$ lb. This force is exerted on the screw, of which the area is 4.9 in.^{2}

Again, F = Area $\times$ psi on the screw, or

$$100{,}000 = 4.9 \times \text{psi}$$

or

$$\text{psi} = \frac{100{,}000}{4.9} = 20{,}400 \text{ psi injection on the material}$$

The reading on the gauge will be less than 20,400 because of friction.

SCREW BACK PRESSURE ON MATERIAL

The pressure of the material in the front of chamber that is exerted on the backward moving screw can be in the range of 500 to 4000 psi. The low end is adequate for most materials and is a requisite for the heat-sensitive ones. The preceding force can be translated into a pressure gauge reading in the same manner as described in the preceding section. Thus, we have

$$F = \text{Area of screw} \times \text{psi} = \text{Area of cylinder} \times \text{psi (gauge)}$$

or with psi on screw of 500 and the screw area of 4.9 in.^{2}, using the example of the preceding section, we have:

$$4.9 \times 500 = 50 \times \text{psi (gauge)}$$

$$\text{Thus, psi (gauge)} = \frac{4.9 \times 500}{50} = 49 \text{ psi on gauge}$$

The back pressure is an important machine setting for repeatability of results.

MELT DECOMPRESS

The pressure generated in the metering chamber while it is being filled with material is called the *back pressure.* The magnitude of the pressure as indicated is on the order of 500 to 4000 psi. When the mold opens, the pressure will tend to cause drooling and to create an objectionable lump of material at the sprue prior to mold closing. This undesirable condition can be overcome by having the screw move backward beyond the feed setting. This depressurizes the metering chamber and to some degree creates a suction at the front of the nozzle. The additional screw movement is controlled by a timer setting, and the time duration determines the distance traveled. The distance of melt-decompress movement ranges from 1/8 to 1 in. long.

MELT TEMPERATURE

Since the heat input into the material is derived from several sources, the temperature of the material should be checked as it leaves the nozzle. This can be obtained by making a shot into an insulated container and using a needle-type thermocouple of a portable pyrometer. Such a shot should be made when machine and running conditions have stabilized. This type of check is also an indication of the accuracy of indicating pyrometers. Readings of the melt temperature can be readily used for thorough troubleshooting.

NOZZLE TYPE

The type of nozzle should be indicated for a particular material, and the opening at the outside that meets the sprue bushing should be about 1/32 smaller than that of the bushing. The pyrometer reading of the nozzle while away from the sprue bushing for about 5 min should not exceed the top limit of the material range and preferably should correspond to the setting of the front zone on the chamber.

SCREW TORQUE

Screw torque and rpm are shown on material data sheets. Since they affect the shear rate of the material, they should be properly recorded.

MOLD TEMPERATURE

Mold temperature uniformity is important in obtaining quality parts. Several areas in the mold should be checked with a hand pyrometer to determine whether the connections of the fluid circuit are correct, in order to produce uniform readings. See Chapter 12 for details.

The range of mold temperatures on the material data sheet can be applied as follows: a mold temperature in the upper range will produce parts with low stress levels, least dimensional change over prolonged time, and high-gloss appearance. The lower end of the temperature of a mold will usually give shorter cycles and is considered the minimum at which no pronounced ill effects of the molded parts will occur. A diagram of the fluid circuit for each mold half should be a permanent part of the setup record. Each circuit should show flow rate in gpm, entering temperature of the fluid, type of connections (size and length of hose), and any information that is deemed to have an influence on the control of mold temperature.

MOLD DAYLIGHT

Mold daylight for safe and easy removal of parts should be recorded along with special information such as mold lubrication, care of removable inserts, and unusual purging information.

SIZE AND NUMBER OF STRIPPER RODS

The length of rods can be determined by adding "*A*" + "*B*" + "*C*" plus stripping length as shown in Fig. 15–3. The number of rods to be used is to be such as to minimize the danger of bending the ejection plates. Whenever feasible, the center rod should be used in addition to those placed under the greatest number of stripping pins.

LIMIT SWITCHES

Limit switches that control the movement of the main ram or the injection screw should be recorded in dimensional relation to the starting point such as daylight opening. The limit switches for "clamp slowdown," "clamp high pressure," "clamp fast return," and "return slow down" should all be located with respect to the parting line of the mold measured with a scale from mold parting line to the respective positions of limit switches and recorded on the setup sheet for reproduction in location. This will aid in expediting the overall setup.

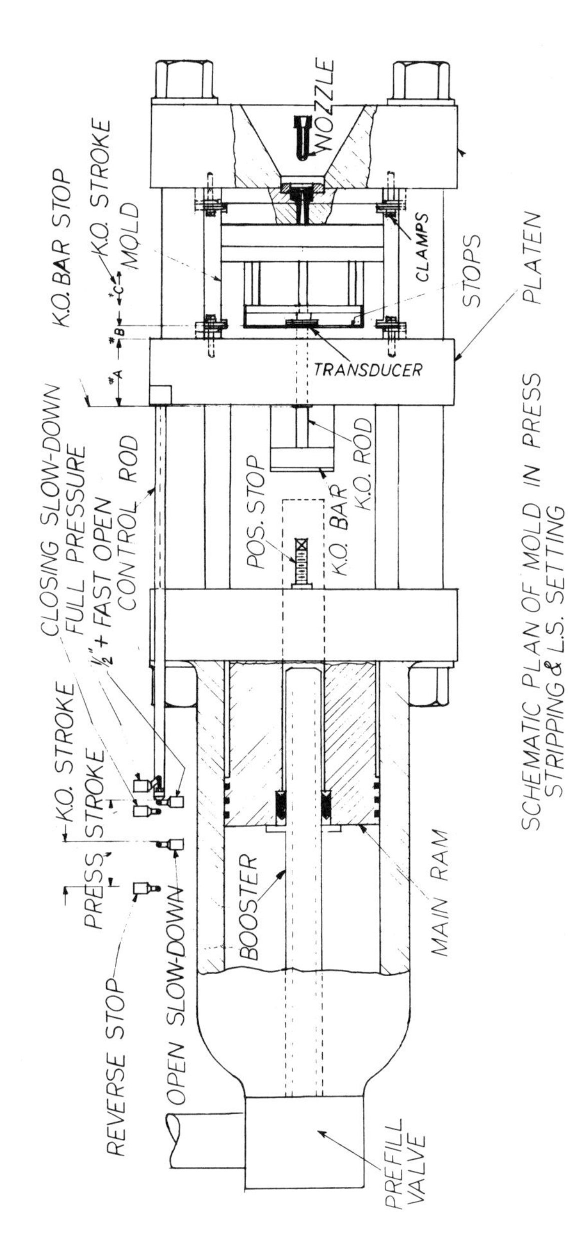

Fig. 15-3. Schematic plan of mold in press stripping and limit-switch setting.

SAFETY CHECK

Before actual placement of the mold into the press is undertaken, it is important to carry out the machine manufacturer's safety procedures. It is especially important to check on the proper functioning of the electromechanical interlocks incorporated in the actuation of the gate. In addition, any required daily attention to maintenance of the press should be adhered to.

Note: the following parameters should be watched during mold transfer from press to press:

1. Clamp size should be adjusted to the value for which the mold was designed.
2. The psi on the material should correspond to that developed during mold proving.
3. The rate of injecting the material–cubic inches per second–should be close to the setup value.
4. The cushion thickness should be per setup.
5. The back pressure on the material should be as specified on the setup.
6. The heating cylinder temperatures should be adjusted, taking into account the capacity of the cylinder under consideration.

16
Mold Placement

The physical placement of the mold should be preceded by a review of setup requirements as well as any other details that will affect the work to be done.

Caution: The prescription for mold handling and machine operation supplied in this chapter is of a general nature and is one of several ways of performing the work of this type. However, each mold and each machine represent individual situations and should be treated accordingly. Each plant may have its own way of machine running, just as each machine has its own instructions for safety, sequence of operations, setting of controls, etc. Obviously, the machine manufacturer's instructions as well as the rules of each shop must be followed so that the warranties are maintained and governmental regulations satisfied.

From the minute the mold is picked off the shelf to the time production is initiated, the main concern should be safety to people around the press and protection of mold and press against damage. The presses are equipped with "setup" selector switches, which put all the functions of the machine into the hands of the setup person. As an added precaution, whenever arms and hands are within a working area of the press, it is best to pull the main power switch to eliminate any accidental operation of a contact switch and with it the possibility of an accident. Other precautions outlined in the manufacturer's manual should be faithfully carried out to prevent malfunction and eliminate hazards.

HANDLING A MOLD

Bringing a mold from storage to the press and back into storage in the majority of cases is done using lifts or slings in combination with lifts. Lifting involves handling the molds around workers and machines that require protection against possible injury and damage from this source. Moreover, the molds are expensive and need to be handled with care. For these reasons, management should assign a person who not only instructs those performing this operation of handling a mold. The designated person should also become familiar with what signs of wear on the lifting parts are critical, so that their use is discontinued and replacement ordered.

There are several sources for the information on slings and the method of weight raising. The supplier of slings for the hoisting method has descriptive literature not only on the capacity of lifting but also on how to judge the approaching end of the sling's useful life. A second source is engineering handbooks under headings such as "strength and properties of wire rope" and "crane chain and hooks." A third source is the federal agency OSHA, which prescribes certain regulations of weight handling and makes the user liable to stiff penalties if its regulations are not complied with. OSHA regulations may be obtained from the Superintendent of Documents, Washington, D.C. This phase of moving heavy molds and placing them in the press has not received the attention it deserves.

CLAMPING A MOLD

How many clamps should hold a mold in a press? Too many clamps waste time, yet too few endanger the bushings and guide pins because of slippage. The most advantageous number can be calculated. Certain assumptions have to be made in order to compensate for conditions prevailing in practice. The assumptions are made regarding the condition of the bolts, the engagement in the platen, and parallelism of the clamp hold-down face relative to the mold slot. The calculations made on the basis of the diagram in Fig. 16–1 give the following results:

Bolt Size (in.)	*Engagement in Platen (in.)*	*Slot in Clamp (in.)*	*Holding Power per Clamp (lb), Safety Factor = 4.5*	*Torque wrench (in. = lb)*
1/2	.75–1	2-13/16	32	210
5/8	15/16–1-1/8	3-3/8	46	340
5/8	15/16–1-1/8	5	36	340
3/4	1-1/8–1-5/16	3-3/8	50	450
3/4	1-1/8–1-5/16	5	39	450
1	1.5–1.75	5-5/16	82	900

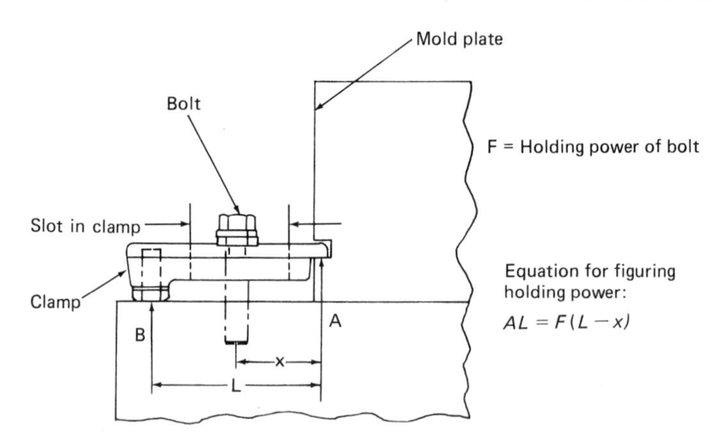

Fig. 16–1. Diagram for figuring number of bolt clamps per mold.

To obtain the number of clamps for each mold, divide the mold weight by the holding power of each clamp. The number of clamps derived in this manner must be divisible by 4 since each mold has four clamping areas. When the result is less than divisible by 4, it should be increased to the nearest multiple of 4. For example, a 300-lb mold will be clamped with 1/2-in. bolts and thus we obtain 300/32 = 9.375. The nearest multiple of 4 will be 12; therefore, each side of the mold will require three clamps per side. These values have been obtained by assuming the least favorable position of the bolt in the clamp slot. According to Fig. 16–1, the closer the bolt is positioned to the mold base, the higher the holding power for each clamp. It is assumed that each clamp will exert the same holding pressure, and for that reason a torque wrench is indicated. In calculating these results, a forged bolt was used with a yield strength of 120,000 psi in shearing, with suitable safety factors. When shearing occurs, it takes place at the root of the thread. In each case the bolt was placed in the least favorable position with respect to the mold. Judgment alone in mold fastening cannot be relied upon since poor judgment can result in slippage and therefore damage to the mold component. For this reason, a torque wrench is strongly recommended for uniform tightening of bolts.

MOLD PLACEMENT

The following steps, not necessarily in the sequence as outlined, can be carried out for placing the mold. These steps would not constitute a preferred method if not in consonance with equipment manufacturer's recommendations. See Fig. 15–3 for schematic outline of mold in press, which shows graphically some of the items described in the following paragraphs, such as setting of limit switches and a stripping system. Other mold placement items are merely explained herein.

1. The positive stripping bars should be adjusted to zero ejecting action.
2. Plasticizing chamber should be in the retracted position.
3. Set clamp opening to the required daylight. For hydraulic clamps, this is accomplished by a limit switch controlling the backward movement of the platen, and, on toggle presses, by the threaded adjustment of the platen.
4. Selector switches on the machine should be placed in setup position. There are three basic selector-switch positions: "mold setting," "hand," and "auto" operation. Some machines have additional selector switches for a variety of purposes. In this chapter, we will be concerned with the three basic ones mentioned.
5. The mold should be lowered from the top or slid from the side so that the locating ring around sprue bushing finds its way into the corresponding opening in the stationary platen. The clamp is now moved slowly forward to hold the mold firmly in position while being clamped to the stationary platen. In the interest of safety, the power-supply switch should be disconnected during

clamping of the stationary half. The preferred arrangement of clamp slots is horizontal in order to avoid accidental slippage of the moving mold half. If vertical slot mounting becomes necessary, an additional bottom support should be provided on the moving platen.

6. Reconnect power, retract moving platen, place ejection rods in position, move platen against mold, disconnect power, and clamp back of mold to platen.

7. Set chamber heat switches to "on" position, and make temperature settings on pyrometers.

8. Set timers for cycle.

9. Water connections for machine and mold should be made.

10. Make setting of stripper bars for proper stripping action. Reconnect power, and, with selector switch in "hand control," open and close press to see that everything is functioning properly.

11. Set limit switches for controlling speed of clamp closing and opening. Closing of mold must be such that no banging or hammering will take place; the parting-line edges of mold halves must be protected against peening if flash-free parts are to be molded. The clamp should start "fast forward," "slow down" at low pressure as the mold halves approach closing, and, finally, close "slow" at high pressure. The opening of the clamp should start slowly until mold halves are separated about 0.5 in., continue "fast," and change to "slow" when stripping starts so that the chance of marking or punching of the plastic is prevented. With the press in the closed position (Fig. 15–3), the "clamp forward slowdown" limit switch should be set for release from the control rod to a dimension of 0.5 in. between mold faces. The high-pressure limit switch should be set to take over at 0.1 in. before mold closing. At the same time, the "clamp fast reverse" limit switch should be set to act 0.5 in. from mold closed position. Fast opening of the clamp should stop 0.5 in. before stripping starts, and in this position the "clamp reverse slowdown" limit switch should be set so that the roller touches the end of the control rod.

12. Change selector switches: (a) "mold set" switch to "off," and (b) "auto-hand" switch to "hand."

With these settings, the approximate limit-switch settings can be checked. Depressing "clamp forward" and holding it will cause the clamp to move in the same manner as in "semi-auto" operation. If the button is released, the clamp will stop. Similarly, depressing "clamp reverse" and holding it will actuate opening of the clamp. Following these steps, final adjustments can be made. The "clamp forward slowdown" with a mold of short stripping stroke can be actuated when stripper return pins contact the opposite mold half. When the stripper plate is spring-loaded or hydraulically returned, the "clamp forward slowdown" can be actuated within 1/2 in. of mold closing. The "clamp return slowdown" should be actuated just prior to stripping action. The "clamp high pressure" limit switch comes into action just prior to clamp closing (about 1/8-

to 1/16-in. spacing). The clamp high-pressure gauge reading should be adjusted so that the tonnage exerted on the mold is 2 to 3 ton/in.^{2} for most materials, and up to twice that amount for highly fluid materials.

13. The "extruder reverse stop" should be set to a position that can be calculated as shown in "setup."

14. The extruder "speed," "torque," and "back pressure" are indicated on the material processing sheet and should be set accordingly.

15. Check cylinder temperatures to determine whether settings have been reached. Also check nozzle temperature. With the extruder unit in retracted position and extruder selector switch in "run-off-on" to "on" position, depress extruder "run" button until "extruder reverse stop" limit switch is actuated, indicating that the shot zone is filled with material.

16. Depress "injection forward" button to purge material into a suitable container, making sure it does not splatter. Repeat this operation until all new clean material is coming through.

17. The needle valve that controls the movement of the heating cylinder by means of the "pull-in" cylinder is opened, and the "seal valve" is moved so that it will cause the "pull in" cylinder to seat the nozzle against the sprue bushing. Depressing the "clamp forward" button will apply the full pump pressure to the "pull in" cylinder and thus bring about a good seat between nozzle and bushing.

18. Set "injection high pressure," "speed of injection," "low pressure injection," as indicated on setup specification.

19. Set "full-semi-auto" switch to "semi." Set extruder switch to "on," and change "hand" to "auto." The press is now ready for normal operation.

20. After a final check of pyrometers to see that they are up to the setting, the press may be operated by opening and closing the gate.

Note: There may be slight variations in designations of switches or preferred sequences on presses of different manufacturers; however, the ultimate objectives are the same.

MOLD REMOVAL FROM PRESS

A mold is a very expensive tool for converting a plastic raw material into a finished product. The vast majority of molds are used for the duration of 10 to 25 thousand cycles per order, after which the usual procedure is to place them in a storage rack. This type of a run for the purposes of this discussion can be called a *short-term run.*

Every press usually has a job in it that requires removal of the mold and, quite frequently, purging of the cylinder. Before removal of a job, any additional information in addition to that shown on the setup sheet should be recorded so that any difficulties experienced will not be repeated. Furthermore, any details in the mold requiring repair or replacement should be noted and called to the attention of supervision for corrective measures. A dated red tag should be wired to the mold indicating the nature of the problem; when the correction is made, it should be marked on the same tag. This will insure that molds producing defective parts will not get into the press. This inspection, judged by the appearance of the last shot, should include cavities and cores, sprue bushing, ejection system, cam action if present, cooling channels, vents, heaters, and wiring if present.

Molds that have been run for about 200 thousand cycles should be subjected to a more thorough examination as enumerated below. Only some suggested areas for inspection are listed; others will depend on judgment of what factors enter into the life of the tool and how to counteract the same. Obviously the schedule for this inspection can vary depending on the abrasive nature of the material. Observations of the mold performance would dictate the time of full examination. Again, it should be emphasized that we are dealing with a very expensive tool, and parts approaching the end of their useful life should be repaired or if necessary replaced in order to keep the mold in active and useful performance.

1. *Cavity of Core.* Any leakage of coolant should be checked. For that reason the mold should be at the operating temperature and the fluid should have the maximum pressure encountered during operation. Any "O" rings should be inspected to be sure that the original shape is intact and the grooves for them are not damaged. Examine closely the cavities and cores for any cracks in the molding area, and for chipped plating if present. A gate should be checked for wear or distortion. Molding surfaces should be checked for loss of polish or design details important for fitting configurations. Nicks or scratches that would spoil the appearance of the product or interfere with withdrawal of the parts should also be removed. Any lands that provide for cores or other purposes, if damaged should be corrected to prevent flashing in the area.

The plates that support cavities and cores should be cleaned of any dirt, grease, or foreign material. All moving parts should be cleaned and measured with suitable instruments to determine that they are still within prescribed tolerance. The same applies to guide pins and bushings. Any rough areas on plates, etc., if not within the accepted limits of good mold workmanship, should be eliminated. All plates and components should be checked for alignment and ground as necessary.

The cavities should be checked for a lip at the edges, which can be present as a result of continuous hammering at the part line during each cycle. If such a lip exists, it should be carefully removed without changing the configuration of the product.

2. *Ejection System.* The heads of an ejection system should be allowed to float on the circumference of the head outline so that the pins will remain in a straight position when engaged in the cavity. This is more significant when there

is a difference in temperature between cavity and pin supporting plates, and clearances should be allowed to suit that difference of temperature.

An examination of the last shot would determine if there is flash around the ejection pins. In that case, a reaming of the hole and substitution of oversize pins (.005 in. larger than nominal) would eliminate this troublesome area. Any bent pins in an ejection system would indicate: (1) a lack of floating of the head; (2) buckling of the pin under the load of ejection (see page 116); or (3) too long an engagement of pin in the cavity, thus requiring more than usual "float" in the plate or better alignment of holes for the pin to start with.

If springs are used, look for broken coils or lack of compresseion. Check the overall length against original specifications.

When support pillars are checked, they should have a .001 to .002 in. greater length than the space provided for them. Flattened support pillars would indicate that the compression on them or on the plates that support them was too great and that additional support pillars are needed (see page 69).

3. *Vents.* The venting at parting lines and around the pins can be plugged by deposits from molding operation. The depth of the vents around the parting line should be checked to see whether they conform to original specifications; if not they should be restored to same. Any deposits around the pin holes should be reamed out to standard clearances. It should be remembered that vents play a most important part in a quality product, and their maintenance during operation is a required feature on every mold. Some circumference vents may call for widening or even slight deepening so as to insure against plugging.

4. *Cooling Channels.* All cooling channels should be checked to make sure that the full flow of coolant is pumped through them. If the cross section of the fluid lines is decreased anywhere in its cooling path, then it should be flushed with a descaling agent for 24 hr under pressure and heat. At the end of this period, it should be cleaned out with processed water for 30 min and then blown out with compressed air and checked out for watertight integrity. A pressure drop on the waterline tester indicates leakage; the likely cause is rusted plugs, which should be removed and replaced with Teflon taped around the new plug. If rusting is severe, consider having the channels nickel-plated by the electroless process to reduce rust and plugging.

5. *Sprue Bushing.* The nozzle seat and possibly irregularity of the entrance hole should be checked out. Also the smoothness of the holes and its correct taper require attention. Two other items to be checked are deformation of the hole at the junction of sprue and runner, and the proper functioning of the sprue puller.

6. *Cam Action.* Each cam-action mold has its own pecularities. The one feature to look for is that the component of the cam embedded in plastic during operation should have nearly the same temperature as the remainder of the mold.

Any looseness in the cam action or binding of components calls for correction of permanent nature to eliminate this from happening. If hydraulically operated, the check should center around leaks or cylinder rod corrosion. It should be normally possible to actuate all the components of a cam by hand to indicate that all parts are working smoothly. Any signs of wear of the pin actuating the slide, the hole in the slide, or the rails that hold the assembly in place should be remachined if possible or replaced with new parts.

7. *Heaters.* Whenever heaters are present, they should be in good working order. Their wiring should be fully protected against short circuits and completely covered against accidental damage. If plugs and sockets cannot be used, be sure to have "strain relief" clamps to prevent wires from being pulled out. A resistance check on the heaters should be carried out so to maintain the initial contemplated wattage.

8. *Lubrication.* Any lubrication on a mold should be so arranged as to keep the lubricant from the plastic and confine it to the area that needs it. Be sure to use only specific-purpose greases since normal lubricants are not satisfactory for the high pressures encountered in molding operations.

9. *Storage.* Most molds run 10,000 pieces, sometimes several times of that amount. A mold should be treated according to its value when stored. As a final treatment, a fully inspected mold should get the same going over as short-run molds receive. In general, storage of a mold should be the same as given to a molding machine.

All accessory jigs, fixtures, and tools should be placed in a container and properly labeled so that repeat orders can be filled without loss of time. The last shot should also be attached to the mold so that any question arising in connection with the last run can be readily settled.

Finally, before the mold is closed for storage, the molding surfaces should be clean and dry and protected against corrosion. The corrosion-proofing materials, in most cases, are best applied to a surface at room temperature which is readily attainable by circulating tap water through a mold.

PURGING

The first step is to shut the slide on the hopper. The molders deal with expensive materials so, to keep waste down to a minimum, they should attempt to use every bit of the material in the cylinder for productive purposes. This can be accomplished by a procedure in which the material in use for producing parts (after the hopper is empty) is followed through the cylinder by a distinguishable purging agent. The parts are run until the purging agent becomes noticeable. The two shots preceding the one in which the purging material is observed should be discarded.

After the purging material starts coming through, the cylinder temperature should be dropped to the low end of the temperature range for the purging agent, but not below the low end of the range for the material being purged. An example is Noryl being purged with styrene. The Noryl minimum temperature is 475°F, and the minimum temperature for styrene is 350°F. The purging temperature will be 475°F.

This step provides greater stiffness in the material being moved through the cylinder; the materials thereby exert better scouring action. The nozzle should now be moved away from the mold. At this point, the injection pressure should be dropped about 25%, and the cylinder purged rapidly until about 1.5 times the cylinder capacity has been flushed with purging material. This purging should cool the cylinder, and a waiting period of 3 to 5 min will permit the absorption of heat for three additional purgings, which should fully clean the cylinder. Now, new material can be introduced, and the cylinder temperatures are set for it. At least two air shots with the new material should be made. It is suggested that purging action be initiated upon completion of the run, if it has been established that the mold being placed in the press will require a different material from the one that is in the cylinder. The reason is that some materials, such as polycarbonate, when maintained at elevated temperature for over 5 min, will degrade and be difficult to purge. If the temperature is permitted to drop below 300°F, the polycarbonate will tend to weld itself to the chamber of screw and damage it. No plastic material can reside at elevated temperatures for too long a time without being degraded, and material in this condition is not prone to controllable action. Materials such as scrap acrylics, polypropylene, and heat-resistant polystyrene are good purging agents.

At this point, it may be well to reemphasize that molders deal with expensive materials; therefore, such materials as that used for air shots and similar purposes should be kept clean (shot into a stainless receptacle) and, upon solidification, placed in a properly identified, clean drum. Materials from air shots, while not up to the standard of virgin material, can be reconditioned at a small fraction of the value of postreconditioned material.

The conservation of costly plastic materials is an important function of those responsible for setup. If setup is carried out conscientiously, the yearly savings in a plant with 8 to 12 machines can amount to several thousands of dollars.

17
Molding Problems and Solutions

The increasing application of plastics to appliances, automobiles, instruments, and general industrial products raises the demand for consistency in quality, which is beyond the present limitations of standard machinery and uniformity of raw materials. The equipment variables contribute to a larger number of rejects, and in some cases overcome the advantages that are expected from plastics. The many applications of plastic parts call for consistency in shape, dimension, and other characteristics that dictate their use, such as resistance to environmental elements and thermal and electrical properties. Additionally, they have to work with parts that are made of different materials so that the complete product is a successfully performing unit.

In discussing molding problems, it must be recognized that there are many sources of defects. The mold plays a very important part in the production of consistent parts. For the sake of this discussion, it will be assumed that a mold is fully debugged and will produce uniform products if the controllable parameters are maintained within a reasonable range. The design of the product will influence how large a variation in pressure, temperature, and time the product will tolerate and still be an acceptable item. The design with wall thickness variation, variety of openings, and complexity of shape will demand a much closer limit of variables in molding than would be the case with a simpler shape, uniform wall, and fewer cores. The plastic raw materials with high shrinkage rates will usually present more problems than the ones with uniform and low shrinkages.

Many products are manufactured economically and with satisfactory properties on existing equipment with its range of variables. By the same token, many other parts demand a closer control over injection pressure, injection rate, mold temperatures, melt temperature, or all of them in combination with each other. Each case will require an analysis of its own to determine what remedy is needed to attain the consistency of a product. As an aid for such an analysis, we will use the experience of raw material manufacturers as contained in their troubleshooting information.

In the review of "Defects" and "Causes," it is taken for granted that the nominal parameter settings on the machine have been properly worked out and recorded on the setup sheet. The defects listed are a result of variables in the parameters.

Before resorting to any one possible solution, it is best to stop and think of the three basic principles in molding–namely, time, temperature, and pressure–and to see which one is the appropriate candidate for adjustment. All three are interdependent, and the side effects of any adjustment should be kept in mind. These factors are:

1. Time-influenced: rate of injection, duration of pressure applications, duration of mold closed, duration of material plastication, speed of screw rotation.
2. Temperature-influenced: mold temperature, cylinder temperature, nozzle temperature, temperature from back pressure, temperature from screw rotation, temperature of melt entering cavity over its entire surface, frictional temperature, temperature causing gas generation.
3. Pressure influenced: injection high pressure, injection hold pressure, back pressure, pressure drop outside of cavity contributed by small gate, long land, small runners, bends without radii, rough finish in runners, small sprue bushing opening, venting, cushion uniformity.

The details for dimensioning of gates, etc. are given for each material in the material processing data sheets. How to optimize machine settings is outlined in Chapter 15. Once the adjustments are begun, it is best to make a written statement about the nature of change and the results therefrom. Obviously, only one change at a time should be made. If the results lead to success, modification of the setup data should be considered so that the experienced difficulty is not repeated.

Changes in pressure respond instantly and therefore provide quick answers. Time adjustments are also quick-acting. Temperature changes usually take anywhere from 5 to 30 min before they reach equilibrium and observations can be properly interpreted.

The following list of defects and their causes covers most of the materials. Wherever there is a deviation, the special problems and solutions are listed in the processing data sheets of each specific material.

- *Surface Imperfections*
  1. Entrapped gas or air in the melt
  2. Inadequate vents
  3. Factors conducive for injection-pressure effectiveness
  4. Hang-up of material between cylinder and nozzle outlet
  5. Contamination and other causes

Appearance defects are the most frequent problems occurring in molding. They are called *splay, streaks, spots, mica, folds,* etc. Fundamentally, one of the listed causes contributes to the occurrence of appearance defects. With the following analysis of each cause, it will be simpler to decide what corrective steps have to be taken to ameliorate the condition.

1, 2. Gas and air present in the melt can be caused by moisture and or by additives to the polymer. The presence of moisture can be readily checked out by a simple system devised at General Electric Plastics, the "TVI drying test." It consists of heating a few pellets to their melting point and compressing them to a 0.5-in.-diameter flatness and observing whether bubbles are present or not. Bubbles indicate moisture, and a relatively smooth appearance indicates a dry resin. Figure 17-1 shows a copy of the TVI drying test. When the melt is at the upper range of temperatures, more gases tend to be generated. The remedy should be to have the melt at the lowest temperature that will produce satisfactory parts. Nozzle temperature should not exceed the front zone reading. While moisture in the material may be absent, a potential source of moisture exists from condensation in the throat that receives the material from the hopper. The temperature of the water-cooled throat should be at the drying temperature of the material.

Every attempt should be made to reduce the amount of air and gases from the melt. Venting of the runner system, as well as placing vents at the gate, will help to dispose of some of the gases. Placing vents at every inch of the cavity circumference will aid in disposing of gases from material in the cavity. Screw rotation at recommended back pressure should be kept going until the press is ready to open. This will tend to homogenize the melt and drive the gases toward its outer surface.

3. Increasing the injection pressure will tend to expel the gas, but the pressure increase should be coupled with a high injection rate, so that the loss of heat from the melt from beginning to end of travel is negligible while the pressure is being applied, thus making it easier to dispose of the gases through softer material.

The gate size, gate land, and runner dimensions should be checked to determine whether they meet the specifications in the material processing data. A larger gate, a shorter land, and a larger runner, as well as a sprue bushing with a larger opening, will make the pressure in the cavity more effective.

Melt not hot enough, mold on the cool side, not enough material in the cavity, injection pressure on the low side, slow injection rate—any one of these alone or in combination with others can create a surface that does not reproduce the smooth finish of the mold. On the other hand, overpacking of a cavity, which is brought about by a high rate of injection and high injection pressure, will cause a product to have a dull surface.

4. Black specks result from material being hung up at some point of the flow path and remaining there until degraded. Degradation causes flaking of particles that get into the stream of the melt. This hang-up can occur at damaged check rings of an injecting screw or at joints between cylinder head and cylinder, nozzle

Engineers at General Electric's Plastics Section, Pittsfield, Massachusetts, have developed a simple, low-cost, foolproof method to determine whether moisture-sensitive thermoplastic pellets are dry and ready for processing.* In brief, this method entails heating a few pellets to their melting point and observing whether bubbles are present, indicating moisture in the resin; or absent, indicating a dry material. Called T.V.I. (Tomasetti Volatile Indicator) after the G. E. application engineer who developed the technique, the system requires little in the way of equipment and calls for just six simple steps.

*Note: This test is not applicable to glass reinforced resins.

EQUIPMENT NEEDED CONSISTS OF: 1) hot plate capable of maintaining surface temperature of 525 F ± 25° F, 2) 75 × 25 mm glass microscope slides, 3) tweezers capable of han-

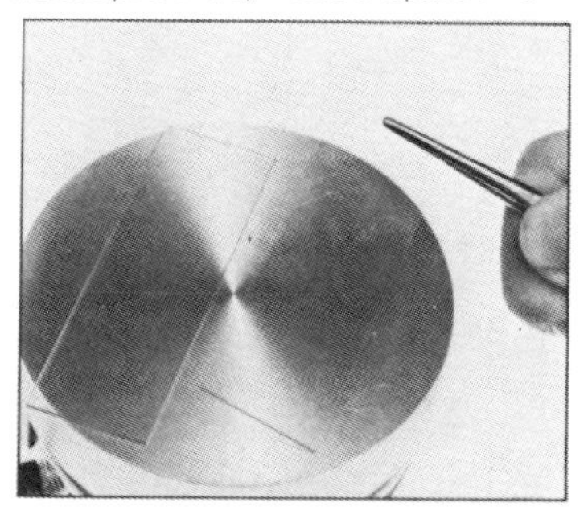

1. Plug in hot plate (be sure surface is clean) and calibrate it to a surface temperature of 525 ± 25° F. Place two glass slides on surface for 1-2 min.

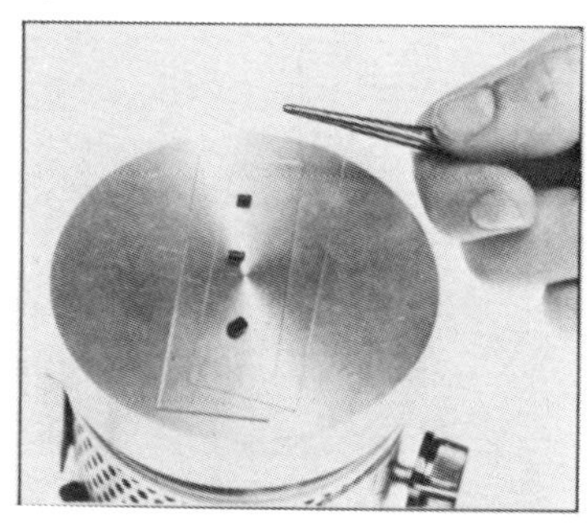

3. Now place a second hot slide over the first one to sandwich the pellets between them.

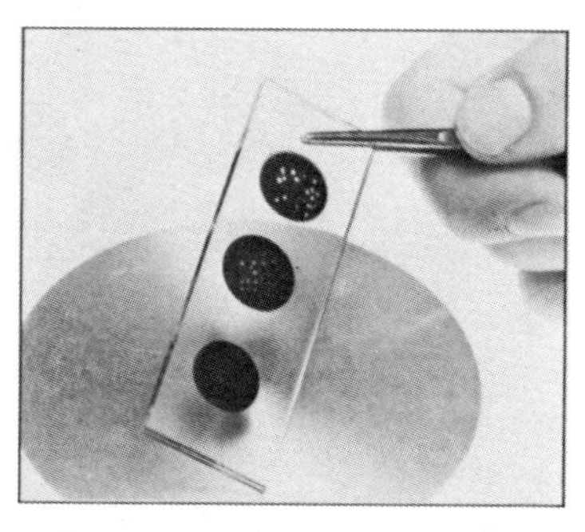

5. Remove sandwich and allow to cool. Amount and size of bubbles indicate percentage of moisture, correlating bubbles with moisture.

Fig. 17-1. Simple test for resin dryness. *(Courtesy of General Electric Co. Plastics Section)*

dling ⅛ in. pellets and 4) conventional wooden tongue depressors. Total cost for this equipment should not exceed $25. Following are the steps to follow in running the test:

2. By this time the glass surface temperature should have reached 450-500° F. Use your tweezers to place four or five pellets on one of the glass slides.

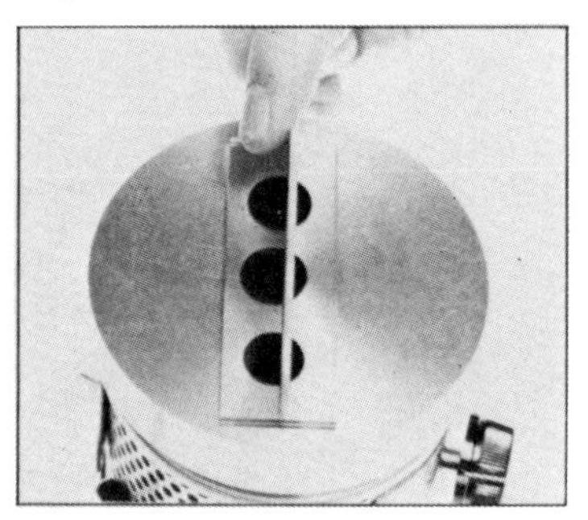

4. Press a tongue depressor on the top of the sandwich until the pellets flatten out to about ½ in. dia.

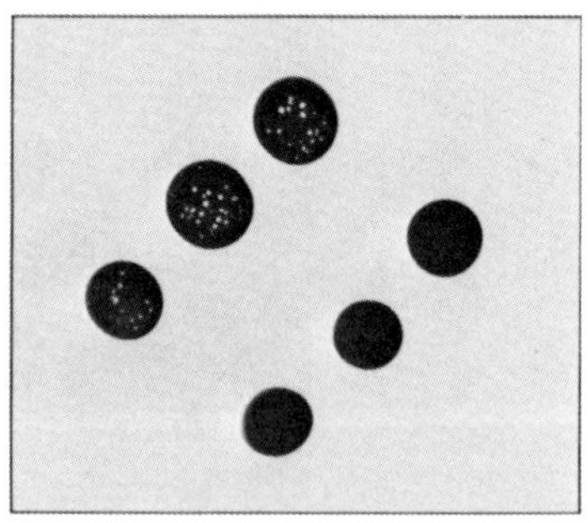

6. Here are typical results. Slide at right indicates dry material; slide at left indicates moisture-laden material. One or two bubbles may be only trapped air.

Fig. 17-1 (*cont.*)

and cylinder head, etc. In brief, anywhere a crack exists or a transition from one part to another is not smooth and where material can thus lodge, there is always a danger of hang-up. Purging with a stiff material can be attempted. However, if this is not successful, physical examination of suspected areas is the only step left.

5. Contamination from dirt in the air, due to the hopper not being covered or fines in the regrind, can cause surface defects. Fines do not heat in the same manner as pellets. They can be trapped easily and in general they may not be clean; therefore, they can contribute to unsatisfactory product appearance. Moisture from condensation on the mold surface or from a water leak, or from excessive mold lubricant can all have a bearing on the poor appearance of a product.

- *Brittleness*

1. Melt temperature is too high or too low.
2. Gate is too small, and land is too long.
3. Mold temperature is too low.
4. Material is not dry or contaminated.
5. Rate of injection needs checking as related to injection pressure and injection timers.

- *Burning*

1. Injection pressure, when higher, will cause a higher cavity pressure and with it a faster rate of filling cavity. This will partially entrap the displaced air, and when compressed will raise the temperature to cause burning.
2. Rate of injection increases due to pump volume fluctuation, which can have the same effect as (1).
3. Resin temperature, when higher, will tend to generate more gases that cannot escape from the cavity. When fully compressed, such gases will increase in temperature and cause burning.

- *Dimensions, large*

1. The injection pressure, when higher, will bring about a lower shrinkage rate because the part is overpacked. This will result in larger dimensions.
2. Mold temperature, when lower, will arrest the molecular chains in the loosened state before they have time to rearrange themselves into the "natural" state. Heated polymers cause a separation between molecules, and when solidified the molecules arrange themselves in a tight interlocked manner. When the molecular chains are arrested in a partially loose state, they will occupy more space and bring about larger dimensions.

- *Dimensions, small*

1. Injection pressure, when lower, will cause a higher shrinkage rate, and thus make parts smaller.

2. Mold temperature, when higher, will cause a higher shrinkage rate because of the expanded condition of mold and plastic material. When removed from mold, the plastic will decrease more and cause smaller sizes.

• *Flashing*

1. Resin temperature, when higher, will cause higher fluidity and require greater clamping pressure to keep the mold closed.

2. Injection pressure, when higher, will tend to open mold at set clamp pressure.

• *Shot-to-shot variation*

1. Injection-pressure variation from maximum to minimum is too great.

2. Resin temperature variation is higher than the product will tolerate. If a high percentage of heat is derived from screw rpm and screw back pressure, their part in the variation may be an important factor.

3. Change of cushion length may cause too large a variation in cavity pressure due to an inconsistent screw travel.

4. Mold temperature variation occurs, which is caused most likely by fluctuation in flow volume of circulating fluid.

• *Sinks and voids*

1. Injection pressure, when lower, will not compress the material to its proper density nor fully expel the air and gas from it.

2. Resin temperature, when higher, will cause an increase in shrinkage. Since molecules at higher temperatures separate more, they occupy more space. Therefore, for the same cavity space, there will be fewer molecules when the material is at a higher temperature. When the material solidifies, the molecules will rearrange themselves from a highly separated state into a tight interlocked manner and thus will bring about a greater shrinkage.

3. Mold temperature, when lower, will cause a lesser shrinkage. The molecules are arrested in a partially separated state by the lower mold temperature, thereby causing larger dimensions or lower shrinkage.

4. When mold temperature is higher, time of "cure" is not adjusted to permit gradual thermal shrinkage. Voids will be formed because the outer walls solidify and permit the inner hot material to shrink toward the outside, leaving bubbles in the middle.

5. Vents, when clogged, will cause entrapment of air and gases within the part.

• *Sticking in cavity*

1. Injection high and injection hold pressures are too high.

2. Injection hold or injection high time is too long, causing overpacking.

3. Cushion length varies.

4. Cure is too short, thus not bringing about full solidification.

5. Cavity temperature is too high, thus not bringing about solidification within set time.

6. Cylinder temperature and nozzle temperature are too high, preventing full solidification during set time.

7. There are possible undercuts, peened edges at parting line, and need of directional polish.

- *Sticking on core*

1. Injection high and injection hold pressures are too high, and injection hold or injection high time is too long, causing overpacking.

2. Cushion length varies.

3. Mold closed time is too long, bringing about excessive shrinkage over the core.

4. Core temperature is too high, not permitting full solidification during set time.

5. Cylinder and nozzle temperature are too high, not conducive to solidification within set time.

6. There are possible undercuts or directional polish needs improvement.

- *Sticking of sprue*

1. Injection high or injection hold pressure is too high. Injection hold or injection high time is too long and does not allow sprue to shrink and pull itself away from walls.

2. Mold closed time is too short, not allowing time for full shrinkage of sprue.

3. Mold temperature in area of sprue bushing is too high, not allowing full shrinkage to take place.

4. Check that hole in sprue bushing is larger than the one in the nozzle and that the fit is such as not to entrap material between them.

5. Check if sprue puller is adequate for the job.

- *Warpage*

1. Injection pressure, when lower, will cause a varying density within the cavity, decreasing from the gate toward the end of the flow. This change in density has a corresponding increase in shrinkage. Different shrinkage rates within a part cause warpage.

2. Resin temperature, when higher, will cause a higher–but not necessarily uniform–shrinkage rate. If the shrinkage rate is not uniform throughout the part, warpage will result.

3. Higher mold temperature will also cause higher shrinkage and react as in (2).

4. The cushion of the resin when on the larger side is caused by inaccuracy of screw travel. This increased cushion absorbs a greater part of the injection pressure, leaving less pressure available to the cavity. The net result will be same as (1).

- *Weak weld lines*

1. Lower injection pressure will not exert enough force on the meeting surfaces to self-weld for the best joint.
2. Lower mold temperature will cause the areas to be welded to be partially solidified, thus making them less conducive to good self-welding.
3. Lower resin temperature will cause the melt in the cavity not to be in favorable condition for welding.
4. Slower rate of injection will cause some cooling of joining surfaces due to the additional time of travel within the cavity. This will make for less favorable welding condition.
5. Partially clogged vents will trap air and gases between the surfaces to be welded. This will detract from the purity of welded areas and cause a weak joint.

- *Process control*

The troubleshooting information supplied by raw-material suppliers is more extensive because they deal with problems encountered in debugging a mold.

For the purpose of controlling consistency in quality of products, the enumerated defects will cover the problem areas that have caused concern to the users of plastic materials. If the defects outlined herein were eliminated, the plastics injection molding could be considered a virtually problem-free process.

The analysis of the causes for defects leads to the inescapable conclusion that the major culprit is the injection pressure variable. Since most of the molding parameters are somewhat interdependent, it appears that by correcting the range of injection pressures, a major step forward would be accomplished. The control of melt temperature seems to be more easily attainable, especially when the heat introduction from screw rotation and screw back pressure are kept to a low level. The new cylinder pyrometers are doing a satisfactory job when the cycle is repeated and when the cylinder is properly shielded against external temperature fluctuations.

The mold temperature uniformity is determined to a large degree, by the flow of the circulating medium through the mold passages. If the rate of flow is maintained to produce the same turbulent condition, the heat absorption should be consistent enough for the purpose. The object is to work the circulating pumps in a pressure and volume range that will be independent from resistance to the flow conditions or to have thermocouples placed in the mold halves that will cause circulation of fluid as dictated by the mold surface temperature.

The question is how to approach the problem of eliminating the defects and assuring product uniformity in properties and appearance. The ideal method would be to establish the degree of variation that exists in injection pressure, mold temperature, melt temperature, and volume delivery. These variables would be recorded concurrently, and the test would be run for a period of at least 24 hr. Considerable expense would be entailed in adapting the recording instruments to the mold and press. How to obtain the necessary instruments is another question. Whether they would have to be purchased, rented, or borrowed from a prospective seller of "Process Controls," the individual circumstances would dictate the best way of obtaining the investigative instrumentation. There is no doubt that getting all the data pertinent to the variables of the enumerated parameters is a sensible and technically sound way of getting to the sources of defects. In many products, the expense connected with such a procedure may appear prohibitive, and thus another approach may be preferred.

Since the advent of instrumentation for more accurate control of the injection process, observations have been made as to which variables contribute the greatest amount of fluctuation in properties and appearance of a molded product. The conclusions have been arrived at on the basis of numerous tests in the field, that the injection pressure, as measured in a strategic point in the cavity, when controlled within a narrow range, will provide the vast majority of improvements needed in a product. The control of cavity pressure within a narrow limit of variation (1% of reading) is attained by means of an instrument designed exclusively for the purpose. The instrument is relatively inexpensive and requires little cost for adaptation to mold and press.

The basic function of the cavity pressure-controlling device is to control the high injection pressure within the cavity to a selected set point for the purpose of filling and packing the part. As soon as this aim is achieved, the holding pressure receives a signal to take over the usual function of preventing backflow from the cavity. In reality, the injection high timer is eliminated from action. The instrument works by placing a pressure-sensitive device known as a *transducer* under the head of a pin, usually a knockout pin. The pressure from the plastic is transmitted to the pin and onto the transducer. The transducer converts the pressure into an electrical signal, which is preset to a value that will produce satisfactory parts. When the set value is reached, the electrical signal causes the secondary or holding pressure to come into play.

The benefits that are being achieved by the cavity pressure control are: close tolerance dimensions, improved and consistent quality, low reject rate, no flashed parts, and no short shots, faster cycles, power saving, material saving, or minimal machine downtime. This respectable list of improvements is provided by controlling only the cavity pressure with a relatively inexpensive process controller.

For some classes of moldings, the cavity pressure control alone may not accomplish the desired result. In that event, a careful review of Chapters 12 and 15 is suggested with the aim of refining existing setup conditions and comparing results with those expected from process controls. If corrective steps resulting from this type of analysis still do not prove adequate, the only solution left is to apply other controls that will insure more precise movement of the screw, maintain a narrower range of cylinder temperature, control back pressure within closer limits, stabilize mold temperature, and provide variation in injection rate to attain uniformity of speed in cavity filling through different sections.

Any one of these additional controls or all in combination should provide the solution to a most complex problem. These controls, by the very nature of their function, are not simple devices and are not easily justifiable for too many applications.

The additional control should be carefully weighed against the benefits they will provide. There are too many proven cases in which too much control can be as detrimental to an operation as too little control. There is a happy medium, and judgment coupled with thorough analysis must be the guideposts.

Figure 17--2 shows one such process control that provides the latest technology of closed-loop servo control for injection molding. It is instrumental in delivering a uniform volume of material at consistent pressure, density, and rate of velocity that will end up in uniform products. All machine variables are corrected with this control to give the molder better products, dimensionally to size, and fewer rejects, and these results warrant the additional expense. Following is an explanation of the numbered control functions in Fig. 17–2. Unnumbered controls are shown in Fig. 17–3.

1. *Visual Display.* Provides a real-time picture of the machine's operations. Shows plastic pressure, hydraulic pressure, ram speed, or shot size. Indication lights show the cycle sequence.

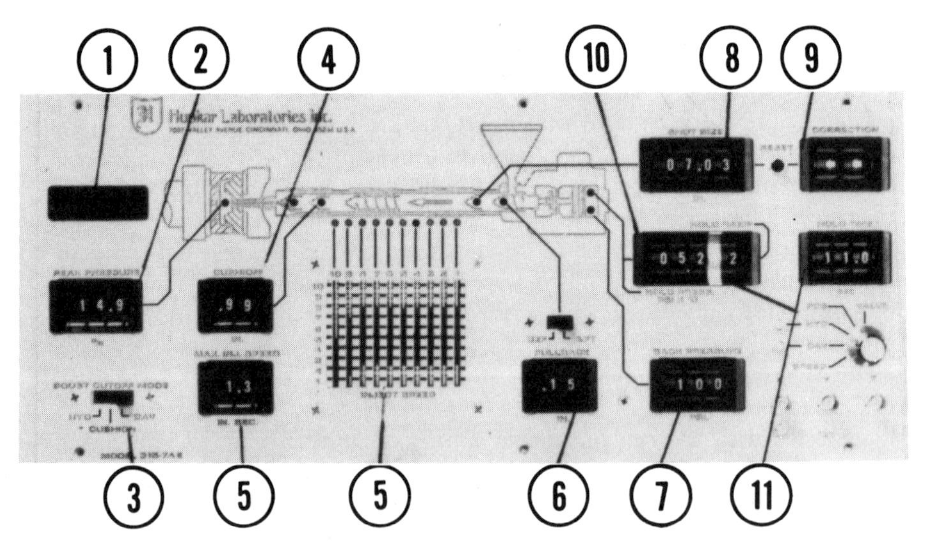

Fig. 17–2. Closed-loop servo control. *(Courtesy Hunkar Laboratories Inc.)*

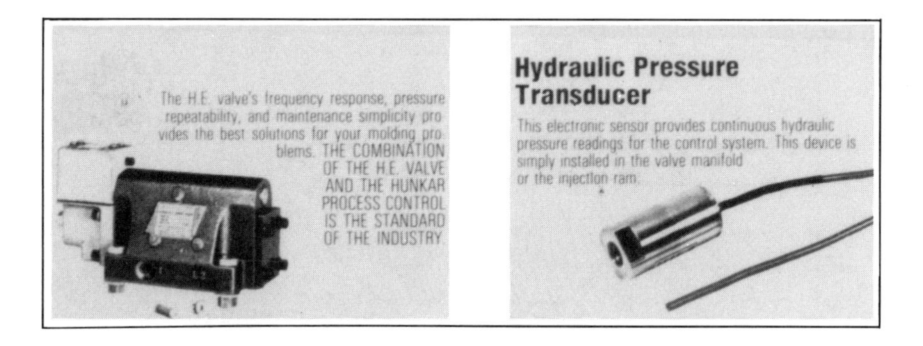

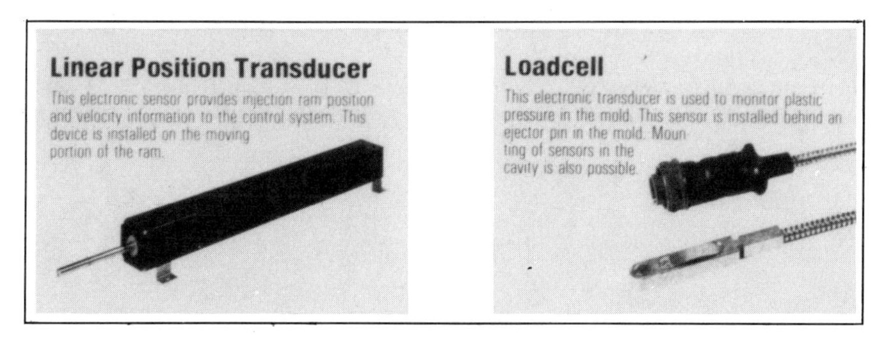

Fig. 17–3. Important parts not shown in Fig. 17–2. These parts are interactive, therefore not always identifiable with each control number.

2. *Peak Pressure on Material.* Problems of mold *flashing, overpacking,* and *underpacking* are eliminated by accurately controlling filling pressures. The injection high timer of the machine is replaced by accurate pressure sensing in the control. This result is *reduced scrap, better dimensional control,* and *shorter cycle time.*

Loadcell. This electronic transducer is used to monitor plastic pressure in the mold. This sensor is installed behind an ejector pin in the mold. Mounting of sensors in the cavity is also possible. (See Fig. 17–3d.)

3. *Boost Cutoff Mode.* Three modes of accurately terminating boost cutoff of hydraulic injection pressure are provided. This gives a flexibility to the molder to assure *consistent initial packing* of the mold. This results in *shorter cycles* and *accurate part dimensions.* Boost cutoff is provided by sensing cavity pressure, ram position, or hydraulic pressure in the injection cylinder, Figs. 17–3d, 3c, and 3b.

4. *Cushion.* The cushion setting is made to suit the material requirements of each mold.

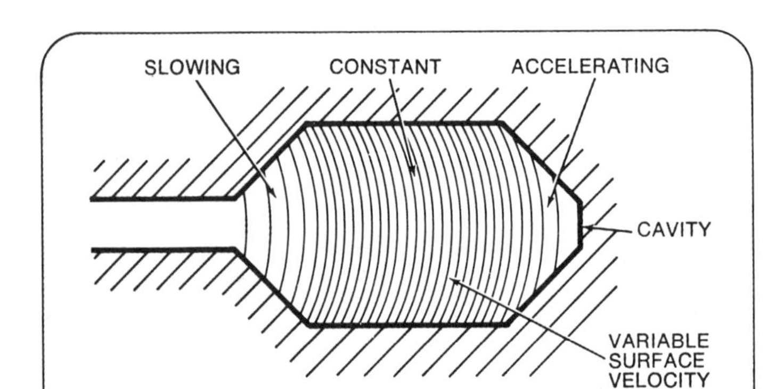

(a) Note how flow of melt at the surface of the cavity varies with mold geometry when filling with *constant ram speed,* indicating need for programmed injection rate.

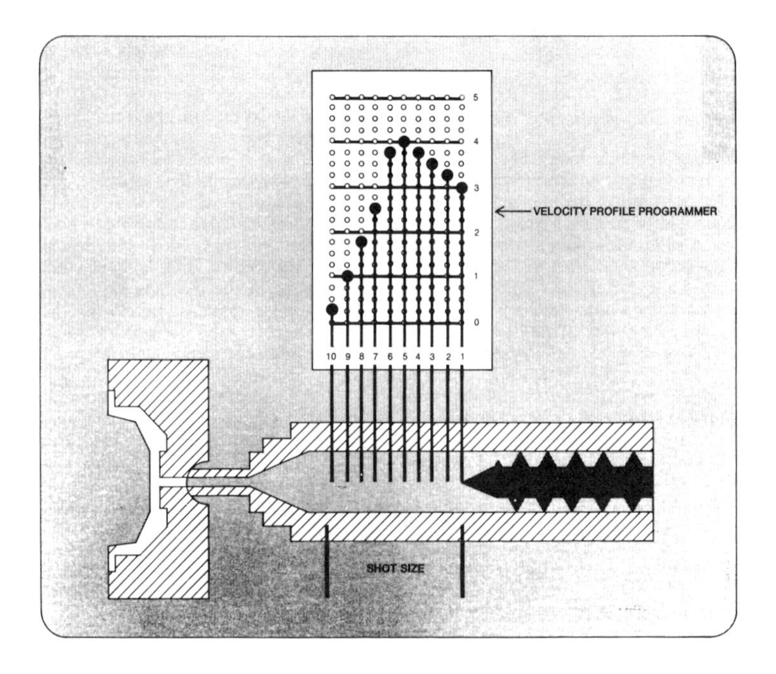

(b) This shows a typical injection *rate programmer* for altering rate according *to geometry of mold.*

Fig. 17–4. Programmed injection speeds. *a,* Constant ram speed (unprogrammed). *b,* Variable ram speed controlled by rate programmer.

5. *Variable-Speed Mold Filling.* The speed of the injection ram is dependent on the material viscosity. Improper speed can cause *jetting, flashing, flow lines, surface stresses, burning,* and *longer cycle times.* Injection velocity control is provided by a closed-loop servo system, Fig. 17–3a. This system maintains a consistent injection speed as programmed regardless of material viscosity, oil temperatures, or changing ambient conditions. (See Fig. 17–4.) Programmed filling will reduce warpage and electroplating problems; reduce surface imperfections and improper location of flow lines; will reduce burning on shear rate–sensitive materials and prevent flashing due to slowness of machine controls.

6. *Pullback.* The pullback position is determined by the condition of the cushion after and before injection.

7. *Back Pressure.* Back pressure is the principal factor in controlling melt conditions. Closed-loop precision control ensures *consistent melt-flow properties.* This is the key factor in controlling *packing conditions* and *surface qualities.*

8. *Precise Shot Size.* Shot size is accurately controlled by a four-position digit switch rather than a mechanical limit switch.

9. *Corrected Shot Control. Material variations* in density are compensated by adaptive shot control. More or less material is added to the shot size as needed depending on the final ram position sensed at boost cutoff. See Fig. 17–3c.

The Gilmore and Spencer equation, which is basic to plastics, reads

$$(P + \pi)(V - \Omega) = RT$$

where P = plastic pressure

V = plastic volume

T = plastic temperature

The other three constants relate to the particular plastic material. The volumetric difference in a material occurs when viscosity changes under the pressure, which is held constant; therefore, the shot volume must be adjusted to satisfy the equations. In a process-controlled mold, there is a loadcell with a constant pressure that is counteracted with a material pressure equal to density times volume. Adjustments have to be made in the volume for the difference in changing viscosity against density that gives a cavity a constant weight.

When a shot is made before the ram arrives at the theoretical position of the cushion and at the same time the pressure in the cavity is achieved, the control stops the screw travel and automatically decreases the subsequent movement by the amount of original stoppage. The reverse is also true; when the volumetric displacement is changed in the opposite direction, the ram may achieve the cushion limit prior to obtaining cavity pressure. In this case the subsequent shot will be increased in order to arrive at a cavity pressure and theoretical cushion limit.

10. *Hold Ramp.* The hold ramp provides a controlled transition ramp from first-stage injection pressure to second-stage injection pressure. It eliminates *ram bounce* and aids in *final packing* of the gate area.

11. *Precise After Packing. Part weight* is accurately controlled from shot to shot by control of the holding pressure. Precise control of the hold pressure and hold time minimizes *sink marks* while assuring accurate *part weights* and *dimensions.*

12. *Hydraulic Oil Temperature.* Variations affect part packing and filling. This process control compensates for oil temperature variations through a closed-loop servo control system. See Fig. 17–3a.

13. Variations in holding pressure and plasticating pressure affect dimensions, weight, and stress in a molded part. This process control provides precision closed-loop control of hydraulic pressures to produce consistent molding conditions. Performance is unaffected by environmental conditions in the plant.

Starve feeding. This technique enables molders to produce parts with lower weight and performance equal to fully dense products. This concept has been practices for a long time, but it has never been popular mainly due to variation in material density and the fact that short shots or severe sink marks were frequently encountered. (See p. 248.)

Adaptive starve feeding of the process control described above refines the basic starve-feeding technique by correcting the shot size in such a manner that the same weight of materials is injected in the cavity with each shot. This is possible if compensation is made for factors that affect material viscosity.

This technique operates as follows. The cushion is set at zero. Precise cavity pressure is to be obtained at the moment the screw bottoms. The constant volume delivered by the screw will result in small variation of cavity pressure as a result of changes in viscosity. The basis for controlling each shot in this operation is thus volumetric. When cavity pressure is higher than desired, the shot size will be decreased, and the inverse is true when the cavity pressure is lower than desired. In practice, changes in cavity pressure will be so small that part characteristics will not be affected. If the density of the part is to be increased or decreased, all that is necessary is to change the cavity pressure in the direction needed. This system will produce parts with dimensionally stable sizes, good appearance, and satisfactory mechanical performance.

The benefits of this technique are (1) fully automatic machine operation without constant supervision; (2) a saving of 3% to 6% of material costs; and (3) appearance and performance of a product meeting all quality requirements.

18
Material Processing Data: Thermoplastics

The final step in the overall plastics picture is to convert the raw material into a product that is shaped by the mold. A mold is built for a specific grade of material with established physical, thermal, electrical, mechanical, and chemical properties. The object of conversion or processing is to see all the properties reflected in the molded product and to obtain from it the expected performance. Each material has its own characteristics as well as its own processing data. The suppliers of raw materials expend considerable effort in evolving manufacturing information that will lead to successful products, and the molder should take advantage of the information and put it to practical use.

The information in this chapter was extracted from materials literature and condensed to a single chart to provide ready reference to the molder and judicious application to each job. It should be emphasized that no amount of perfect information will overcome a poor product design, a mold lacking in necessary features for moldability, or a press that is not properly maintained. Therefore, it is believed desirable to review the highlights of frequent omissions in part design, moldmaking, and machine conditions before getting into the details of material processing data.

Plastics processors, if they are to be considered qualified and knowledgeable manufacturers, will assume responsibilities for the product even when the problem raised is beyond their scope of activity. Since they are last in the process of bringing a plastic product into the commercial world, any cause for complaint will be directed to their attention with the expectation that they instigate the corrective steps. It matters little whether the defect is due to poor design, bad tooling, or variations in processing; the consumer expects the manufacturer to deliver a product that will perform in accordance with anticipated requirements. Whether or not this attitude is justified, the merchandiser and the consumer have the notion of manufacturer's liability on their minds.

Processors, therefore, must protect themselves against liabilities not of their making and be fully cognizant of all the elements that go into the manufacture that can adversely affect product quality. They should be aware of the design features that detract from the performance of the product, the tool design characteristics that are responsible for a weak mold as well as poor moldability, and, finally, what molding variables contribute to the unpredictable product behavior. With such an awareness, the processor can in the early stages of production call to the attention of all concerned the deficiencies encountered and instigate proper corrective measures until a well-performing product is obtained.

Let us now review the main and frequent omissions and inattentions from a design, tooling, and processing point of view.

DESIGN FACTORS

1. Many drawings call for close-tolerance dimensions in places where repeatability of size is the real need. If correctly indicated, the toolmaker could concentrate on the features that are essential to the performance of a part. A drawing used in quality control should be marked with notations as to the significance of each close-tolerance dimension. This will insure the proper function without concern for specifications that are part of a drawing but are not of a consequence from the use point of view.

2. Material designations should indicate equivalent grades. In many instances, an equivalent grade may cause fewer rejects and produce a more consistent quality. It is important to evaluate several equivalent grades, keep records of them, and concentrate on the use of the most favorable one.

3. Shrinkage of dimensions that may be responsible for possible warpage should be closely followed in relation to material grade and molding parameter settings. There are combinations of machine control settings that will generate the most desirable dimensions and quality. The mold temperature is considered one of the vital control elements and should be included in the term of machine controls.

4. Molding stresses. Very few specifications–if any–indicate the tolerable level of stresses. Admittedly, this is a difficult problem to handle. Where close dimensions have to be maintained, this could be treated by specifying an annealing operation of the part 24 hr after molding time and tying down the dimensions of the annealed part in comparison with dimensions of a part that was not annealed. Needless to say, performance tests would have to be carried out prior to agreeing upon what is tolerable.

5. Parting line. There is usually a slight sharpness at the parting line, and its height should be agreed upon whether it be 0.002, 0.005, or even 0.010 in.

6. Weldline. When strength of a part is a factor, the weldline is a point where checking is needed. A test that is repeatable and indicative of quality should be devised and used at certain intervals on each operating shift. The appearance of the weldline can be agreed upon by visual observations and a sample, properly countersigned, retained as a standard of comparison.

7. Tapers. When parts are stressed during ejection, it frequently indicates that there is insufficient taper on the vertical walls. The permissible amount should be checked out, and, if the maximum amount tolerable is on the part, then the only assistance that can be obtained for withdrawal is from a smooth directional

polish. Lubricants should be frowned upon since they are usually a source of undesirable side effects.

8. Deep ribs and projections. The air being displaced from the spaces of ribs and projections should have a way of connecting to the parting line, so that it can escape. Ribs and projection are normally used for supporting some component and should have good strength. In that case, provision for venting should be made by means of auxiliary K.O. pins, etc.

9. Square corners, thick and thin sections. Inside sharp corners act like a notch, causing the material to be brittle. A radius of 0.020 to 0.025 in., while visually sharp, reduces the brittleness to a large degree.

Thick and thin sections cause variation in shrinkage rate and with it stresses, warpage, and related problems.

In analyzing warpage, the following observation should be kept in mind. If we consider a flat part molded under conditions in which each mold half has a different temperature, we find that the cooler side will generate a larger dimension, while the warmer side will be smaller. The larger side will cause bending toward the warm side, and the part will not be flat. This example can be a valuable aid in determining temperature adjustments of cavity and core for minimal distortion and reduced stresses.

10. Miscellaneous. Molded-in metal insert should present no outside sharp edges to the surrounding plastic. Inside threads should have the outside of the thread form rounded to minimize brittleness. In general, a thorough review of Chapter 4 will call the attention to product design details that merit attention.

MOLD FACTORS

1. The passages for the plastic from the nozzle up to the cavity should be checked for sizes, smoothness of finish, radii in bends and smooth transitions, adequate size of cold well to take care of cool nozzle material, and gate size. All of these have an impact on the pressure drop outside the cavity and indirectly affect the amount of pressure left for densifying of parts.

2. Cavity surfaces need to be checked in quality and hardness to insure long-lasting life. If excessive wear is observed anywhere on the mold, the cause should be investigated and corrected.

3. Temperature control of cavities should be established, and a determination made whether there is a variation of temperature within a cavity and whether turbulent flow is prevailing in the cooling passages. Temperature uniformity throughout the cavity can provide many beneficial results. Chapter 12 will call attention to many items frequently overlooked during operation.

4. The rigidity and strength of the mold base and cavities should be examined to see that deflections are not present and that support items embedded into plates are not noticed.

5. Are precautionary notes about limit of press clamp and limit of stripping force stamped on the mold for safeguarding the mold and press?

A review of Chapter 5 will call attention to other details related to processing.

6. The calculations for various aspects of molding and moldmaking were introduced in Chapters 5 and 15 for a dual purpose. First, for individuals so inclined, calculations are step-savers in some instances; in others, they verify strength conditions where judgment alone has frequently been found to be faulty. Second, the calculations point out how the variables will influence the result. Thus, for example, if excessive deflection is experienced, it becomes evident from the formula that

- a harder or higher tensile steel will not correct the difficulty, but to reduce deflection it will take a higher dimensional adjustment.
- when an item fails in tension or compression, the formula shows that substitution of a higher-strength material will solve the problem.
- when heat dissipation in the mold becomes of concern, the formulas show that, not only is the coolant temperature a factor, but also the gpm or volume delivered to the passages that create a turbulent flow can be a factor in increasing the heat-absorption capacity by a multiplier of three in comparison with a low-velocity flow.
- when machine capacity in terms of ounces per shot is in question, the calculation shows how the hopper drier can be used as an aid in increasing this capacity.
- when a small mold is placed in a high tonnage press, the formula points out how to protect mold and press against damage.

As a matter of record, it should be noted that each calculation was made in order to point out the factors that may have a detrimental effect on a specific problem, and the steps that have to be taken to provide positive solutions.

MACHINE FACTORS

Maintaining a consistent pressure in the cavity is most important in the quality of a product. The most frequent cause of variation in injection pressure is the working order of screw check rings. Leakage in these rings will limit the application of the pressure by the degree of leakage. A simple indicator of such leakage is the inability to maintain a uniform cushion.

The injection-pressure gauge is another item that requires checking in order to insure its correct indication. Other maintenance questions are elaborated in Chapter 14.

PROCESSING FACTORS OR "MATERIAL PROCESSING DATA"

The material processing data chart was conceived to serve as a convenient, concise, and practical source of information for those involved in the molding operation. It occurs in this chapter on pp. 320–330. In almost every chapter so far, some reference has been made to the material processing data charts. These references indicate that the listed data can be applied to almost every phase of injection-molding thermoplastics. The processing data can be used by

- Mold designers, to find information for dimensioning material passages.
- Molders, to find the range of molding parameters for making setups to run a mold.
- Molders, in conjunction with mold designers, to predetermine desired parameter settings for a new mold. Designers can use such information as the basis for designing a mold-cooling system.
- Processors, to become aware of the prominent characteristics of a material and to exert every effort to convert such characteristics into part properties.

The chart is followed by notes for each material, in which unusual processing requirements are found. These include special purging instructions, unusual start-up and shut-down procedures, maximum residence time of a material in cylinder at melt temperature, and similar requirements that may vary for a specific material. Additionally, the same note may contain troubleshooting information regarding a certain material, but not covered in Chapter 17. In the category of "material characteristics and uses" in these notes, statements such as "good electricals" and "good impact strength" are made. These definitions should be interpreted relative to the cost of material and not as an absolute comparison. Finally, when process information is needed for a grade of material not listed herein, a blank sheet can be sent to the supplier of material to be filled out.

Following is a description of each heading on the chart.

Material contains the generic description and any other data that differentiate it from another grade; for example, "polycarbonate, glass-filled 40%." In shop practice, the category would also include the name and number of one or more manufacturers, arranged in order of preference based on performance experience.

Mold shrinkage. In./in. is used not only for mold-design purposes but also for checking dimensions of parts against the mold to insure that molding parameters are within correct limits to produce anticipated shrinkage values.

Specific gravity. Used to compare machine capabilities of a material with that of polystyrene, such as screw travel, rate of injection, etc.

Melt temperature. The range of heat within which good fluidity of the polymer is attained with a degree of safety. Melt temperature range provides ample margin to manipulate pyrometers to suit each individual condition.

Drying temperature. Some materials require drying prior to introduction into the machine. The drying temperature is a set value for a specified duration. It is also an indication of the temperature at which the plastic granules will not adhere to each other and cause difficulty in dropping onto the screw.

Mold temperature. Another range of temperatures that permits heat absorption for favorable properties in the product. It is measured in most cases by means of a portable pyrometer.

Back pressure psi on material. The pressure that indicates the work of mixing and, to some extent, of shearing the polymer in preparation for injection. An important setting, it therefore should be carefully watched so that shear-rate-insensitive materials are not harmed, i.e., their molecular structure is not disturbed. This setting in psi is comparable to that of injection and should not be confused with the "pump pressure" scale of the gauge.

Specific heat, Btu/lb/°F. This information makes it possible to figure the amount of water needed to cool a mold or to estimate machine capacity in relation to polystyrene.

Runners, vents, gates, land. All dimensions of these passages are dependent on the viscosity of the material at melt temperature and are established by the material supplier.

Screw speed (rpm). This column is related to the amount of work that is put into the polymer for plastication. In this case, also the shear-rate-insensitive materials are limited in ability to absorb mechanical energy; therefore, the numbers indicated should be carefully followed.

Screw torque. This column is governed by the nature of the polymer when fluid. In practice, it was found that a high torque will work satisfactorily on most materials, provided the desired screw speed is available to suit this setting.

Nozzle. If specifically indicated by material supplier, the type of nozzle will be specified.

Heat deflection temperature @ 264 psi. A useful temperature for stress relieving and checking for stresses at changing operating conditions.

Remarks. Any unusual requirement or precaution for a material that appears to be of special interest to the processor will be recorded under "Note." For example, if a material is acetal, the note would read: "If temperature is not properly controlled, material may decompose, causing objectionable odors. Should not be permitted to be idle at melt temperature for more than 15 min."

MATERIAL PROCESSING DATA SHEETS

The materials selected for listing in the data sheets are representative. They contain the kind of information that is necessary for successful conversion of raw material properties into performance properties of a product.

The subdivisions in each material type are numerous, frequently being modified, and the processing data may or may not deviate from the "typical" resin. It is impractical to list all the special grades here. For instance, nylon 6/6 has 25 subdivisions not including grades with fillers. If a special grade is used, it is easy to obtain data sheets and processing information from the supplier of the material by requesting that a blank material processing data sheet with the heading given in the chart be filled out as well as to supply additional useful processing information.

The information in the data sheets is important, and each material grade used in a particular establishment should have its own material processing data sheet completely filled out. Such sheets should serve as a permanent record should a question of setup information arise.

"Material Processing Data" Sheets

Material	Mold Shrinkage*	Specific Gravity*	Melt Temperature (°F)*	Drying Temperature (°F) and Time	Mold Temperature (°F)*	Back Pressure on Material (psi)	Specific Heat (Btu/lb/°F)
ABS	.006	1.04 (27.5 in.³/lb)	475–525	170–190 (2–4 hr)	100–200	500–1000	.3
Acetal	.020 av.	1.41 (19.7 in.³/lb)	380–420	185 (4 hr/surf)	150–260	500	.35
Acetal, glass-coupled 25%	.002–.008	1.61 (17.4 in.³/lb)	400–440	185 (4 hr, moist surf.)	220–255	500	–
Acrylics	.002–.008	1.19	400–490	170–200 (2 hr)	180–220	400–800	.3
Cellulosic compound	.003–.010	1.15–1.22 (24–23 in.³/lb)	340–480	170–190 (1-1/2 hr)	120–160	500–2000	.3–.4
Fluorocarbons	.03	2.12–2.17 (13–12.8 in.³/lb)	620–750	–	200–320	500	.28
Noryl, modified PPO	.005–.007	1.06 (26 in.³/lb)	475–575	225–240 (2–4 hr)	180–230	500	.32
Noryl, glass-filled 20%	.001–.004	1.21 (22.9 in.³/lb)	480–600	225–240 (2–4 hr)	180–250	500	–
Nylon 6/6	.010–.022	1.14 (24.3 in³/lb)	520–700	140–160 (2–4 hr)	70–210	500	.4
Nylon 6/6, glass-filled	.002–.006	1.38	550–590	140–160 (2–4 hr)	210–250	500	.4
Polyaryl ether	.007	1.14 (25.2 in.³/lb)	540–590	–	160–200	1500–3000	.35
Polycarbonate	.005–.007	1.24 (25.2 in.³/lb)	520–620	250 (4 hr)	175–230	500	.3
Polycarbonate, glass filled 40%	.0005	1.51 (18.2 in.³/lb)	520–620	250 (4 hr)	200–250	500	–
Thermoplastic polyester, unfilled	.015–.018	1.31 (21.1 in.³/lb)	450–500	250 (2–4 hr)	100–140	250	–
Thermoplastic polyester, glass-filled 30%	.006–.010	1.53 (18.1 in.³/lb)	475–520	250 (2–4 hr)	150–230	250	–
Polyethylene, high-density	.012–.022	.925–.965 (30–28.6 in.³/lb)	360–550	150–160 if needed	50–200	500–1500	.55
Polyethylene, glass-filled 30%	.002–.006	1.16	450–480	–	80–200	500	–
Polypropylene	.010–.025	.905 (30.5 in.³/lb)	450–525	180–200 (1 hr if needed)	60–150	500–1000	.46
Polypropylene, glass-filled 30%	.004–.008	1.2 (24 in.³/lb)	450–550	180–200 (1 hr if needed)	60–200	500	–
Polystyrene	.002–.006	1.01–1.09	325–500	170–180 (2 hr if needed)	50–100	500–1500	.32
Polystyrene, glass-filled 20%	.001–.002	1.20–1.33 (21 in.³/lb)	350–520	170–180 (2 hr if needed)	50–150	500	.25
Polysulfone	.007	1.25 (22.3 in.³/lb)	575–750	250 (5 hr)	200–325	500	.31
Vinyl, rigid	.002–.005	1.35	340–420	–	150 max	min possible	–
Urethane	.010–.020	1.25	375–425	200–230 (3 hr)	100–150	500	.4

*Values will change with some grades.

(continued)

Vents Depth*	Gates (in.)*	Land (in.)*	Screw Speed (rpm)	Screw Torque	Runners (in.)*	Nozzle	Heat Deflection Temperature (°F)*	Remarks
.001 or water blast parting line	diam same as part thickness	.030–.040	75–100	Low	Main 1/4–3/8 Branch 3/16–5/16	ABS type	200–218	See note 1
.001–.003 (see Remarks)	.04 min diam, 40–60% of wall	.035–.050	40–50	High	1/8–3/8 diam	General purpose	230	See note 2
.001–.003	.05 min diam, 1/2 of wall	.035–.050	40–50	High	3/16–3/8 diam	General purpose	325	See note 2
.003	1/2 of wall, × 2 for width	.060	40–60	High	1/4–3/8 diam	General purpose	166–198	See note 3
.001	.030–.125, × 2 for width	.050–.070	50–300	Low	1/8–3/8 diam	General purpose	113–202	See note 4
.002–.004	2/3 part thickness, × 2 for width	1/8–1/16	50–80	High	1/4–3/8 diam	General purpose	–	See note 5
.0015–.003	1/2–full thickness, × 2 for width	.020–.040	25–75	High	3/16–3/8 diam	General purpose	212–265	See note 6
.002–.003	2/3 full thickness, × 2 for width	.030–.045	25–50	High	1/4–3/8 diam	General purpose	270	See note 7
Blast parting line line with water	1/2 full thickness, × 2 for width	.040–.060	30–50	High	1/8–3/8 diam	Nylon type	220	See note 8
.002	2/3 full thickness, × 2 for width	.035–.050	30–50	High	2/3 of thickness, × 2 for width	Nylon type with 3/16-in. throat	470	See note 9
.002	1/2–2/3 part thickness, × 2 for width	.020–.045	60–100	High	1/4–3/8 diam	General purpose	300	See note 10
.002–.003	1/2–2/3 part thickness, × 2 for width	.020–.050	30–50	High	1/4–3/8 diam	General purpose	265–285	See note 11
.002–.004	Full part thickness, × 2 for width	.030–.050	30–40	High	1/4–3/8 diam	General purpose	300	See note 12
.0005–.0075	1/2 part thickness, × 2 for width	.020–.040	20–60	Low	1/8–3/8 diam	General purpose	130	See note 13
.0005–.0075	1/2 part thickness, × 2 for width	.020–.040	20–60	Low	1/8–3/8 diam	General purpose	415	See note 13
.001	.020 min, 1/2 part thickness	.030–.040	40–120	Low	.060–3/8 diam	General purpose	100–130	See note 14
.001	2/3 part thickness, × 2 for width	.030–.040	40–60	Low	1/8–3/8 diam	General purpose	–	See note 15
.001	.025 diam min, 1/2 of thickness	.040 max	50–150	High	Min 1/4–3/8 diam	General purpose	125–140	See note 16
.002	.050 min. 1/2 of thickness	.040	50–75	High	1/4–3/8 diam	General purpose	230–300	See note 17
.001	1/3 part thickness, × 2 for width	.045	50–200	Low	1/8–3/8 diam	General purpose	200	See note 18
.002	1/2 part thickness, × 2 for width	.050	50	High	1/4–3/8 diam	General purpose	195–220	See note 19
.002–.004	1/2 or equal part thickness, × 2 for width	.040–.060	30–50	High	1/4–3/8 diam	General purpose	345	See note 20
Water blast parting line	2/3 part thickness, × 2 for width	.060	15–25	High	1/4–3/8 diam	General purpose	163	See note 21
.001 or less	.030–.080 diam	.030–.125	40–60	High	1/4–3/8 diam	General purpose	–	See note 22

NOTES FOR MATERIAL PROCESSING DATA

Note 1—ABS

• *Molding Problems*

Blush at gate: Increase injection pressure, decrease injection hold, check land, use nozzle with larger opening, decrease cylinder temperature, increase mold temperature, check vents, check material if dry, or decrease injection speed.

Jetting: Use nozzle with larger opening, decrease cylinder temperature, increase mold temperature, check material if dry, or decrease injection speed.

Laminations and color streaking: Check land, use nozzle with larger opening, decrease cylinder temperature, increase mold temperature, check material if dry, or check to see whether material is contaminated.

• *Material Characteristics and Uses*

Acrylonitrile butadiene styrene is a tough material with outstanding impact strength. Its good dimensional stability, temperature resistance above boiling water, chemical resistance, and electrical properties are also desirable.

Applications: Helmets, automotive and refrigerator parts, appliance parts, carrying handles, etc.

Note 2—Acetal

Shrinkages should be obtained from Nomograph's or actual sample information as supplied by material manufacturer. Prior to starting a machine for acetal molding, the cylinder should be at 400°F and purged with low melt index polyethylene or polystyrene. The nozzle should be heated also to 400°F to prevent accidental plugging. Precautions should be taken to prevent any chance of acetal decomposing and generating gas pressures that may blow out material through either end of the machine. Acetal should not lay idle in the cylinder for more than 15 min without making air shots. The cylinder temperature should be brought up gradually and the material moved through it. When starting and finishing a job with acetal and when stopping for less than 15 min, complete purging is necessary.

There should be no material left in the measuring chamber (the screw or plunger must be in full forward position). When the material starts decomposing, it becomes uncomfortable to the eyes and nose. (See supplier's literature for more details.) Purgings should be dropped into a water bucket.

• *Molding Problems*

Splay, brown streaks, unmelted particles: Temperatures of melt above limits tend to cause decomposition and with it generation of gases. Gas bubbles bursting at the mold surface cause splay or brown streaks. Low melt temperature or short exposure to heat causes unmelted particles.

Pits, orange peel, and wrinkles: These are caused by insufficient cavity pressure or low mold temperature.

Smear, blush, frost, and folds: These are caused by the formation of a thin skin as the material enters the cavity. Later on, the skin is distorted by incoming material causing these defects. Skin formation can be eliminated by raising the mold temperature and decreasing the rate of cooling.

Jetting: The first resin entering the cavity is in the form of a twisted worm. If the surrounding resin does not remelt it, the result is bad appearance. Directing the flow toward a wall or a nearby core will change the flow pattern, or flaring the gate and reducing the rate of injection or occasionally increasing the mold temperature should eliminate jetting.

Improper venting of cavities may cause a white deposit to form that has to be removed from the mold. Heat from a brazing torch or heavy-duty cleaning compounds will dissolve the substance.

Stress relieving may be done in an air-circulating oven at 320° ±5°F for 15 min for each 1/8-in. thickness.

• *Material Characteristics and Uses*

Acetal is one of the more rigid thermoplastics, has a low coefficient of friction, is resistant to most chemical solvents and is odorless, tasteless, and nontoxic. It has good toughness, fatigue life, and tensile strength. It is one of the better creep-resistant materials and has a relatively high specific gravity.

Applications: Small gears, bearings, bushings, carburetor parts, door handles, plumbing components, and parts for business machines and appliances.

Glass-coupled Acetal

See above for operating conditions and precautions, and for troubleshooting and other information pertinent to the resin.

• *Material Characteristics and Uses*

This material not only has the advantages of the unfilled grade but also possesses unusually high-notched impact strength for a glass-filled material. Its outstanding rigidity and increased heat-use range make it a most useful addition to the plastic materials.

Applications: It is used extensively for appliances and for automotive and instrumentation purposes.

Note 3—Acrylics

For long cycles and thick sections, use lower cylinder temperature and slow rate of injection. Molding stresses can be uncovered by brushing 5% toluene in

2-B denatured alcohol. Keep part wet by repeated brushing; if craze cracks appear in 30 sec, this indicates a stressed part. Annealing temperature *180°F for each 1/8 thickness 2 hrs.

• *Molding Problems*

Crazing (minute surface fractures) may use adjusting of injection hold time, injection pressure, mold temperature, mold closed time, or ejection slow down.

Breaking or cracking during ejection may call for adjusting injection hold time, injection pressure, mold closed time or ejection slow down.

• *Material Characteristics and Uses*

Acrylic is an outstanding material with excellent optical properties, rigidity, weather resistance, and colorability. It is odorless, tasteless, and a good insulator.

Applications: Lenses for automotive and industrial use, airplane canopies, dentures, surgical instruments, signs for outdoors and indoors, nameplates, skylights, etc.

Note 4—Cellulosic Compounds

Areas around machines should be ventilated because of the tendency to liberate gases.

• *Molding Problems*

Moisture will cause flow lines and streaks in gate area, a scaly surface, blisters in which bubbles have expanded after ejection, a depression, or a rough shallow crater where a bubble broke the surface.

• *Material Characteristics and Uses*

Cellulosics are resistant to household chemicals, oils, gasoline, and cleaning fluid. They are clear and scratch-resistant. The butyrate and propionate are good for outdoor use and very tough.

Applications: Acetate—Glass spectacle frames, vacuum cleaner parts, combs. Butyrate—Steering wheels, pipes, tool handles, telephone handsets, pens, etc.

Note 5—Fluorocarbons

Specific molding information on any one grade should be obtained from the supplier. The high cost of the material makes it imperative that all the details be carefully studied before undertaking a mold design and part manufacture.

• *Material Characteristics and Uses*

Fluorocarbons have high-temperature resistance, excellent chemical and electrical properties, no-stick properties, and zero water absorption.

Applications: Valve seats, high-voltage insulation, no-stick appliances.

Note: There is a very high material cost ($4 to $10 lb.) and a high specific gravity.

Note 6—Noryl, Modified PPO

Tab or fan gates are preferred; cushion 1/16 to 1/8 in. When molding is interrupted for a prolonged period (2 hr or so), drop cylinder temperature to 350°F. Material has excellent flow properties and outstanding repeatability of sizes. Machine capacity may be utilized within 40% to 80% of rated capacity. Nozzles should have a minimum opening of 3/16 in diam. Hot runner systems should be checked with supplier. General molding parameters for each grade of material should be carefully reviewed from suppliers literature.

• *Molding Problems*

Excessive shrinkage is caused by a gate too small, injection pressure too low, injection hold time not long enough, or overall cycle too short.

• *Material Characteristics and Uses*

Modified PPO has good physical properties, dimensional stability, electricals properties, and low water absorption.

Applications: Air-conditioning housings, computer and typewriter components, electrical components, and coil forms.

Note 7—Noryl, Glass-filled

Tab or fan gates are preferred. Cushion should be 1/8 to 3/16 in. When molding is interrupted for a prolonged period (2 hr or so), drop cylinder temperature to 350°F. Material has excellent flow properties and outstanding repeatability of sizes. Machine capacity may be utilized within 40% to 80% of rated capacity. Nozzles should have an opening of at least 1/4 in. diam. Hot runner systems should be checked with supplier. General molding parameters for each grade should be carefully reviewed from suppliers literature.

• *Molding Problems*

Excessive shrinkage is caused by a gate too small, injection pressure too low, injection hold time not long enough, or overall cycle too short.

• *Material Characteristics and Uses*

Good physical properties, dimensional stability, good electricals properties, and low water absorption.

Applications: Water-meter and other industrial components, and as automotive parts that require good wear properties, temperature resistance, and dimensional stability.

Note 8—Nylon 6/6

Mold temperature has a bearing on toughness and dimensions. Cylinder temperature profile should be (1) the selected melt temperature at the front zone, (2) 5% less at the rear zone, and (3) the average of the two in the middle zone. The nozzle temperature should be that of the front zone or adjusted to keep out drooling and prevent freezing. Fill rates should be fast, but within limits that would not cause brittleness or dimensional instability.

• *Molding Problems*

Drooling: Can be controlled by decreasing nozzle temperature, decreasing melt temperature, decreasing cycle, making sure resin is dry, using melt decompress, or applying correct nozzle.

Screw either does not retract or is erratic in its movement: Caused by low back pressure or low cylinder temperature.

Accurate mold shrinkage can be estimated from nomographs from suppliers literature.

• *Material Characteristics and Uses*

Nylon moldings are strong and long-wearing, resistant to extreme temperatures, and limited in chemical resistance.

Applications: As filaments, slide fasteners, gears, bushings, bearings, fishing lines, etc.

Note 9—Nylon 6/6, Glass-filled

Cycles may be faster than plain resin. Ejection from hotter molds is possible. Other pertinent notes are the same as Note 8.

• *Material Characteristics and Uses*

The use temperature is raised considerably over the unfilled resin and the applications broadened for such uses as portable tool housings and parts that require reasonable creep resistance.

Note 10—Polyaryl Ether

Nozzles should be as short as possible unless accurately controlled in temperature. The material is shear-rate-sensitive; therefore, work input from back pressure, screw rotation, gate size, frictional heat, etc., may be higher for better plastication.

• *Material Characteristics and Uses*

It stands out as a high heat deflection–resisting thermoplastic with very high impact strength and excellent chemical resistance. It is readily platable.

Applications: Automotive, plumbing, electrical, and appliances.

Note 11—Polycarbonate

Material must not be exposed. A dehumidifying hopper drier will dry virgin material in the hopper at 250°F for 2-1/2 to 3 hr. Profile of temperature should be for specified melt temperature in front, 10% lower in rear, and the average of the two in the middle. Nozzle shutoff valves, nozzles with reverse taper and nozzles with openings less than 3/16 in. diam. should not be used. Minimum pinpoint gate is .040 diam. For long cycles, the melt temperature should be closer to the low end of the range. Periodic checking of nonreturn valve is recommended.

• *Material Characteristics and Uses*

Polycarbonate is one of the highest impact materials, has good electrical properties, and weatherability, is transparent, and has good colorability.

Applications: Helmets, eye protection guards, portable drill housings, parts for automotive and aircraft use, appliances, gauges, street shields, and many industrial uses where impact resistance is of utmost importance.

Note 12—Polycarbonates, Glass-filled

Material must not be exposed. A dehumidifying hopper drier will dry the virgin material with a residence in the hopper at 250°F for 2-1/2 to 3 hr. Nozzle opening should not be less than 1/4 in. diam. Other details are the same as Note 11.

• *Material Characteristics and Uses*

A very strong material with a high degree of creep resistance.

Applications: Used in pressure vessels for water meters and other instruments where strength and creep resistance are important considerations.

Note 13—Thermoplastic Polyester

Clamping pressure is in the range of 3 to 5 tons/in.^{2} of projected molding area. Material should not be permitted to lay idle in the cylinder for more than 10 min. Air shots should be taken at this interval. If shutdown is needed for a longer period, cylinder should be fully purged and temperature dropped to 250°F. For accurate shrinkage determination, the supplier literature shows relation of shrinkage to part thickness, injection pressure, melt temperature, and mold temperature.

• *Material Characteristics and Uses*

Material possesses good lubricating properties, colorability, toughness, chemical resistance, and dimensional stability, and it lends itself to fast cycling.

Applications: Instrument components and miscellaneous industrial parts in which the above characteristics are a factor.

• *Material Characteristics and Uses (Glass-filled)*

The filled grade has surprisingly good antifraction properties as well as excellent impact strength. The potential applications for industry are very large to replace small metal parts for all sorts of mechanical and electrical devices.

Note 14—Polyethylene, High Density

Shrinkage is slightly different in direction of flow and perpendicular to it.

• *Molding Problems*

Screw does not return: Can be caused by cylinder temperature in rear too high, screw running too slow, or low back pressure.

Gate brittleness: Can be due to mold temperature being too high, inadequate cooling in the gate area, or nonuniform resin temperature.

Silver streaks: Caused by melt flow too fast or too slow.

• *Material Characteristics and Uses*

High-density polyethylene is strong, flexible, rigid, and resistant to heat and cold. It has good electricals properties, is odorless, tasteless, and shows good resistance to chemicals. Flexibility can be varied.

Applications: Ice cube trays, containers of every type, squeeze bottles, pipe and tubing, and insulation.

Note 15—Polyethylene, Glass-filled

Screw back pressure and rpm should be on the low side to prevent breakup of glass. Otherwise, same as for Note 14.

• *Material Characteristics and Uses*

Good rigidity and appearance. With mold temperature on the upper end, the surface will hide the glass appearance.

Applications: Gas-meter parts and automotive applications under the body where rigidity is required.

Note 16—Polypropylene

Parts made for maximum shrinkage and properly vented will usually be void-free. Stress relief at 275°F for 15 min. for each 1/8 thickness.

• *Molding Problems*

Post shrinkage: 90% of the total shrinkage occurs 6 hr after part leaves mold. The remaining shrinkage takes place in 10 days. Injection rate on the slow side will bring lower shrinkage and less tendency to warp.

• *Material Characteristics and Uses*

Polypropylene is one of the lightest thermoplastics, low in price, and has the best heat resistance of low-priced thermoplastics. It has good electrical properties and low water absorption.

Applications: Pipe and pipe fittings, heat sterilizable bottles, refrigerator parts, household articles, battery cases.

Note 17—Polypropylene, Glass-filled

Screw back pressure and rpm should be on the low side to prevent breakup of glass fibers. Otherwise same as Note 16.

• *Material Characteristics and Uses*

The glass reinforcement improves the rigidity and doubles the heat deflection temperature. These two properties should provide considerable increase in applications.

Note 18—Polystyrene

• *Molding Problems*

Cloudiness can be caused by contaminated material, melt too cold, excess lubricant on mold, or volatiles in material.

Surface dullness: Due to cold material (melt), cold mold, or foreign matter on mold such as water, oil, grease, dirt or excess lubricant.

Location of stresses: Can be established by the dip-test method, in which a part is immersed into kerosene for 2 min. Parts and kerosene are at room temperature. When parts are withdrawn, they must fully drain and be permitted to air dry. Within 10 min stressed areas will show up as cracked or crazed surfaces.

• *Material Characteristics and Uses*

Polystyrene is a very low priced material and is used in large volume in combination with other resins. It is also used for throw-away utensils and containers.

Note 19—Polystyrene, Glass-filled

Screw back pressure and rpm to be on the low side to keep the glass fibers from breaking up. Otherwise same as Note 18.

• *Material Characteristics and Uses*

The glass reinforcement decidedly improves many physical properties. Large parts have an appearance of quality that compares favorably with much more expensive materials.

Note 20—Polysulfone

Material shall not stay in cylinder over 1/2 hr at temperatures above mean of melt range to avoid degradation. If lubricant is required, use only zinc stearate, as others may cause stress cracking.

• *Material Characteristics and Uses*

Polysulfone is a high-strength, high-temperature-resistant, low-creep thermoplastic with low water absorption and good electrical properties.

Applications: Electronic and industrial control uses, water-meter components and, generally, parts with requirements for outstanding physical characteristics.

Note 21—Vinyl, Rigid

Injection screw machine requires a smear tip. The nozzle should be very short, with a minimum opening of 5/16 in. diam., and the land should be no more than 1/2 in. All passages to the cavity should be on the large side to prevent velocity burning. The cavity and core should be chrome-plated on .001-in. thickness. The speed of injection at the start should be low and gradually increased as needed, but preventing the overheating of the material. Back pressure and screw speed can be utilized to increase melt temperature if needed. During shutdown, the machine should be purged with polystyrene. For other details, obtain processing literature from supplier.

• *Material Characteristics and Uses*

The large applications are in pipe fittings, conduit fittings, and related parts.

Note 22—Urethane

Do not use lubricants because of their degrading effect. Mold shrinkages for the following thicknesses are: less than 1/8–.008 to .011; 1/8 to 1/4–.010 to .015; over 1/4–.015 to .020. Drying to be done with a dehumidifying hopper-dryer with a residence time of 2 hr prior to use. Post curing may be done at 230°F for 8 to 24 hr, in an air-circulating oven and controlled temperature. Flow pattern in the cavity must be such to avoid possible gas pockets. Cores should be anchored to prevent shifting. Cold-slug wells should be at the end of the runner. Plate or air ejection should be used wherever possible. Good temperature control of cavities and core is essential. Cavities and cores should be vapor-blasted for easy release. The blasting should include runners and sprue bushing. The nozzle land should have a reverse tapered opening. For other details, see suppliers processing literature.

• *Material Characteristics and Uses*

Urethane is tough and shock-resistant as well as resistant to chemicals and abrasion.

Applications: Abrasive wheels, automotive parts, industrial tire wheels.

19

Foamed Engineering Plastics Molding and Molds

Foamed engineering plastics are also known as *engineering structural foam,* and the term is usually applied to foaming of engineering thermoplastics. The foaming process produces a product that has an internal cellular structure as well as an external integral tough skin. The term *structural foam* is intended to convey the thought that the material has a high strength-to-weight ratio and that it lends itself to larger products where structural integrity and appearance of rigidity as well as that of strength are important considerations.

The specific gravity of the foamed material usually ranges from 70% to 80% of the solid material. This shows a considerable saving of raw material, at the same time providing other benefits that are of special interest to larger part sizes. For instance, sink marks are eliminated since the outer skin is held firm against the mold surface by the expanding gas until cooling is completed, at which time the skin is rigid and retains the surface that was formed during foaming.

Molded-in stresses are absent since the cavity is not filled by high injection pressures. The gas-resin mixture loads the cavity by expansion pressure, which is not high enough to cause stressing or warpage of parts.

For the same weight of material, foamed plastic will display a considerable increase in rigidity because it increases with the third power of the thickness. This is partially offset by the reduced density, since both rigidity and density are also direct functions of the flexural modulus.

Equipment that is specially designed for structural foamed material will be less expensive to purchase as well as to operate. The cycle time will be longer due to the cooling needs of a better heat insulator and thicker part. All other properties will be somewhat affected; these will have to be fully evaluated for each material and each product to determine their influence on the performance of the completed unit.

From a review of some specific parts, indications are: parts weighing up to 1.25 lb are preferably made by conventional injection molding; for parts from 1.25 to 12 lb, the advantages of conventional vs. foamed molding should be fully analyzed and compared; parts between 12 and 40 lb are most likely candidates for foamed materials.

One major disadvantage of the structural-foam molded parts is the cost of postmolding surface improvement by painting and the necessary preparatory steps associated with it.

Extensive research and development is being carried out to improve the surface finish in processing, to eliminate the cost of painting for many applications, to reduce the cost of operation through equipment design, to improve the control factors in upgrading the resins and blowing agents, and finally, if painting is essential to improve the painting system, to minimize the steps in the operation and keep the cost at a reasonable level. The potential applications are for business machine housings, computer component housings, automotive parts, material handling aids, etc.

With this formidable list of possible uses and the high activities involved in those fields, the impetus for improvement of every phase of the process is great and is being followed through with intensity and dedication by raw material suppliers, equipment manufacturers, and to some degree by prospective users of products.

DESCRIPTION OF PROCESS

Fundamentally, foaming is achieved by introducing an inert gas into the molten polymer or by compounding a chemical blowing agent with the polymer, which will release the inert gas at an appropriate temperature and cause it to spread itself throughout the molten polymer. As the resin and gas mixture is injected under pressure into the mold, the gas will expand, forcing some of the melt against the mold faces and thus forming an outer skin. The space between the skins is formed into a cellular mixture. The pressure exerted by the gas against the skin prevents sink formation and provides good contact of the plastic material with the mold for heat transfer during cooling.

Two basic processes are used in structural foam molding: the high-pressure and the low-pressure method. In the high-pressure process, material and blowing agent are injected into the mold at customary thermoplastic injection pressure where it is held for a few seconds; the mold then opens in the amount that the volume increase is desired. In the low-pressure system, the mold is partially filled, and thus pressure cannot be exerted in the cavity. With the partially filled cavities, the blowing agent forces the material against the walls where the skin is formed, and the remaining space is occupied with cellular material.

With the general concept in mind, we can proceed to a more detailed description of each system.

HIGH-PRESSURE SYSTEM

The pressures used in the high-pressure process are in the range of 2000 to 20,000 psi. Other processing considerations, such as clamp requirements, material flow, and general mold strength and configuration needs, follow the pattern of injection molding. Foaming is achieved by increasing mold volume. The problem of increasing the volume of the mold becomes involved and frequently expensive depending on the direction in which the enlargement takes place; this in turn depends on part configuration. If a part is to be foamed in the direction of thickness only, then the mold can be made to have the core entering the cavity by the amount of volume increase and opened for the additional space when actuated by press movement. When side walls are involved in addition to the depth of the part, it becomes necessary to provide a means of moving the walls of the cavity with the aid of cams, cylinders, or similar devices. The problem of proper support of the moving sides during injection deserves careful consideration (see Chapter 5).

The high-pressure system is used in conjunction with a single-shot method as developed by USM Corp. and a double-injection method known as the *sandwich process* and developed by ICI Americas, Inc.

In the single-shot method, the material is preblended with a blowing agent and fed into a standard screw injection machine where it is plasticized and brought to a temperature at which the blowing gases are released and foaming is completed. At this point, the material is injected into the mold at a high pressure. The mold is kept closed long enough to form a skin thickness by cooling (usually around 5 to 8 sec). The skin has a surface, which reproduces the mold finish. The formation of skin is followed by enlarging the volume of the cavity, allowing the molten material between the skins to foam and cause expansion into the new space. The thickness and rigidity of the skin will determine whether any of the gas bubbles will break through it and possibly mar the appearance generated

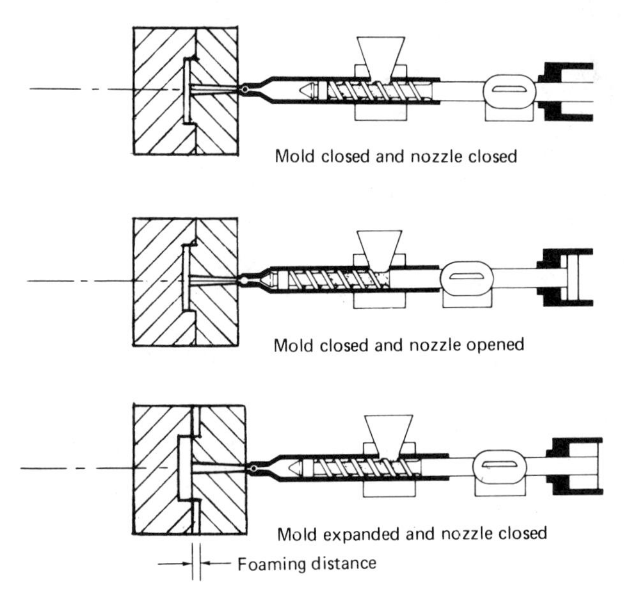

Fig. 19–1. High-pressure injection of foamed material.

during injection. (See Fig. 19–1.) When the right combination of parameters is used, such as length of time and mold temperature for skin formation, a smooth-appearing product can be produced with this system.

In the double-injection system, the first injection cylinder charged with solid plastic material (without a blowing agent) partially fills the mold and is immediately followed by the second injection unit charged with a foaming material through a fast-acting valve and completes the loading of the mold. The foaming mixture locates itself inside the solid material injected from the first unit and pushes the skin-forming material to the walls of the cavity. Since the last material to enter the cavity is the foaming mixture, the product may have a cellular-like surface exposed to the eye at the point of entrance to the cavity. If such an appearance is undesirable, a small amount of solid material can be injected for covering purposes. When the materials enter the cavity in a smooth nonturbulent flow, there will be no tendency of the solid skin material to mix with the foaming mixture. The foamed mixture will tend to push the solid material ahead until all mold surfaces are covered with it to form the skin. After a few seconds, for skin buildup, the cavity space can be increased, and the molten-foamed material between skins will expand to the controlled density that is established by mold opening (Fig. 19–2).

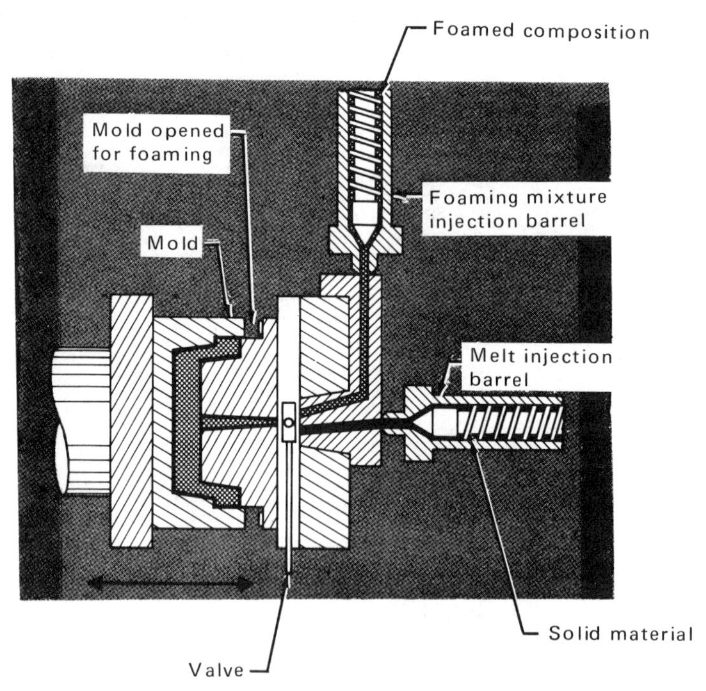

Fig. 19–2. Sandwich process.

LOW-PRESSURE SYSTEM

In a low-pressure system, a volume of material with blowing agent smaller than the space in the mold is injected into the mold. At this stage, the material–blowing agent mixture expands and fills the mold to its full volume. The material that is forced against the mold surfaces by the gas forms a skin against each mold half, and the space between the skins forms a cellular core structure. During injection, as stated before, the mold is partially filled; therefore, it cannot provide the resistance that would permit exertion of any high pressure in the cavities. The pressure that develops in the cavity as a result of injecting the material, compresses the air in the mold and increases the pressure of the gas from the blowing agent to about 500 psi. This pressure value can be used for determining clamp capacity of the press. At this low cavity pressure, the molded part is free of molding-in stresses, and the parts will not distort a feature that is very important for large parts with intricate configurations.

There are two subdivisions to the low-pressure system:

1. *Chemical blowing agents* are blended with the resin and fed into the screw machine, where the mixture is plasticized and heated to a temperature at which the blowing agent will decompose and provide the gas needed for foaming. The blowing agent of necessity has to be matched to the melt temperature of the resin so that decomposition and melt heat will be in reasonable balance.

The same foaming results can be attained by the use of concentrates mixed with plain resin and molded. The concentrates have a proportionately large amount of blowing agent in relation to volume of resin, but when properly applied they will work in the same manner as color concentrates do in regular injection molding. The advantage of chemical blowing agents as well as concentrates is that they can be used on standard injection machines and thus afford an easy introduction to the foaming field. To avoid potential problems, inquire of the material suppliers whether any side effects from a particular blowing agent might be experienced with mold, machine, or surrounding atmosphere.

2. *Nitrogen gas blowing agent,* when introduced into a molten polymer, requires specialized equipment. Such equipment was developed by Union Carbide Corp. and consists of a continuously running extruder, a gas inlet into the cylinder, one or more accumulators to hold the foam mixture, and a mold. All of these are connected by suitable pipes and one or more injection nozzles that feed the mold. The multiple nozzle arrangement is necessary due to the limited flow length of the polymer and blowing agent mixture, and facilitates the use of multicavity molds and the making of large objects. The extruder thoroughly mixes the gas and material, and feeds a prescribed volume of material and foam mixture into one or more accumulators, where it is kept under pressure to prevent premature expansion. When the proper volume of the mixture is reached,

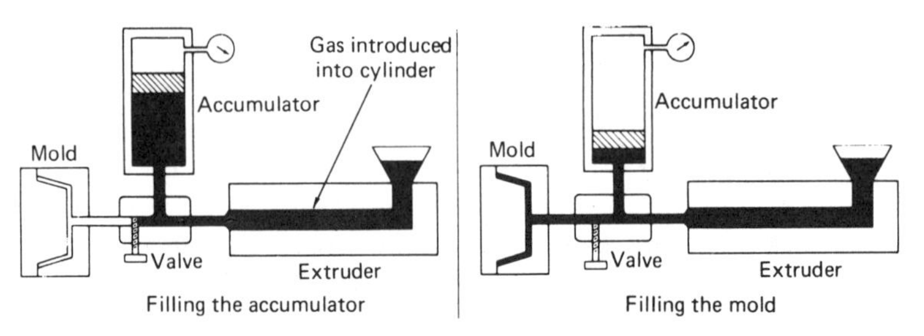

Fig. 19–3. Union Carbide system.

a valve opens, and a piston in the accumulator quickly forces the material into the mold. The stroke of the piston determines the volume of material delivered to the mold. The mold is only partially filled. At this point, the valve closes, and the expanding gas fills the mold and exerts pressure on the forming skin to prevent sink marks. With a high melt temperature of the polymer, rapid delivery of the material to the mold, and 25% of the circumference of the parting line devoted to equally spaced vents, a smooth surface finish can be attained (Fig. 19–3).

The system is relatively inexpensive as fár as materials and blowing agents are concerned. It is effective in performance because the screw breaks up the large gas bubbles and disperses them into the polymer. We now have a mixture of material with small and uniform gas bubbles well dispersed throughout. When this mixture is quickly introduced and partially fills the mold, the expanding bubble under the low cavity pressure will produce a part with uniform cell structure, lower density, and a favorable surface appearance.

There are other processes either in the development stage or in use for specialized applications. Most of them, including those described herein, involve patents, and the owners of such patents look for licensing arrangements. The patent question is another aspect of structural foam molding that requires attention and analysis before making a move toward the application of the system of structural foam molding.

PROCESSING

1. *Equipment.* The injection-molding machines of the screw plastication type can be used in conjunction with chemical blowing agents. If we recognize the limitations of the standard machine with respect to structural foam molding and compensate for them during product design and process prescription, foamed parts can be successfully produced.

The rate of injection on standard machines is about 2.5% of that of special-purpose machines (25 vs. 1000 in.^{3}/sec). Therefore, the material takes longer to arrive at the outpoints of the mold. This in turn calls for larger passages in the flow system and a possible increase in part thickness so as to overcome the problem of short flow properties of foamed materials. The shot size is limited to the metering section in the cylinder.

Nozzle drooling is a problem that has to be overcome since conventional drool control by means of melt decompress control cannot be applied to this material due to the possibility of premature foam.

Finally, the platen area is hardly ever large enough to accommodate a mold with a projected molding area that is within the clamp capacity to keep the mold from flashing during injection; here again, this is because the pressures developed in the cavity are roughly 2.5% of those in conventional molding.

The special-purpose machines are designed to overcome all these limitations. In addition, they may also be equipped with several nozzles that can extend the length of flow to very large objects. All the piping and connecting parts as well as manifold, valves, etc., must be designed to provide streamlined flow throughout the system in order to properly process the foamed products.

2. *Mold Design.* The strength requirements of mold components should follow the pattern outlined in Chapter 5, except of course the appropriate pressure values will have to be substituted. The low cavity pressure will obviously tolerate materials of much lower strength except that good polishability will be a requirement. For intermediate activities (10,000 parts per year), cast aluminum, cast beryllium copper, and Kirksite are used. Machined molds are usually made of Alcoa aluminum alloy No. 7025-T73 or AISI No. 4130, or prehardened P-20 and P-21 steels.

In connection with aluminum molds, the possibility of galvanic cell formation should be checked to avoid rapid deterioration of any aluminum mold part.

Let us now follow the path of material flow, and see how alterations in design details have to be made to fit into the structural foam processing needs.

Nozzle. In structural foam, the nozzle has to perform two additional duties, namely, permit fast flow and prevent drooling. On standard injection machines, the nozzle should have a 3/16-in. land or straight portion, and 1/4-in.-diam minimum opening. If a long nozzle is needed, it should have a 1-in.-wide tip heater to control drooling and another heater in back to control melt temperature. If available, a shutoff nozzle that can control drooling without interfering with injection speed is recommended. On the special machines with high-speed material flow, the nozzles should have openings to suit the high rate of flow. The nozzle tips in those cases are usually of Teflon or other heat-insulating materials that will aid in preventing freeze-off.

Sprue Bushing. The standard sprue bushing with 3/4 taper per foot and an "O" dimension 1/32 in. larger than the nozzle opening is satisfactory. The bushing should be short (not more than 2.5 in.) so that it does not adversely affect the flow and speed of injection. Its outlet opening and transition to the runner or part should be well rounded, with a radius equal to the opening diameter.

Runners. A full round or trapezoidal runner cross section is preferred. They should range in diameter from 3/8 to 3/4 in. depending on length of flow, speed of injection, and volume of part. (See Chapter 7.)

Gates. The recommended gate, whenever applicable, is the direct sprue gate; when this is not feasible, the edge gate is preferred. Tunnel gates or other small gates should be limited to small parts with short flow lengths (less than 3 in.). The gate should at no time be so small that it will interfere with a high rate of injection. When a runner has a 0.75 in. diam, the starting point for the gate size could be 0.25-in. diam and the land 0.25-in. long. Of this 0.25-in. land length, 3/16 in. should be used for a radiused transition from 0.75 to 0.25 in. diam. If the restricting action to the rate of flow is noticeable, increases in gate size could be made in steps of 1/16 in. The application of the flow formula ("Melt Rheology," Chapter 7, p. 000) could aid in establishing a reasonable gate size.

Ideally, the location of gate should be such that the length of flow to the extreme points will be about equidistant. The area of degating should be accessible for removal of the gate and finishing if necessary. The flow pattern should be such that entrapment of air is eliminated (see Chapter 7). The location of the gate should not cause weld lines to be formed in areas that are subjected to loading or any stress in use.

Vents. Disposing of air and gas during molding of structural foam takes on added significance. Not only are we dealing with more air because of greater thicknesses, but we also have to contend with the large volume of gas to be expelled, which is inherent in the operation. For these reasons, venting must be more extensive than it is with conventional molding. Vents should be placed at 2-in. centers of part circumference and should be 1/2 in. wide by 0.003 in. deep for a length of land of 0.25 in. and cleared beyond that point to 0.010 in. deep. At the extreme corners of the part (furthest from the gate), the depth of the vent should be 0.010 in. deep. A certain amount of trying for ultimate depth of vents is necessary, and this has to be done in relation to speed of injection with the aim of attaining an acceptable surface with a uniform density.

Mold Temperature. Controlled and uniform mold temperature can contribute to good surface appearance, a low level of molded-in stresses, proper cell structure, uniform skin, and good flow. This is an important list of properties, and the design of the cooling system deserves close attention. For example, the material used for the mold will determine the distance of passages from the molding face, and its heat conductivity will dictate the spacing of lines and possible use of heat-diffusion plates such as copper to avoid cold spots. If steel is used, the distance of passage opening to the face of the mold can be 3/8 in. instead of the

9/16 in. for conventional materials molding. The passages should be of a size that will create turbulent flow. Dimensions of hose connections for the cooling media should be such as to cause little pressure drop. Other details pointed out in Chapter 12 should be analyzed as to the application for the structural foam. Where large molds are involved, special attention must be given to the problems of expansion, contraction, and insulation.

PROCESSING PARAMETERS

Molding parameter settings for melt temperature, mold temperature, and drying (if necessary) should follow the recommendations for the base resin from which the foamed product is made. The only exception is that the rear zone temperature should be about 10% lower than the middle and front zones in order to prevent premature decomposition of the blowing agent. Once the parameters are set, they should be closely monitored and controlled so that the following characteristics are maintained: surface quality, cell structure, skin thickness, good properties, and low stress levels.

The time of "cure" depends on part thickness, efficiency of heat transfer, complexity of part design, and the mold material. It is usually established by determining the point during mold closed time when satisfactory parts are produced.

Rate of injection is important because flow properties of the structural foam are adversely affected after a certain injection time. The rate of injection should be such that the extreme length of flow is reached within a time most favorable to a particular polymer. Thus, for example, 10 in.^{3}/sec may be satisfactory for a 12-in. flow length, whereas a 30-in. flow may require 500 in.^{3}/sec etc. The optimum flow time for each polymer must be established, and the rate of injection must be adjusted to suit.

Whenever required, mold release agents should be selected for compatability with the material used and effectiveness against a particular mold material. Another important characteristic of the mold release agent is to assure that it does not clog vents. Open vents are very important for making a consistent product with consistent properties.

At the end of a run, the cylinder and flow system should be thoroughly purged with a recommended purging material. In most cases, general-purpose polystyrene should do a satisfactory job.

What are the criteria for a good structural foam product? In general, they are

1. Surface finish
2. Skin thickness
3. Cell structure
4. Part weight and density
5. Discoloration

6. Fully filled-out part
7. Voids and sink marks absent
8. No warpage
9. Dimensions
10. Paint adhesion

These and other requirements are the basis for a troubleshooting guide, in which the cause of a trouble will indicate the corrective measures to be taken, and in most cases the trouble will point to a closer control of a specific parameter setting.

TROUBLESHOOTING GUIDE

A troubleshooting guide for foamed engineering plastics with corresponding defects and their causes are enumerated herein.

- *Defect:* Brittle parts

Possible causes: High stock temperature
wet material
contaminated blowing agent
contaminated regrind

- *Defect:* Discoloration, part

Possible causes: High stock temperature or long residence time
nonuniform material heating
nozzle too hot
trapped material in areas of restriction

- *Defect:* Extruder amperage high

Possible cause: stock temperature too low

- *Defect:* Flash

Possible causes: Low clamp pressure
shot size too large
parting line in poor condition
low stock temperature
nozzle not fully open
mold not adequately vented
cylinder heaters need checking
flow length too long

- *Defect:* Slow injection speed

Possible causes: Low stock temperature
nozzle not fully open
low injection gas pressure
mold and melt temperature not in proper relation

• *Defect:* Nozzle freeze-off
Possible causes: Low nozzle temperature
too much heat taken from nozzle to sprue bushing
improper nozzle shutoff

• *Defect:* Post blow
Possible causes: Cooling cycle too short
overpacked shot
high gas pressure or too much blowing agent
manifold temperature not uniform
high mold temperature

• *Defect:* Surface, poor appearance
Possible causes: High gas pressure
inadequate vents
low stock temperature
slow rate of injection
excessive flow length

• *Defect:* Large, glossy voids
Possible causes: Inadequate vents
high stock temperature
low gas pressure
excessive flow length
small shot size
inconsistent stock temperature

• *Defect:* Heavy part
Possible causes: Amount of blowing agent is small due to
1. more solid material in the shot
2. slower fill time
3. low stock temperature

• *Defect:* Weld line, poor strength
Possible causes: Low stock temperature
inadequate venting
excessive flow length
slow rate of injection

Most defects are due either to stock temperature, mold temperature, or venting. A portable pyrometer with a thermocouple for reading or checking melt temperature and another for determining mold surface temperature will prove a valuable aid in diagnosing molding problems. Depth micrometers for checking vent openings are a necessary tool for those responsible for a smooth functioning operation.

20 Reaction Injection Molding and Molds

In the processes of injection molding of thermoplastic, injection-molding thermosets, structural foam molding, and expandable polystyrene molding, we are dealing with materials that are chemically completed compounds, ready for conversion into a finished product. The materials are received from suppliers with certain properties based on test-bar information and recorded in material processing data sheets. The processors are expected to convert these materials into products with similar mechanical, electrical, and environmental characteristics, as indicated on the data sheets. The processors are also furnished with a range of molding parameters that should be optimized to attain the desired product properties. In brief, they are given a material along with guidelines for its conversion, but can do little to change the processing behavior of the material since they are dealing with a finished raw material that is fully prepared for conversion into a finished product by application of time, temperature, and pressure.

In reaction injection molding (RIM), the starting point for the conversion process is liquid chemical components. These components are metered out in proper ratio, mixed, and injected into a mold where the finished product is formed. In reality, it is a chemical and molding operation combined into one system of molding in which the raw material is not a prepared compound but rather chemical ingredients that will form a compound when molded into a finished part. The chemicals are highly catalyzed to induce extremely fast reaction rates. The materials that lend themselves to the process are urethane, epoxy, polyester, and others that can be formulated to meet the process requirement.

The system is composed of the following elements:

1. formulating the chemical composition that will produce a material of desired physical and environmental properties. Normally, this consists of two liquid chemical components that have suitable additives and are supplied to the processor by chemical companies.
2. chemical processing setup that stores, meters, and mixes the components ready for introduction into the mold.
3. to facilitate smooth continuous operation, a molding arrangement consisting of a mold, mold-release application system, and stripping accessories.

The success of the overall operation will depend on the processor's knowledge of

1. the chemistry of the two components and how to keep them in good working order.
2. how to keep the chemical adjunct in proper functioning condition so that the mixture entering the mold will produce the expected result.
3. mold design as well as the application of auxiliary facilities that will bring about ease of product removal and mold functioning within a reasonable cycle, e.g., 2 min.

The potential applications for the products made by the RIM process are very large in volume and dimension. These applications include automobiles, transportation in general, furniture, building components, housings for electronic apparatus, and recreation. The potential applications have created a tremendous interest on the part of machinery and material suppliers, which in turn have instituted intensive research-and-development programs for the purpose of producing a quality product while minimizing the amount of labor required and keeping material waste at a low percentage of part weight. Several systems are offered, each varying from the other in such areas as materials feeding and mixing the chemicals—but all trying to end up with the same result.

Two actively used systems will be described that can be used as a basis for evaluation of others. Most systems are patented subject to a licensing arrangement. The materials that are predominant in the use of the RIM process are polyurethane compounds that consist of a polyol with suitable additives and polyisocyanate. When molded, the material has a solid skin on the surfaces that contact the mold; in between the skins, there is a microcellular structure. The skin faithfully reproduces the mold surface, without sink marks, and lends itself readily to finishing operations. The raw material ingredients are considerably less expensive than a conventional molding compound with comparable properties. The formulations are available in flexible, semiflexible, and relatively rigid compositions.

Modern controlling means are applied to every stage of operation to insure consistency of the materials and least variation in the processing elements so that the end result is an acceptable product with few rejects and little waste of raw material.

THE PROCESS

Figure 20–1 shows the schematic outline of the Mobay Chemical Corp.'s RIM process, and Fig. 20–2 is a diagram of the Cincinnati-Milacron process components. Essentially, the process consists of these elements: storing and conditioning of the component chemicals, controlled delivery of volume from each chemical, maintaining the proper ratio of components, complete mixing of the chemical components, delivering the mixture at the proper rate to the mold,

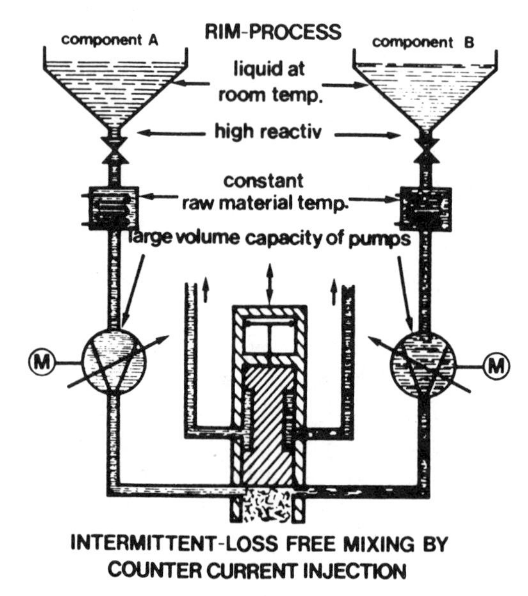

Fig. 20–1. Schematic of the RIM process.

reaction curing or polymerizing in the mold, and removing the finished product from the mold.

Let us now follow the flow of material through every step of the process to determine what some of the requirements are.

Monomer or Component Tanks

The tanks are usually pressure-rated, and the liquid is normally subjected to some pressure so that the rate of feeding the inlet of pumps or the rate of filling the chemical chambers will be repetitive and conducive to consistent metering. Viscosity and temperature controls are incorporated since they are important parameters for carrying out the overall functions of the process. A mixing mechanism is provided as a means of maintaining the homogeneous state of the chemical. Fluid-level sensors insure that the liquid level does not fall below limits and at an appropriate position calls for replenishment action.

Heat Exchanger

The feeding of the liquid chemicals with pumps or pressurized tanks is continuous, and the exact needs per shot are determined by the mold. The excess is recirculated into the tank. The bypassed liquid picks up heat, and thus it has to

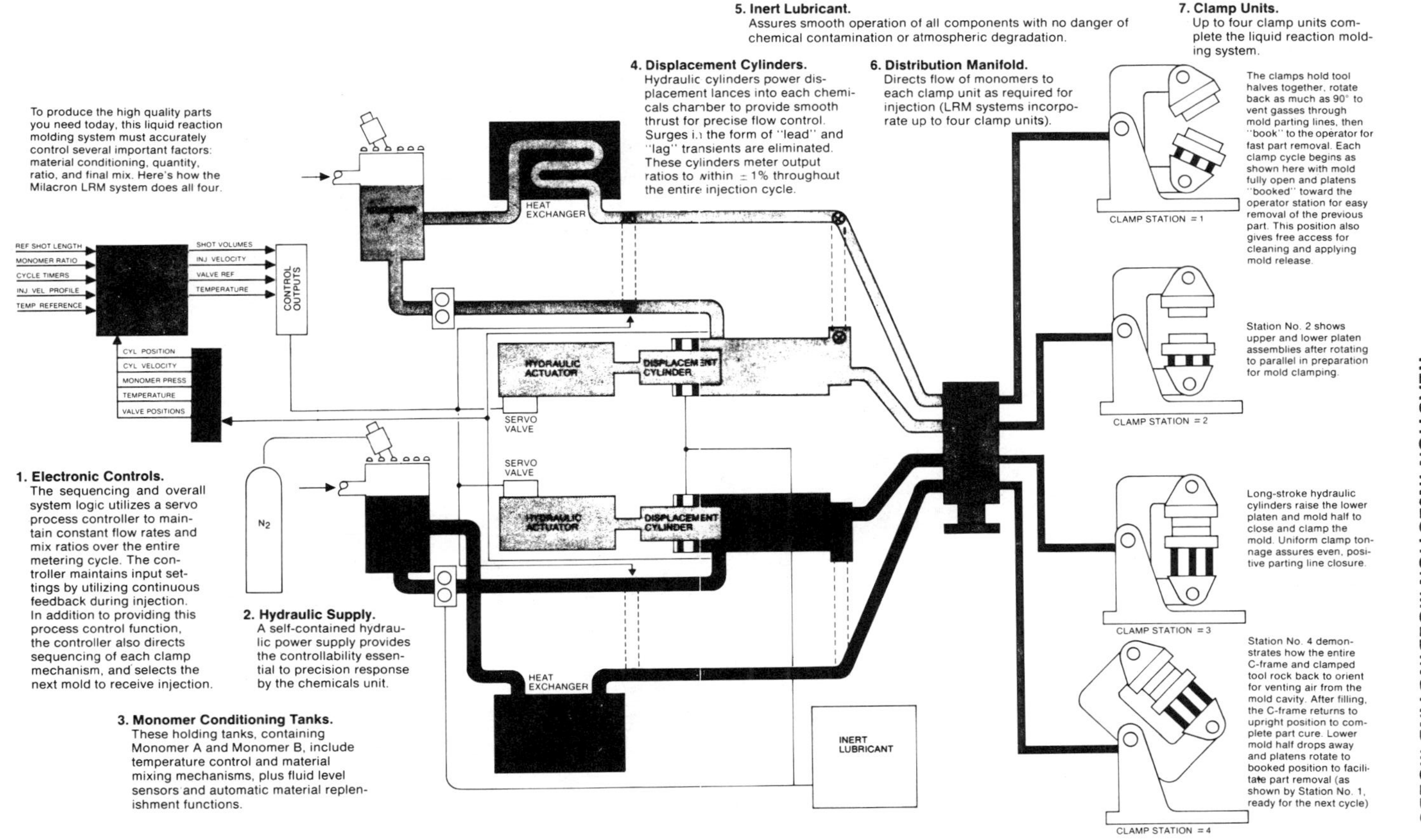

Fig. 20–2. Control of material conditioning, quantity, ratio, and final mix *(Courtesy of Cincinnati-Milacron, Plastics Machinery Division system).*

pass through a heat exchanger so that a constant temperature of the chemicals is maintained prior to mixing. In the interest of having the components homogeneous and the system in equilibrium, temperature of the materials entering the mixing head must be closely controlled. The water from the heat exchanger is recirculated through a temperature-controlled water unit.

Metering System

Normally, the metering system is large enough to cover the needs of four presses in which the molds are mounted. High pressures of 1500 to 3000 psi are employed in the metering system in order to utilize the highly reactive chemicals and bring about the shortest possible cycle time. For accurate metering, viscosity, temperature, and specific gravity of the component chemicals must be consistent within narrow limits of ±1%.

The metering units in the Mobay system consist of axial-positive-displacement pumps with adjustable cylinder stroke to deliver varying volume per revolution. They are highly precise devices made to close tolerances to enable them to deliver uniform and repetitive volumes with each stroke of the cylinders over prolonged periods of time. They are made in a variety of sizes and are direct-motor driven.

The other system uses hydraulic cylinders to displace a desired volume of material from the filling chamber. The movement of the cylinder is controlled by servo-valves with the aim of attaining uniform speed of piston movement and consistent distance of travel. This system has a self-contained hydraulic power unit equipped for precision response to the needs of the hydraulic cylinder metering method.

In both systems, a prerequisite to the success of the operation is that there be no "lead" or "lag" of one component against the other throughout the cycle.

Mixing Head

The mixing head is attached to the mold so that all of the polymer up to flush with the mixhead can be removed with part ejection. The features incorporated in the mixhead are as follows:

1. Accurate temperature control is achieved by recycling both chemical components through the head and their respective passages.
2. The mix chamber and mix ports are wiped clean each time the material is delivered to the mold.
3. The simultaneous and rapid shift of the mixing piston eliminates lead or lag problems and accomplishes the change from recycle to pour position.
4. The high pressures (1500 to 3000 psi) with which the chemicals enter the

mixhead and which force them through small variable orifices creat a condition that is conducive to favorable mixing.

The Cincinnati-Milacron mixing head shown in Fig. 20–3 has two lines for each monomer, which recirculate the chemicals from each tank to the head and back to the tank. This makes conditioned chemicals available for immediate needs. Displacement cylinders deliver the fluid through mix ports in the head. After achieving the shot volume, the mixhead ports close, the mix chamber is wiped clean by the piston, and the system is ready for the next cycle. The mixhead utilizes four mix ports located 90° apart around the mix chamber. Each chemical is fed through two sets of ports opposing each other and impinging on each other for effective mixing. The orifice openings of the mix-port adjust to maintain optimum mixing pressure. Each mixing head contains an electronic insert for the chemicals unit and an electronics unit for control at the clamp.

The signals for the mixhead controls and adjustments are set at the chemical unit, and the sequencing controls originate with the clamp.

In Fig. 20-2, one line is shown connecting the distribution manifold with each clamp-mixhead station. This single line represents two supply lines and two recirculating lines for each clamp station.

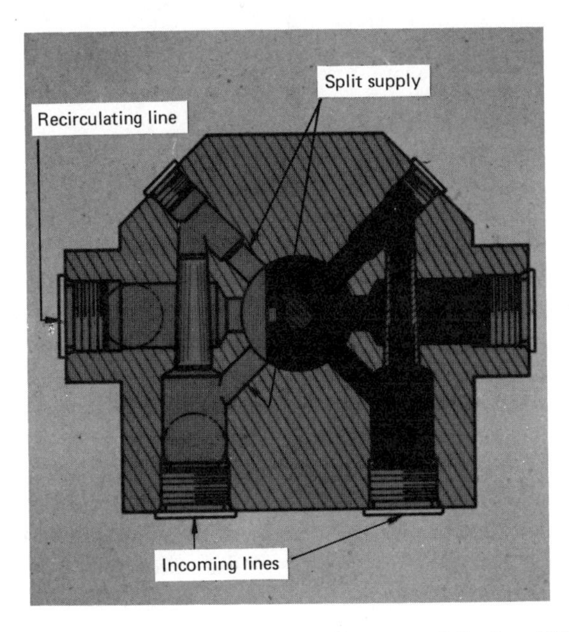

Fig. 20–3a. Mixhead view showing connections. *(Courtesy of Cincinnati Milacron, Plastics Machinery Division)*

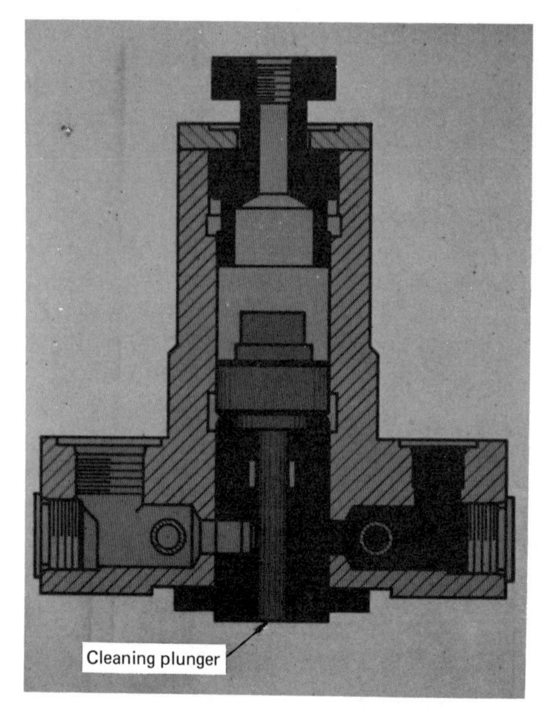

Fig. 20–3b. Mixhead: View showing cleaning plunger. *(Courtesy of Cincinnati Milacron, Plastics Machinery Division)*

THE MOLD

Schematically, the process components appear to be not too complicated. In reality, however, they represent complex units with controls and electronic signaling devices that perform intricate functions to enable a chemical operation to accurately feed and control a monomer for polymerization in the mold. Proper mold construction should make it possible to meet the following requirements:

1. Filling of the mixed monomers must be rapid—that is, a little over 1 sec even for such large parts as auto bumpers.
2. The rate of filling should create a laminar flow (nonturbulent).
3. The mixing head should be attached in an area that will be the lowest point from a filling consideration so that the air from the mold is driven toward the last portion to be filled. At this point, vents are provided to permit free exit of air.
4. The mold is filled to 90% of its volume.
5. The parting line must be placed so that all the molded configuration is pro-

duced at and below the parting line. Any molding taking place above the parting line will tend to entrap air and produce defective parts.

6. The pressures in the mold are 50 to 100 psi max.

7. The temperature of reaction is about 230° to 250°F and has to be brought down to 110° to 150°F in a uniform manner over the entire molding surface. The heat of reaction is about 200 Btu/lb.

8. Molding surfaces must be in condition for easy release of the moldings.

Mold Materials

The molded polyurethane faithfully reproduces the surfaces of the mold and tends to stick to them. The application of mold-release agents becomes necessary with each cycle at the present state of the RIM technology. After polymerization, if the mold is not covered with mold release, the part will adhere to the mold, making it difficult to remove from the mold. In addition, a film will remain on the mold surface, which will impair the appearance of the product. In view of these occurrences, the mold material should be highly polishable and platable with nickel since this coating has proven to be most effective.

The strength requirements for the mold material are very low, about 50 psi, but the ability to provide a surface with a smooth and high polish is very important.

For production runs of 50,000 parts per year, a P-20, P-21, or H-13 steel would be most appropriate not only because of their homogeneous nature, but also because of their excellent polishability and adaptability for a good plating job. The prehardened grades of 30 to 44 RC are preferable because of the degree of permanency that they impart into a tool. After machining, a stress-relieving operation is very important in order to avoid possible distortions or even cracking.

Nickel shells that are electroformed or vaporformed when suitably backed up and mounted in a frame are also excellent materials for large volume runs. For activities of less than 50,000 parts per year, aluminum forgings of Alcoa grade No. 7075-T73 machined to the needed configuration will perform satisfactorily. They have the advantage of good heat conductivity, an important feature in RIM.

Cast materials are used for RIM molds with reasonable success. One such material is Kirksite, a zinc alloy casting material described in detail in Chapter 11. Kirksite molds are easily castable, are free from porosity, will polish and plate well, and have been used with favorable results. This zinc alloy, when treated by a hardening process known as "IOSSO," is claimed to produce a skin of up to 0.020 in. deep in a hardness ranging up to 70 RC. It is also claimed that the coefficient of friction of the surface is decidedly improved, which could enhance the ability to release the moldings.

Cast aluminum molds have the potential of porous surfaces, and so are a secondary choice as a mold material. Release problems from a porous surface become severe, and, unless a foundry can produce castings with dense surfaces, the cast aluminum can become expensive in terms of poor productivity and maintenance.

Mold Features

The mold design should accommodate all of the features listed in the following section and should incorporate a shrinkage of 0.014 in./in.

Parting Line. The part geometry will to a large degree determine the location of the parting line. If it is possible to have more than one position for the mold parting line, then the choice should be made so that the part will be molded at or below the parting line. The most favorable filling positions are at the parting line and at the low point of the part. Since the expanding reaction mixture displaces the air from the cavity, it is imperative that the displaced air have an escape outlet at the highest point of the mold, which means that the parting lines pass through the same point and venting outlets be provided there. The parting-line position determines the location of the mixhead, the sprue, runner, gate, venting, and removal of the part with all the auxiliary adjuncts that polymerize and are attached to the molding. Figure 20–4 shows a part design with a favorable parting line. In Fig. 20–5, the top of the mold has to be made with two plates, with venting provided between them in order to end up with a filled-out part.

When bosses, ribs, or other internal projections are present in a part, venting must be provided or porous and unfilled sections will result due to the entrapped air in the blind spaces (Fig. 20–6). The width of the face at the parting line where mold sealing takes place should be:

For steel, 0.25 to 0.5 in.
For aluminum, 0.5 to 0.75 in.
For Kirksite, 0.5 to 0.75 in.
For nickel (shells), 0.5 in.

Vents. The number and location of vents are dictated by the level position of the mixture during filling and the rate at which the material enters the cavity. Vents should be deep enough (0.008 to 0.016 in.) to permit easy exit of air and their width, anywhere from 0.25 in. and up depending on part thickness. After the sealing face of 3/8 in. length has been cut through, the remainder of the parting line should be cleared away by a 1/32-in.-deep groove. The best location for vents would be opposite the gate and parallel to it. Some venting problems were described above, under "Parting Line."

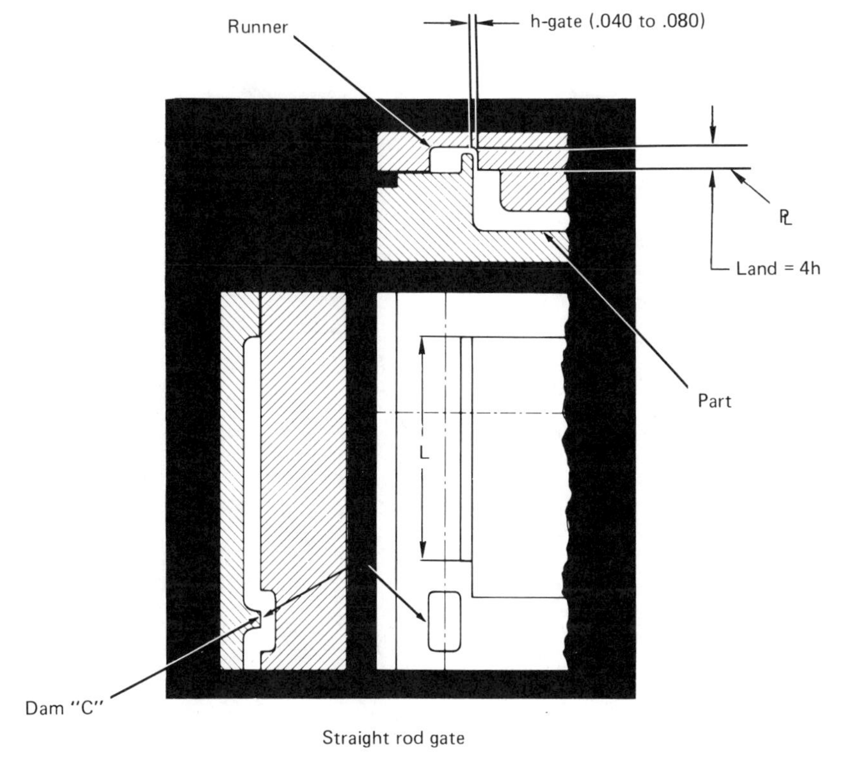

Fig. 20–4. Favorable parting line. *(Courtesy of Mobay Chemical Corp.)*

Sprue and Gate. The location of the sprue should be such that the reaction mixture enters the cavity at the lowest point as viewed from the cross section of the part. There should be no projections or changes in the section in the front of the sprue that would restrain the free flow of the reaction mixture. Thin and complex sections should be under the level of the liquid; therefore, these factors should also be a consideration for sprue location. The physical placement of the sprue can be either in the mold itself or a separate block attached to the mold. If adjustments in the sprue are anticipated, a separate block would be preferred for larger molds.

In order to attain a laminar flow that will not entrap air in the material as it fills the cavity, the velocity through the sprue and gate has to be within certain limits. For flexible urethane parts with a thickness of 0.120 to 0.160 in., the velocity should not exceed 15 ft/sec. For thicknesses of 0.240 to 0.400 in., the velocity should be within 6.5 ft/sec. For rigid urethane, the velocity should not be greater than 5 ft/sec. These velocities are for currently popular compositions.

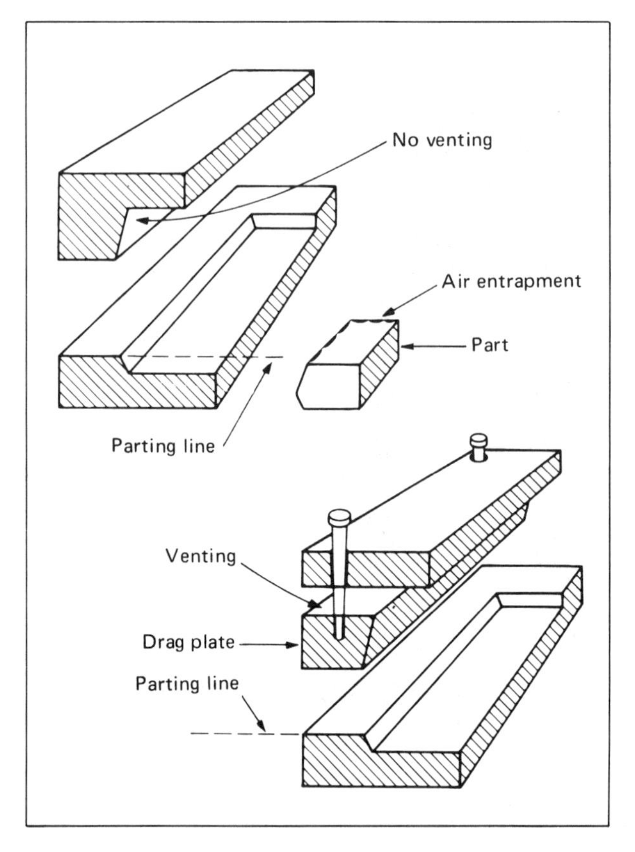

Fig. 20–5. Venting for unfavorable part geometry. *(Courtesy of Mobay Chemical Corp.)*

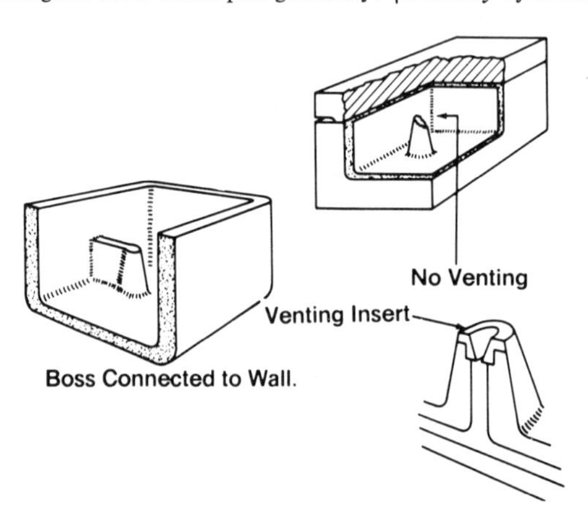

Fig. 20–6. Venting bosses, etc. *(Courtesy of Mobay Chemical Corp.)*

As the viscosity of the reacting mixture changes, so will the velocities have to be adjusted accordingly. The gate depth is from 0.040 to 0.080 in. depending on part thickness and volume of material being fed to the cavity. The length of the gate may run up to 24 in. and is a function of velocity and volume of material as well as of part configuration. The land of the gate should be at least four times its depth.

The length of the gate can be calculated as follows:

$$\text{Area} \times \text{Velocity} \times \text{Density} = \text{Weight/sec}$$

Substituting the appropriate letters according to Fig. 21–4, we have $h \times L \times V \times d$ = Weight/sec. As an example, we use $Q = 0.75$ lb/sec, Density = 70 lb/ft^3, $V = 5$ ft/sec, and $h = 0.060$.

$$\frac{0.060 \times L}{144} \times 5 \times 70 = 0.75$$

$$L = \frac{0.75 \times 144}{0.060 \times 5 \times 70} = 5.1$$

The weight per second is determined by the metering capacity of the machine. Figure 20–7 shows a fan gate and related dimensions of the cross sections.

The fan gate is limited to small outputs (0.75 lb/sec), since at larger outputs there would be too much material wasted. Under certain conditions in the fan gate shown in the top of Fig. 20–7, the liquid will lose contact with the wall at the transition from sprue to fan. The lower part of the figure shows an arrangement that eliminates the problem and in fact permits the increase of the fan angle to 150°. At every section, the proportions of areas in the fan portion must be equal.

Figure 20–8 shows a runner and gate system in which it is pointed out that the area of the runner is equal to the cross section of the gate. Figure 20–9a shows the importance of directing the reaction mixture parallel to the lower cavity wall, by incorporating a land that equals at least four thicknesses of the gate. Figure 20–9b shows an incorrect method that is being used. In order to prevent splashing of the reaction mixture as it first enters the runner, a dam "C" in Figure 20–4 is inserted to change the direction of flow.

Aftermixer. The mixing of the chemicals by high-pressure impingement takes place at the mixhead. To insure that the mixing is at its best condition, an aftermixer is cut into the runner system that divides the flow and reimpinges the mixture. Figure 20–8a shows one such system, and Fig. 20–8b is a view of the molding from aftermixer with runner gate and part.

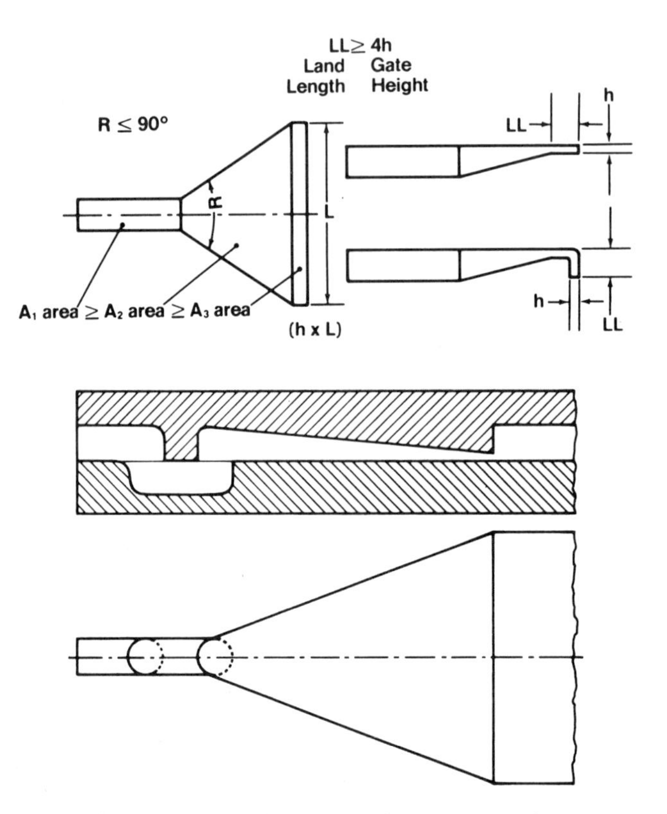

Fig. 20–7. Fan gates. *(Courtesy of Mobay Chemical Corp.)*

Mold Temperature Control. For consistent quality and molding cycles, the mold temperature should be maintained within ±4°F. The mold temperatures will range from 101° to 150°F depending on the composition being used.

The cooling lines should be so placed with respect to the cavity that there is a 3/8-in. wall from edge of hole to the cavity face. The spacing between passages should be 2.5 to 3 diameters of the cooling-passage opening. These dimensions apply to steel; for materials with better heat conductivity, the spacing can be increased by one hole size.

A review of Chapter 12 will aid in recognizing other details of mold-temperature control. Figure 20–10 shows a mold design for shoe soles and indicates the practical carrying-out of some or most of the features described.

Molding Press. The design of the mold will, to a large degree, be influenced by the type of press in which the mold will be used. All the detail requirements of

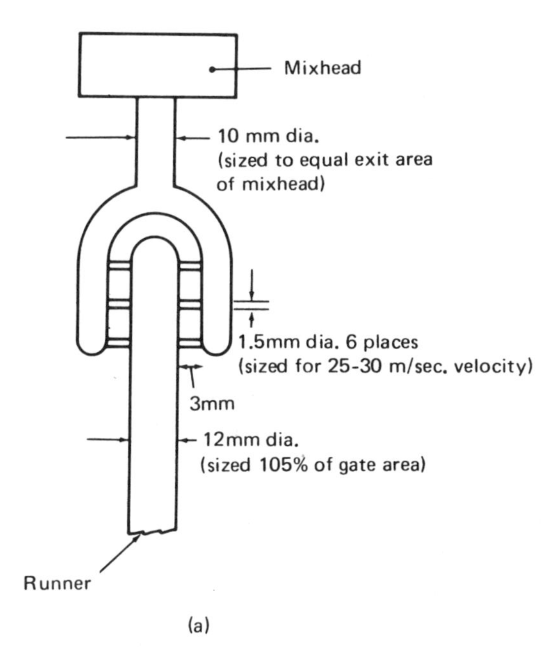

(a)

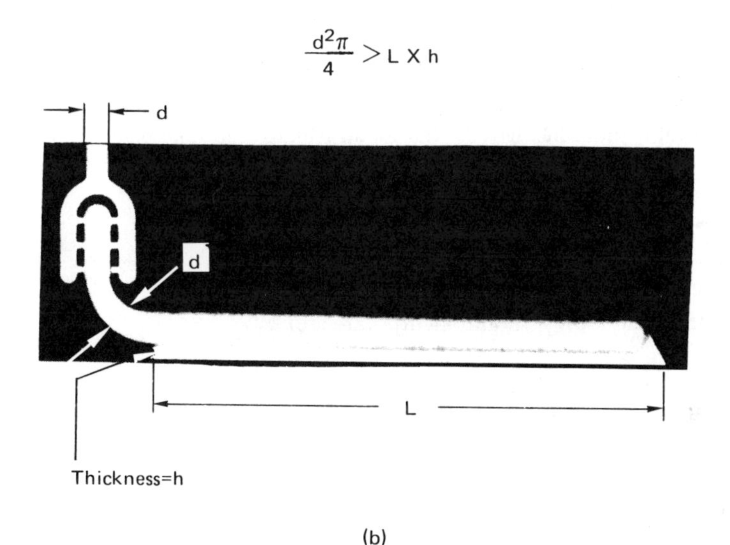

(b)

Fig. 20–8. *a,* Typical aftermixer. *b,* Typical runner-film gate. *(Courtesy of Mobay Chemical Corp.)*

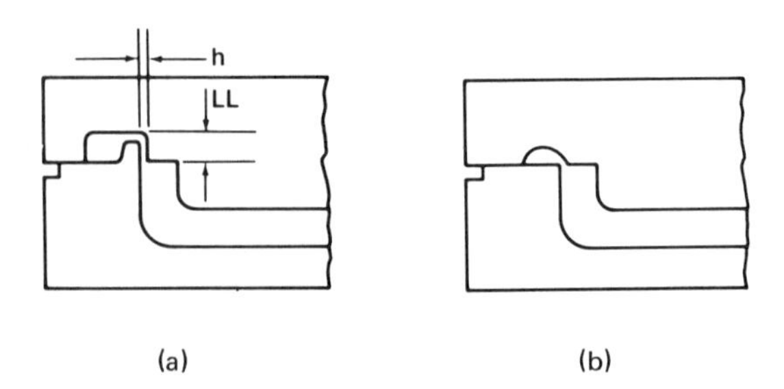

Fig. 21–9. Gate and land transition from runner to part. *a,* Correct method. *b,* Incorrect method. *(Courtesy of Mobay Chemical Corp.)*

the press and the manner in which it operates should be obtained and thoroughly reviewed; only then should mold design be undertaken.

Part Removal. Stripping of the part from the mold is one of the major time-consuming elements along with the application of mold release. These two elements are responsible for about 2/3 of the overall cycle time. The problem of part stripping is being approached with compressed-air ejecting disks actuated by hydraulic or air cylinders and to some extent by mechanical gripping devices that help remove the part from the mold. The degree of stripping difficulty depends on the compound, the mold surface finish and configuration, and the type of mold release employed.

Mixing Head. The attachment to the mold should be such that its exit opening and entrance to the mold be properly aligned and of the same cross section. The attaching joint must be tight so that no air leakage can take place into the mold. The mixing head should preferably be connected to the stationary half of the mold.

PRESS FEATURES

The features that a press should incorporate are:

1. Adequate pressure to assure tight parting-line closure.
2. Efficient attaching arrangement for mold halves.
3. Capability of rotating the mold up to 90° for orienting to fill and for effective venting to take place. After filling, the mold should return to horizontal position for cure.
4. After cure, the press should open the mold and rotate the halves to "book" position for fast part removal, cleaning any flash or adhering material from the mold, and applying mold release.

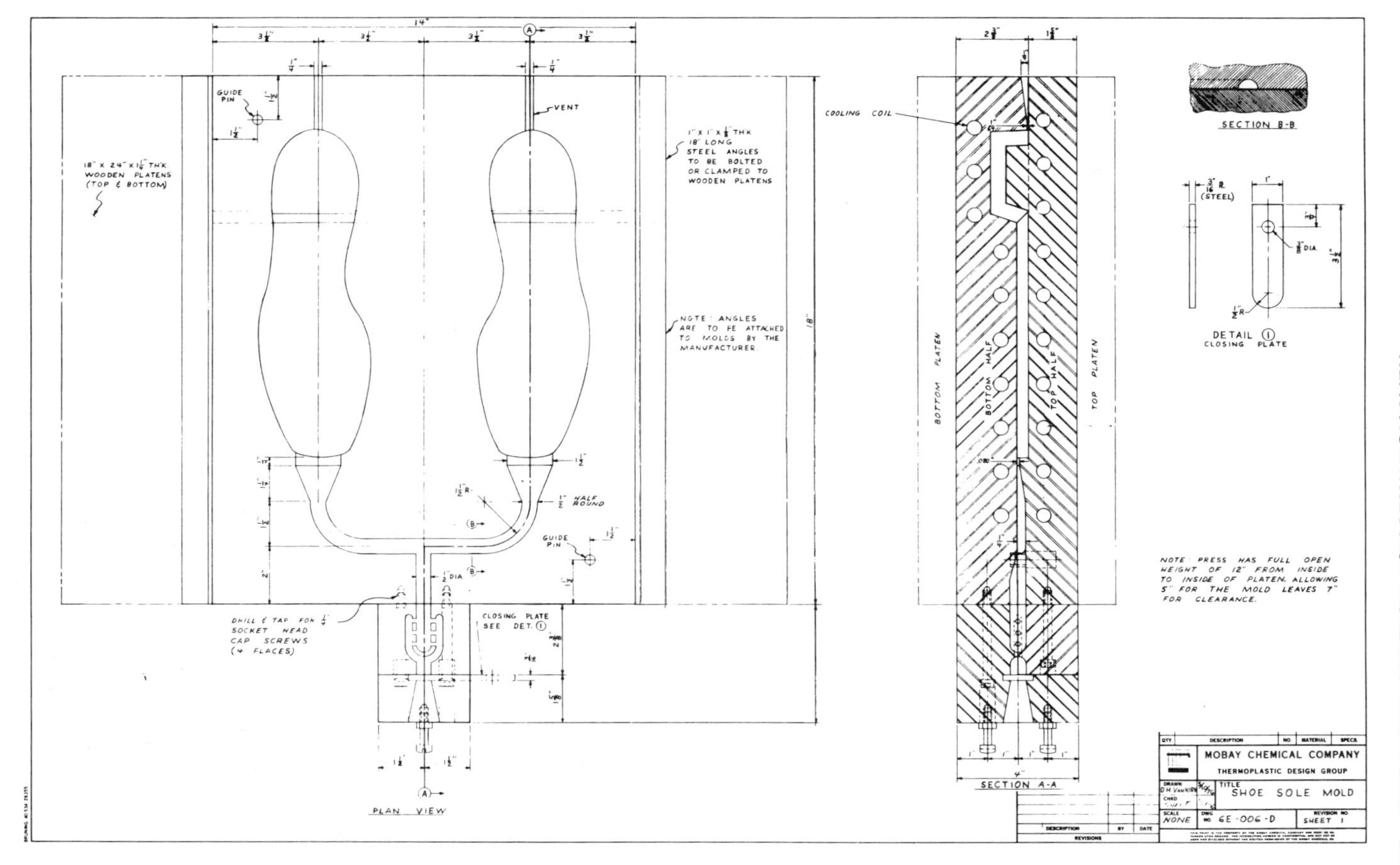

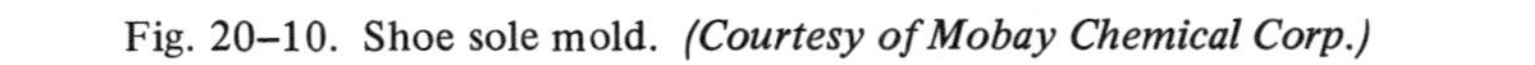
Fig. 20-10. Shoe sole mold. *(Courtesy of Mobay Chemical Corp.)*

PRODUCT PROPERTIES

All the accurately controlled processing factors have been pointed out and their importance stressed for the attainment of required part properties. These include monomer conditioning, quantity of each precisely measured out, maintenance of correct ratio of monomers, entrance of the monomer mixture to the mixing head without lead or lag, thorough mixing, delivery to the mold at correct velocity, venting of air from the mold, mold temperature within certain limits, curing, and removing the molded product from the press. Each type of industry will have its own requirements, and formulations of material are or will be prepared to suit the needs of the specific products in an industry. Thus, the automotive industry is looking for some of these characteristics in materials that they consider for application:

1. Good stiffness at 70°F and a flexural modulus of more than 20,000 psi.
2. At elevated temperatures, the material should retain sufficient stiffness and resiliency to prevent noticeable sagging.
3. The tensile strength should be greater than 2000 psi.
4. Impact resistance at -20°F should be such that there would be no damage during a prescribed impact test.
5. The surface of the molded product must lend itself to economical paintability.
6. The strength of the die "C" tear test should be greater than 400 pli.

Many materials formulations can fall within this list of specifications. In addition to these requirements, some of them are expected to provide other properties that may be important for a particular application. Thus, for example, one part may call for a low coefficient of expansion, another may demand greater elongation, and some other end use may look for greater or lower shore hardness. Within the broad range of specifications, there are special needs that have to be met, and a material selection has to be made on the basis of circumstances that are pertinent to each application.

From a processing point of view, the materials must also display certain properties, such as:

1. Good storage life and stability.
2. At processing temperature, the monomers should have easy flow properties, i.e., viscosities below 1500 cps.
3. The components should mix readily and thoroughly under conditions of mixhead design.
4. After mixing, the polymerization (even partial) and change to solid should be within 5 to 9 sec.
5. The solidified material after mixing should have sufficient strength for re-

moval from mold, without tearing or distortion, within 1 min or less of residence in the mold.

6. The molded part should release easily from the molding surface and not tend to deposit a buildup.

Because RIM is a relatively new process, the chemistry of the materials is constantly undergoing modifications to improve the properties, the process, or both.

In view of the circumstances, exact production records should be maintained regarding the variation of properties and the causes for same. By so doing, it will be possible to establish the most favorable composition and whether the process control limits are adequate for the needs. The potential of the process is great, and everyone along the line, including chemists, equipment manufacturers, moldmakers, and processors, can contribute their part to make it the success that it deserves to be.

MOLDING PROBLEMS OF RIM PROCESS

1. *Surface Defects*

- Delamination and blisters usually show up after parts are subjected to a postcure treatment in the range of 250°F. These defects are a result of insufficient mixing and/or lead-lag condition of components.
- Nonuniform appearance is caused by entrapped air during filling. This is brought about by turbulent flow, i.e., velocities above prescribed limits.
- Shot-to-shot appearance variation is due to inconsistent mold surface temperature.
- Quality of surface finish is roughly a reproduction of mold surface provided that no buildup is allowed to take place.

2. *Skin thickness and quality* at their optimum determine favorable mechanical properties and ultimate surface quality.

- High mold temperatures can produce thin and porous skins, which when painted will cause a reject; they can also be responsible for sink marks.
- Mold temperature on the low side will inhibit the reactivity of the system, prevent the development of full green strength, and may cause delamination.

3. *Shrinkage.* The polymerization in the mold is incomplete and is carried out to a degree that permits removal of the parts from the press and subsequent handling. At the time of part removal, the shrinkage is not stabilized until the time and temperature schedule have completed the cycle of full cross-linking, at which time the product reaches stability and ultimate shrinkage. Reaching ultimate shrinkage may mean overnight postcure at ambient temperature or whatever the specific composition may require. The shrinkage varies from 0.013 to 0.015 in./in. according to the type of RIM system.

21
Injection Molding and Molds of Thermosets

Thermosetting plastics are among the oldest industrially useful plastic materials. In fact, they have always been considered the workhorse of the plastics industry.

When molded, these plastics display properties that in most respects outperform thermoplastics. They possess superior heat and chemical resistance, they are rigid, their resistance to creep is better than glass-filled thermoplastics, and their electrical properties and general stability under use conditions are outstanding.

The thermoset plastics are chemical compounds consisting of a resin, filler, and additives that will aid in moldability and/or provide color, lubricity, or other desired features. Most of these compounds are supplied in granular or nodular form, and, when subjected to heat and pressure, they will soften and flow to conform to the shape of a mold. With the continuation of the heat and pressure in the mold, a chemical change known as *polymerization* or *cross-linking* of the molecular chains takes place. This cross-linking results in large interconnected molecules, becomes irreversible, and forms a new chemical composition. It may be compared to bread, which once baked will not revert to dough.

The most frequently used resins are phenolics, polyesters, melamines, ureas, epoxy, and silicones. The fillers generally used are wood flower, mica, alpha cellulose, cotton flock, paper, macerated fabric, fiberglass, carbon fibers, talc, asbestos, graphite, etc. These fillers impart such properties as impact resistance, electrical properties, heat resistance, flexural strength, rigidity, toughness, lubricity, etc.

The resin when combined with filler and additives is partially polymerized and formed into granules or nodules for use by the molder. When stored at recommended temperature, the resin will have a good shelf life of approximately several months. Each compound has its own shelf-life limitation.

In the past, most thermosetting materials were formulated to work advantageously under compression or transfer molding. Some of those materials in their softer flow range may also work satisfactorily in injection moldings. The requirement for an injection thermoset material is that, at plasticating temperature and residence time in the preheating chamber, the polymerization rate be so slow that it will not affect the flow characteristics nor initiate any noticeable setting-up. With this type of formulation and the resin capable of feeding from hopper to screw, almost any thermoset material can be successfully injection-molded.

INJECTION MOLDING OF THERMOSETS

When we think of injection molding thermosets, the first reaction is to compare it with the same operation of thermoplastics. It is true that the thermosetting material is plasticized in a cylinder from where it is injected into the mold until solidified. When the shot is rigid, the mold opens, ejects the part, and a new cycle begins. This is where the similarity stops.

The plasticizing in the cylinder is done at temperatures from 200° to 230°F. The low heat is necessary in order to minimize the polymerization of the material and its hardening during the working of the screw and while the material is residing in the cylinder. The screw has to be so dimensioned that its work input into the material is accurately controlled, and the cylinder temperature must be evenly maintained. As a matter of fact, most of the thermosetting materials are so formulated that the kickover point of rapid polymerization takes place at a temperature well above the preheating or softening point that exists in the cylinder. In production terms, this means that plasticized material will maintain its consistency for minutes, whereas the polymerization or cure takes place in parts of a minute.

It must be recognized that thermoset materials continue to polymerize very slow, even at the low temperatures. Therefore, the material must be protected at every stage of storing and processing against the possibility of unnecessary heat absorption. In addition to controlling the plasticizing heat input, the duration of exposure even to the plasticizing heat must be limited in terms of length of barrel or residence time.

The mold temperature is maintained at from 330° to 400°F and is thermostatically controlled. The cure time may range from 10 sec in thin sections up to 1.5 min in heavy walls.

The transition from cylinder to mold is through a nozzle, which is normally maintained at a heat slightly below that of the cylinder. The contact between mold and nozzle should be held for the duration of injection hold pressure to prevent excessive heat absorption by the nozzle and the possible curing of the material in it. When injection-hold pressure is completed, the injection head should be retracted to prevent heat transfer from the mold.

We now turn to the overall operation of injection molding thermosets. Material is fed into the hopper (normally in granular form), and from there it drops onto a screw. The screw brings the material into its heated housing or cylinder. The screw is driven by a suitable power source, and, while moving the material forward, heat is absorbed from the cylinder and from the frictional work of the screw. As the heated material works its way to the tip of the screw, it brings about a backward movement of the screw for a distance that provides enough material for the shot at the front of the cylinder. At this point, screw rotation stops, and a hydraulic cylinder forces the screw forward. This move injects the plastic into the mold through the nozzle, which is in close contact with the mold. In the mold, the material passes through the sprue bushing, runners, and gates and into the cavities. The heated mold supplies the necessary energy for cure; upon completion of the hardening process, the mold opens and the parts are ejected. Upon completion of the injection hold time, the cylinder retracts to break contact with the mold and starts preparation of the material for the next shot. Upon ejection of the parts, the mold halves have any loose flash blown off. Following this, the mold closes, and the next cycle starts.

Obviously, unlike compression and transfer molding, no preforms nor preheaters are required, and the handling of material in the in-between steps is eliminated.

The injection process is a relatively simple, automated method. Thermoplastic and thermoset injection are similar enough that some press manufacturers supply interchangeable screw and cylinder assemblies that can be interchanged in a few hours, and the machine can be used for either thermoplastic or thermoset purposes, depending upon the type of injection assembly that is in place. The main difference is in the screw design. The thermoset screw has a compression ratio of nearly 1:1. Thus, the flights are of the same depth for the full length of the screw. Furthermore, no check rings exist nor any other places where material can hang up and cure itself in place and thus form a permanent obstruction. The screw with uniform flights and no obstructions provides a smooth path for the flow of material. The cylinder in which the screw rotates is in most cases heated with hot water by units similar to the circulating units used for mold-temperature control of thermoplastic jobs. Diagrammatically, the machines will appear the same for the two basic materials (thermoplastic and thermoset) except for the details of construction of screw and cylinder, which are not too clear in the reduced scale drawing of Fig. 2–2.

For high-volume production, the analysis of the overall production cost of other processes, including molds, machines, and labor, leads in most cases to a conclusion in favor of injection molding of thermoset plastics. So far, the limitation has been that not too many thermoset materials as presently formulated lend themselves to this process.

For the injection machine outline, see Fig. 2–2.

ELEMENTS OF THE PROCESS

Let us now follow the path of material flow from hopper to the mold to see what specific attention is required at each point.

First of all, the material amount stored on the floor in the vicinity of the machines should not exceed the amount used during one shift unless the temperature in the area is 70°F and the air is dry. Material is introduced into the hopper of the machine. From there, it travels through the various machine elements. These include:

1. *Hopper.* The shape of the hopper should be conducive to free flowing of the powder. The taper per side should be 15° to 20°, and the volume of material in the hopper should not fall below the half mark in order to insure a pressure on the material that will bring about a consistent flow to the screw. From the hopper, the material enters the throat.

2. *Throat.* Throat temperature should be checked with a portable pyrometer and maintained at midpoint between room temperature and the rear zone setting of the cylinder pyrometer.

3. *Cylinder and screw.* Before starting a run, the cylinder and screw should be checked out to insure that no material from a previous job has been left behind. With the machine controls for manual operation, the barrel and screw assembly is moved away from the mold, and the screw is put in forward position. The hopper slide is closed, the nozzle is removed, and the screw is rotated until it reaches the out position. At this point, the screw is put into injection position so that any material left on the inside can be purged. This procedure is repeated to clear out all material possible. Incidentally, this is the correct procedure for shutting down a job.

The cylinder temperature should be maintained to suit each material's specifications. Excessive temperature can cause advancement of the cure in the cylinder, as well as slow injection time, short shots, cushion increase, and improper filling of cavities. The higher temperatures can be caused by higher back pressure, higher speed of screw, or any cause that adds to the frictional heat. It is most important to check out what the optimum temperature range is for a specific material and to make settings for heater, screw speed, back pressure, injection pressure, and injection speed that will ensure the desired preheat temperature of the material.

The screw itself is built so that the amount of heat it adds to the softening of the material is limited. The compression ratio is usually 1 to 1 and the L/D ratio is 12 to 15:1. These construction features mean that the time of residence of the material in the cylinder is on the low side. The tip of the screw matches the internal shape of the cylinder head and is fluted for smooth flow of the material, so that after injection there is a minimal amount left behind.

The heat profile in the cylinder calls for the front zone to be at the optimum setting, while the rear zone is 30 to 50° cooler, and the middle zones are in proportion to the difference of the two.

4. *Screw Rotation.* Screw speed should be in the range of 50 to 100 rpm. The important part is the coordination of the screw rpm with mold opening. The rotation should stop just at the time of mold opening. If the material is per-

mitted to lie in the cylinder for too many seconds prior to shooting time, the preheated material can start curing and adversely affect the shot. If for any reason it is not feasible to have the screw rotation time fit in between the pressure hold timer and mold opening timer, then the delay part should be after the pressure hold times out—that is, the screw rotation is delayed so that it finishes at the time of mold opening.

5. *Back pressure.* The back pressure on the material should be 500 to 750 psi (50 to 75 gauge pressure) during the production run. It can be close to zero during start-up and increased as running conditions are adjusted. Excessive back pressure can cause compressing of the material over a longer distance of the screw flights, making the injection more difficult. The back pressure contributes to temperature rise of the material due to the added work input. When filled with higher pressures, the material in the measuring chamber is denser, making the shot larger. If back pressure is adjusted upward, it should be accompanied with a decrease in size of feed volume.

6. *Feed volume.* The machine specifications indicate the volume for full screw travel. When the weight of the shot is known, it can be converted into cubic inches, and the travel setting can be proportioned to the full length of screw travel. The conversion is

$$\text{in.}^3 = \frac{\text{oz.} \times 27.7}{16 \times \text{Specific gravity}} \text{ or } \frac{\text{Grams} \times 27.7}{453.4 \times \text{Specific gravity}}$$

Thus, a machine with a shot size capacity of 20 in.^{3} and a screw travel of 5 in., when making an 8-oz shot with a specific gravity of 1.4, will travel:

$$\frac{\text{in.}^3}{20} = \frac{X}{5}; \text{ in.}^3 = \frac{8 \times 27.7}{16 \times 1.4} = 9.89$$

We have

$$\frac{9.89}{20} = \frac{X}{5}; X = \frac{5 \times 9.89}{20} = 2.47\text{-in. setting}$$

When adjusting for the final setting, the back pressure can be used for making minute adjustments.

7. *Nozzle.* Material enters the nozzle from a warm cylinder and exits into a hot mold. Conditions in the nozzle have to be such that curing of material is prevented. For this reason, the nozzle temperature should be such that while contacting the mold, it will not get hot enough to cause cure of material. A water jacket with controlled coolant flow is a desirable feature. The opening of the nozzle should

not cause a restriction to the flow, which would thus eliminate the possible frictional heat. The contact area between nozzle and sprue bushing need not exceed a 3/16 to 1/4 in. wide ring as measured on the sprue bushing from the opening out. This would keep the heat-conducting area to a minimum. Finally, retracting the cylinder away from the mold will, on the average, keep the contact time with the mold to about 33-1/3% of the cycle time. In many cases, only some of the heat-insulating steps might be necessary; in others, all of them may be required.

8. *Mold.* Before any shots are attempted into the mold, parting line, molding surfaces, runners, and gates should be carefully inspected to see that polish is in good condition, no material nor flash is adhering anywhere, and the vents are open. After the condition is found satisfactory, the mold temperature should be checked in several molding areas with a portable pyrometer.

The heaters are normally calculated to bring the mold to proper temperature in 1 hr or less. The mold should be insulated from the platens and clamped in a manner that will permit expansion due to the heat.

9. *Injection pressure.* The cavity pressure, which is a result of the injection pressure, should range from 10,000 to 15,000 psi. The high cavity pressure will usually mean better plasticized material, which is brought about by additional frictional heat associated with the higher injection pressures. The duration of the injection high pressure is about 5 sec. With a shorter time, there is a danger of gas entrapment or even irregular curing surface.

10. *Injection hold.* Injection hold pressure can be reduced to about half of the injection pressure since its sole function is to prevent backflow of the material from the cavity and possible sinks in the gate area.

When uniform part density is important, a small cushion (not more than 1/16 in.) may be desirable, and the holding pressure would be closer to the injection high pressure in order to supplement any additional material to the cavities. The duration of this pressure is between 5 and 10 sec depending on gate size and its speed of sealing.

11. *Speed of injection.* The rate of injection should be the highest the machine can deliver, but should be within the 5-sec range of the high injection pressure application. At the high rate of injection, there is a surge of heat when the material passes through the restricted openings. The surge comes from frictional energy and aids in accelerating the cure.

12. *Observation of screw travel.* After all the parameters have been set and shots are being made, the smooth travel of the screw to the full distance will indicate good performance. Stopping short of the destination may call for adjusting speed of injection and/or pressure, or lengthening of injection time. Jerky screw movement will call for more preheat in the material by a slight increase in the back pressure. If this step does not correct the condition, an increase in the front zone of the cylinder in steps of 10° is appropriate. It takes about 6 to 10 shots for cylinder temperature adjustment to become effective.

13. *Cure of part.* The thickness of the part determines the length of time a part has to be cured and to be rigid enough for ejection. A thin (0.060-in. wall) piece may take 10 to 15 sec; a thick piece (0.3-in. wall) may take 1 min or more depending on the compound. An undercured part will blister on the surface, which is caused by gas pressure formed during polymerization. If the part is not hard enough from cure, it will lack the rigidity to contain the internal pressure. This difficulty can be overcome by applying more heat into the material, longer time, or both.

14. *Opening and closing of press.* This part of the operation is identical with that of the thermoplastic machine. The machine approaches closing at high speed and slows down prior to the touching of mold halves, when full clamping pressure is applied. Opening starts slow, followed by high speed until the start of ejection; it then slows down during ejection. Mold protection, when included in the machine, works in same manner as that of thermoplastics.

MISCELLANEOUS

1. *Mold blowing.* In thermoset molding, there is usually a film of flash at the parting line, especially in the vent openings, that requires removal. This is done by blowing compressed air across the faces of cavities, cores, and parting line to make sure that all loose flash is removed from the mold surfaces. The time of blowing and volume of air should be controlled, so as not to affect the temperature of the mold to any noticeable degree; above all, the blowing time and volume should be consistent from cycle to cycle. The compressed air should be filtered and dry so that no foreign substances would have a chance to bake themselves to the hot mold.

The compressed air regulators are usually equipped to add controlled amount of lubricant to the air; in that event, substituting the lubricant with a mold release material would aid in maintaining a clean and good working mold.

2. *Heat shield of mold.* The heated mold should be protected against drafts and blowing air to avoid temperature variation and, with it, molding variables. Voltage fluctuation, if too great, can also affect the mold heating system, so corrective measures should be taken.

3. *Mold heating.* Heating a mold with electrical cartridge heaters is advantageous for injection molding because it provides great flexibility and ready response to controls. Steam and hot oil are used with success in many applications, but, unless the circumstances that dictate the use of these heat sources are very convincing, the cartridge heaters are preferred.

4. *Mold clamp.* The size of the clamp needed is determined by the projected molding area–including runners–and the fluidity of the compound. A pressure from 3 to 5 ton/in.^{2} of projected area has been found satisfactory. The lower figure is applicable to harder flowing materials, whereas the 5 ton/in.^{2} will be needed for the soft-flow compounds.

TROUBLESHOOTING OF THERMOSETS

The following list of injection-molding problems of thermosets and appropriate solutions is considered representative of most thermosetting material, although they were developed principally in conjunction with phenolic materials. Even with phenolic compounds, we may find some resin and filler combinations that will cause problems not shown in the list. Generally, all troubleshooting problems can be resolved with the aid of three principles of molding–i.e. time, temperature, and pressure. A little thought and analysis of these elements will lead to a logical solution of the problem. Incidentally, the time, temperature, and pressure approach should also be used in connection with the list of problems and solutions.

If we consider the parameters involved with time-, temperature-, and pressure-related factors, it becomes easy to pinpoint a parameter that could cause a defect that is not listed in injection-molding problems. Thus, if we list the factors involved, the problems can be resolved with greater ease.

Time-related factors are cure time, rate of injection, mold open time, time of air blowing, residence time in the cylinder, duration of high- and holding-pressure application.

Temperature-related factors are mold temperature, cylinder temperature, frictional heat from gate and other passages that interfere with the smooth flow of the material, surrounding atmosphere and its effect on mold heat, roughness of runners, sprue diameter, back pressure, screw rpm, nozzle temperature, frictional heat from rate of injection.

Pressure-related factors are injection high and/or injection hold pressure, clamp pressure, back pressure, vents, pressure drops up to gate, cavity pressure.

Injection-Molding Problems of Thermosets

• *Defect:* Blistering, gas entrapment

Causes: 1. Mold temperature low or uneven throughout molding surface. *Cure:* increase in steps of 10°F for 12-shot duration in each setting. If necessary, consider additional heaters in mold to even out surface temperature.

2. Incorrect venting. Vents should be open and clean with each shot. Additional vents may be necessary.

3. Mold temperature may be too high. Decrease in steps of 10°F.

4. Trapped gases. Decrease injection speed to give gases more time to escape and or increase back pressure to force gases out of the melt.

• *Defect:* Distortion and warpage

Causes: 1. Temperature differences within the mold. Check that all heaters are working.

2. Mold temperature too low.

3. Cure time too short.
4. Holding pressure too low.
5. Moisture in material.
6. Cooling or heating not uniform in the mold or outside after ejection.

• *Defect:* Dull surface

Causes: 1. Mold temperature too high.
2. Screw back pressure too high.

• *Defect:* Flash

Causes: 1. Mold not fully closed due to retained flash or distorted fit of halves.
2. Shot size too large.
3. Injection pressure too high.

• *Defect:* Flow lines

Causes: 1. Mold temperature too high.
2. Screw back pressure too high.
3. Injection pressure too high.
4. Gate may need fan shape.

• *Defect:* Slow injection

Causes: 1. Flow-control valve setting may be low for flow of injection.
2. Injection pressure too low.
3. Material from cylinder not plasticized correctly due to low heat input.
4. Mold temperature too low.

• *Defect:* Nozzle plugged

Causes: 1. Excessive exposure to mold heat, or material from cylinder too hot.
2. Pressure hold time is too long, and space between nozzle and mold, after pressure hold, is too small to prevent heat absorption.
3. All the work-inducing elements related to the screw, such as screw speed, back pressure, and injection speed, are high.

• *Defect:* Porosity or unfilled part

Causes: 1. Insufficient material, mold not closed due to flash or other causes, vents not open, nozzle not fully open due to partial pressure, screw is worn and therefore incapable of delivering full pressure to the cavities. Some of these causes could be overcome by increasing injection pressure, increasing shot size, increasing back pressure, checking material preheat temperature, and adjusting for best flow.
2. Mold temperature is high. If so, it should be adjusted downward in steps of 10°F.

• *Defect:* Screw retraction, not uniform

Causes: 1. Material from hopper not flowing evenly; could be bridging due to excessive throat temperature.
2. Screw speed too slow for consistent plastication.
3. Screw is worn unevenly.

- *Defect:* Staining

Causes: 1. Mold kept open for long period of time (one shift or more) with heat on. *Cure:* polishing.
2. Vents not open, or mold temperature or injection speed too high.
3. Vents are too open or mold does not close properly, which prevents the application of full pressure to the material and thus the desired wiping action.
4. High heat-resistant formulations tend to stain a mold.
5. In most cases, a commercial oven cleaner such as Easy-Off will remove stains and restore the original mold surface.

- *Defect:* Sticking part or sprue

Causes: 1. Hold pressure too high or too long.
2. Mold temperature too high.
3. Parts not fully cured.
4. Condition and taper of sprue bushing need correction.

It is to be noted that about half of the heat needed for polymerization is induced into the material outside of the mold and the other half from the mold. This would lead to the conclusion that both heat sources have their contribution to the good functioning of the operation, and troubleshooting should be carried out with this in mind.

The basic processing data such as shrinkage, molding temperature, flow number, and pressure of molding are available from the supplier of the raw material. Because resin, filler, additive, and flow numbers are so great, the specification sheets alone would fill a 2-in.-thick book.

The injection-molding formulations are not too numerous, but it has been found that most of the transfer grades are suitable for injection and even those are large enough for listing.

During start of a new compound, it is best to set all the parameters with the exception of curing time at the low end of the range, and after a few shots one can start to fine-tune each setting for optimum production.

STARTING AND ADJUSTING MOLDING CONDITIONS

The following is a starting setup from practice for a general-purpose phenolic compound as the processing material. The starting setup conditions were:

Mold temperature	330°F (both halves)
Cylinder temperature	200°F (front zone)
	150°F (rear zone)
Nozzle temperature	180°F
Screw rpm	40

Back pressure	50 or less
Injection pressure	15,000 psi
Injection time	8 sec
Hold time	10 sec
Cure time	Add 20% to value from Fig. 21–1

Figure 21–1 shows cure time in relation to part thickness for a general-purpose phenolic preheated in an injection cylinder to about 220°F with mold temperature of 330° to 350°F.

The first step is to determine the approximate screw travel distance, so that cavities are filled with material during the adjusting shots. From the machine data, we obtain the cubic-inch capacity of the measuring portion of the chamber that corresponds to the full screw travel. The cubic inches are found to be 25 for a machine that has a corresponding 5-in. screw travel. The estimated shot weight including runners is 12 oz, and the specific gravity is 1.4.

$$\text{in.}^3 \text{ of shot} = \frac{27.7 \times \text{oz}}{16 \times \text{Specific gravity}} = \frac{27.7 \times 12}{16 \times 1.4} = 14.84$$

Therefore,

$$\frac{14.84}{25} = \frac{X}{5} \text{ or } X = \frac{14.84 \times 5}{25} = 2.96 \text{ in. of screw travel for the shot.}$$

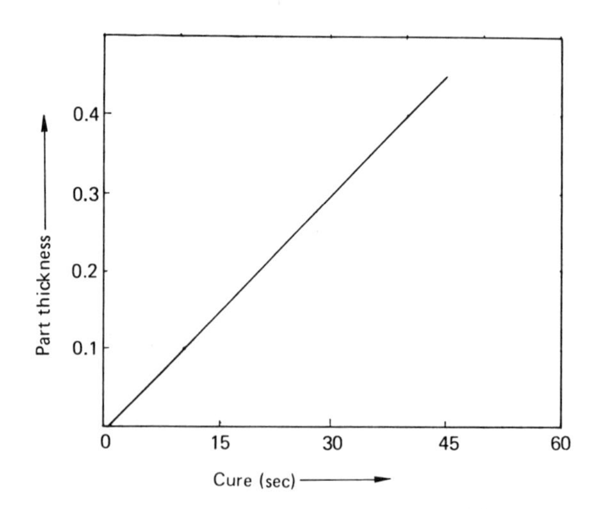

Fig. 21–1. Relation of cure time to part thickness.

To be on the safe side, we would start with 2-7/8-in. travel and make very few shots to see how close we are to filling of the cavities. We adjust the volume of the shot as indicated by observation. The adjustments are made in small increments of volume change, and a few shots are molded with each setting. For the ultimate fine adjustment, the back pressure should be utilized.

We have now established the needed shot size, and we start observing the screw movement to make sure that it's smooth, steady, fast, and completes its stroke to the full forward position. Should the screw stop short of its destination, this means that the speed of travel, or injection pressure, or time of injection need to be increased.

If the screw travel is not smooth, it is usually necessary to increase the material temperature by adjusting the front zone of the cylinder temperature in steps of 10°F or slightly increasing the back pressure. With the smooth traveling screw going to the full distance of the setting, timers can be set for most favorable cycle.

Injection speed timer should be around 5 sec. A faster speed may entrap gases or cause precuring in spots. Injection hold timer should only be long enough to stop backflow, which causes a slight sink at the gate.

Cure time should be long enough to eliminate blisters and rigid enough for good ejection. A few additional seconds above these requirements should be added to the cure time as a safety measure. In all the adjustments of parameters, one important objective should be kept in mind—the quality of the product at the lowest possible cycle.

THERMOSET INJECTION MOLDS

Mold Materials

The thermoset molding materials are predominantly composed of a resin and filler. A larger number of the resins by themselves have an abrasive action on the mold, but, when combined with the fillers as a molding compound, their wear of mold surfaces becomes pronounced. In comparison with unfilled thermoplastics, the abrading action of thermosetting compounds can be considered severe. Added to the abrasive action are high pressures (8000 to 15,000 psi) that force the material into the cavities and high heat at which the mold is maintained (around 400°F), yielding a combination of factors that demand a steel with outstanding wear-resistance properties.

These molding conditions are severe and apply to a large number of materials. On the other hand, sound mold-design practice would dictate analyzing the molding characteristics of the material for which a mold is contemplated and deciding on mold material selection on the basis of the facts at hand. The preceding molding conditions call for steels in the mold that are harder and more abrasion-resistant; possess higher thermal shock resistance; and will not crack readily when subjected to concentrated and uneven pressures. It is not unusual to find a piece of flash adhering to either half of mold, having it prevent the mold from closing and thereby concentrating the full clamping force on a small area of the parting surface. If a tool steel does not have springlike characteristics, after a few concentrated compressions it will crack, and so a very expensive tool will be damaged beyond repair. Using the correct tool steel for thermosetting plastic is a very important part of moldmaking. Before deciding on a specific grade for a mold, review Chapter 10.

For example, if a hobbed cavity is being considered, the steel should not only be hardenable, but should have a tough inner core that can give necessary reinforcement to the hardened outer shell to minimize the tendency of cracking. Or when other processes of fabricating are used, the stress-relieving operation takes on more importance because of the potential of experiencing concentrated stresses on top of the built-in stresses in a mold component. For a tool with long life, the best steel has been found to be the H-13 group, heat-treated to 54–56 RC, polished, and hard-chromeplated to a thickness of 0.001.

For high-activity cavities (60,000 to 100,000 pieces per year per cavity), it may be desirable to make high wear sections such as the gate section as a replaceable insert. As soon as the plating of the inserts begins to wear off, it can be replated and replaced for additional use. If this type of replacement insert does not last to satisfy the needs, tungsten-carbide inserts might be considered; they cost up to three times as much as steel ones but they will most likely last 10 times as long.

Cavity Design

The injection molds for thermoset or thermoplastic materials are very similar except for heat exchange. (See Fig. 21–2.) The problems that prevail in thermoplastic injection molds will also be found in thermoset injection molds. For this reason, it is advisable to review Chapters 3 through 11. Whenever deviations exist, they will be outlined.

The overall thickness of the cavity block will be lower since there is no need for cooling passages and it is not feasible to mount heaters of correct capacity in them. The thickness under the lowest point of the cavity can be equal to the width of cavity face. A cavity can be looked upon as a pressure vessel with ends closed and uniform wall thickness that meets the strength requirements pointed out in Chapter 5, pages 49–51.

Let us consider the example used in Chapter 5 for making the part from wood-flower-filled phenolic, and calculate the face width. The projected area of the four cavities, including runners, is 28.6 in.^{2} The injection psi for thermoset

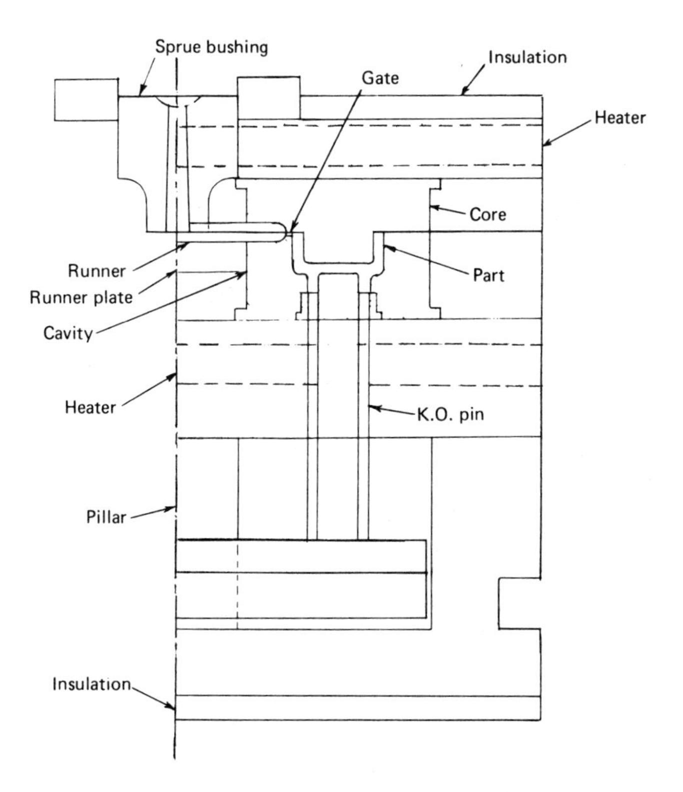

Fig. 21–2. Outline of thermoset injection mold.

materials will be counteracted by a clamp force of 3 to 5 tons per projected square inch. For woodflower phenolic, we will use 3 ton/in.^{2} Thus, the clamp force will be $28.6 \times 3 = 85.8$ tons. The nearest press capacity available that will accommodate this mold is 100 tons. The cavities will be made of H-13 steel and heat-treated to 54–56 RC with an allowable safe compressive stress of 5 ton/in.^{2} (Chapter 5). The area of cavity faces subjected to compression will be:

$$\text{Area} = \frac{\text{Tons of clamp}}{\text{Allowable stress}} = \frac{100}{5} = 20 \text{ in.}^2$$

for each cavity, it will be $20/4 = 5$ in.^{2} The area of the face will be

$$0.7854\,D^2 - 0.7854\,d^2 = 5$$

or

$$0.7854\,D^2 - 6.6 = 5$$

$$0.7854\,D^2 = 11.6$$

and

$$D^2 = \frac{11.6}{0.7854} = 14.77$$

$$D = 3.84$$

$$d = 2.9 \text{ (from Chapter 5)}$$

and

$$\text{face width} = \frac{D - d}{2} = \frac{3.84 - 2.9}{2} = 0.47$$

In the case of thermosets, the venting needs are greater, and in this case we will provide eight vents, 0.25 in. wide and 0.003 in. deep. We will therefore increase the area of compression by the amount we have subtracted for venting, namely,

$$8 \times 0.47 \times 0.25 = 0.92$$

Thus,

$$0.7854\,D^2 = 11.6 + 0.92 = 12.52 \qquad D^2 = \frac{12.52}{0.7854} = 15.94$$

$$D = 3.99$$

$$\text{and the corrected width of face} = \frac{3.99 - 2.9}{2} = 0.545$$

Let us now check if the thickness will satisfy the needs of a pressure vessel. According to Lame's equation (*Machinery's Handbook,* "Strength of Cylinder") for "brittle" material, the thickness is

$$t = \frac{d}{2}\left[\sqrt{\frac{s + p}{s - p}} - 1\right]$$

$$5 = \frac{300{,}000}{5} = 60{,}000\text{-psi tensile allowable stress}$$

p = 15,000-psi maximum cavity pressure at injection.

Therefore,

$$p = \frac{2.9}{2}\left[\sqrt{\frac{60{,}000 + 15{,}000}{60{,}000 - 15{,}000}} - 1\right] = \frac{2.9}{2}(1.29 - 1) = 0.42$$

By considering the compressive strength of the cavity face and by viewing the cavity and force as a pressure vessel, we find that the dimension 0.545 will meet both strength requirements.

Since the full clamp pressure is absorbed by the cavities by having them protrude 0.005 above the plate on one half of the mold only, the mold should be stamped "100 Ton Max. Clamp" in the proximity of the eyebolt so that no one handling the mold will overlook this safety precaution.

Sprue Bushing

In the interest of keeping frictional heat up to the gate to a low value, the nozzle opening should be 0.25 in. diam, which would make the sprue bushing "O" dimension 9/32 in. The inside dimensions of the nozzle should lead to smooth transition to avoid generating frictional heat and with it the possibility of setup of material in this area.

A cold slug well should be provided at the end of the sprue to assure that any material from the front of the nozzle, which is not in best flowing condition or is being partially set up, will be trapped and thus will not enter into the path of flow to the cavity. Sprue pulling can be incorporated in the cold well (see Chapter 7), but all transitions from diameters should be rounded to avoid breaking while extracting the sprue. Transition from sprue to runner should also be rounded. Nozzle openings for hard-flowing materials may have to be over 0.25 in. with a corresponding increase in sprue bushing opening. Sprue bushings over 2.5 in. long should be avoided, not only because of additional waste of material but also because of increase of flow length. A recessed sprue bushing with a temperature-controlled extended nozzle should be considered when sprue length exceeds reasonable dimensions (2.5) (see Chapter 7).

Runners and Gates

The size of runner depends on such factors as softness of material, distance of flow, molding temperature, and volume of material. Since thermoset runners are wasted material, the tendency should be to start out on the smaller side and adjust to requirements. Figure 21–3 shows a relationship between runner diameter and volume of material flowing through it, based on average conditions of molded phenolics. In thermoset materials, runner plates are an important design feature not only because of wear but also because of possible heat insulation from cavities. Thus, for example, the runner plate could be resting on machined rails against the heater plate so that it would absorb the minimum heat. Needless to say, the rails would have to have sufficient area to prevent embedding into the heater plate. The face of the runner plate on which the runners are incorporated requires attention so that the flow of material is smooth and encounters little resistance. The surface finish of the runner should be around 15-μ-in. profilometer reading (RMS). Turns if present should have a generous radius at the bend, i.e., no less than runner diameter at the bend. Anything that can be done to improve the smooth flow of the material to the cavity will aid to keep the runner on the small size.

Whenever the runner size is established for production operation, the runner plate should be properly polished and chromeplated with 0.001 in. thickness. The most advantageous runner cross section is the full round shape since it results in faster cavity fill at lower injection pressure. The trapezoidal shape, if essential, should have well-radiused corners and be otherwise dimensioned as shown in Chapter 7.

Gates with full round cross section are also preferred for easier filling, and the transition from runner to gate should be well rounded. The land of the gate should be around 0.060 to 0.090 in. Rectangular gates are thinner in depth than an equivalent round gate and therefore will present a more attractive appearance after breakoff. Should the appearance be a consideration, the gate should be 75% of runner width. The cross-section area from which the depth is derived should be obtained from Fig. 21–3 where the relation between gate area and material volume flowing through it is shown. Here again, the transition from runner to gate should be smooth, and the land should also be around 0.060 to 0.090 in.

For automatic degating, the submarine gate is being applied successfully to thermoset material. The submarine gate size should correspond to a round gate (Fig. 21–4).

The dimensioning of the gate will depend to some degree upon the behavior of the material while being ejected. If the material shows some flexibility during ejection, the dimensioning can be similar to a thermoplastic design. Thus, the centerline of the gate with respect to a vertical side of the cavity can be between 25 and 30°. The included angle of the gate may also be about 30°. The transition of runner to gage should be smoothed out. The K.O. pin that aids in shearing the gate and ejecting the runner should be of the same diameter as the runner and should be placed one to two runner diameters away from the end of the runner (Fig. 21–4).

Some materials are fairly rigid during ejection. Gate dimensions as shown in Fig. 21–4 would bring about breaking of the gate and plugging of the cavity for the following shot. This type of a problem can be overcome by providing a hold-down slug for the runner placed in parallel to the gate and having the K.O.

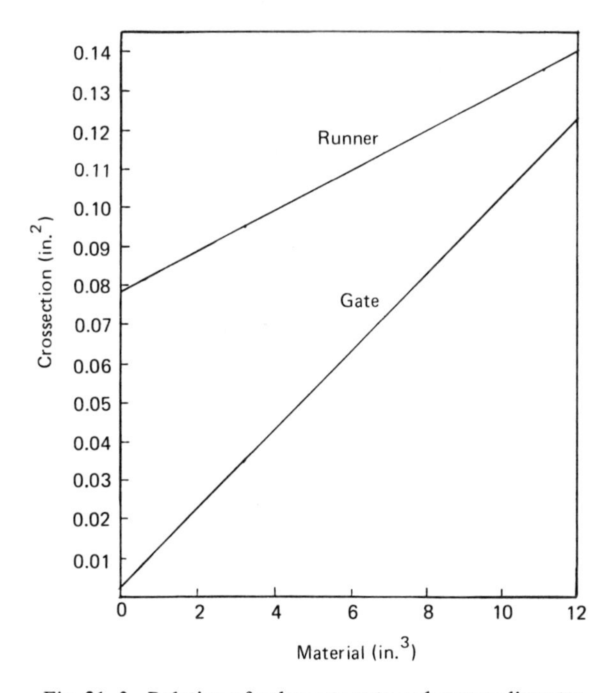

Fig. 21–3. Relation of volume to gate and runner diameter.

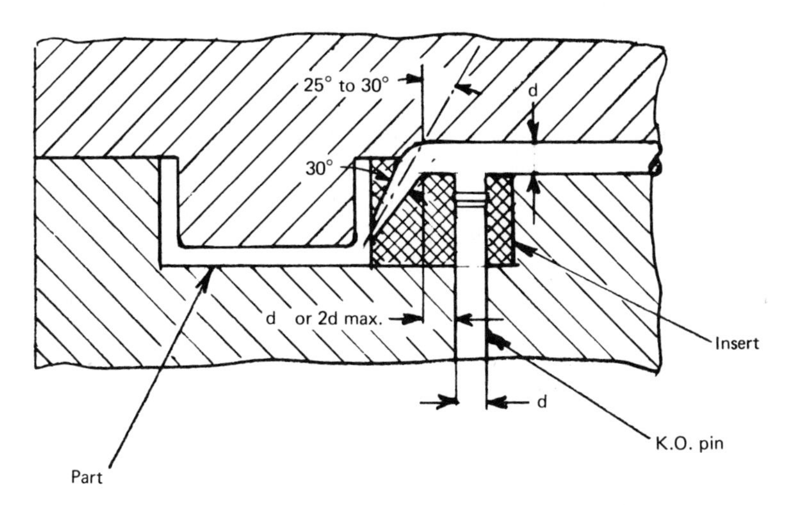

Fig. 21–4. Subgate for material that is not too brittle at ejection.

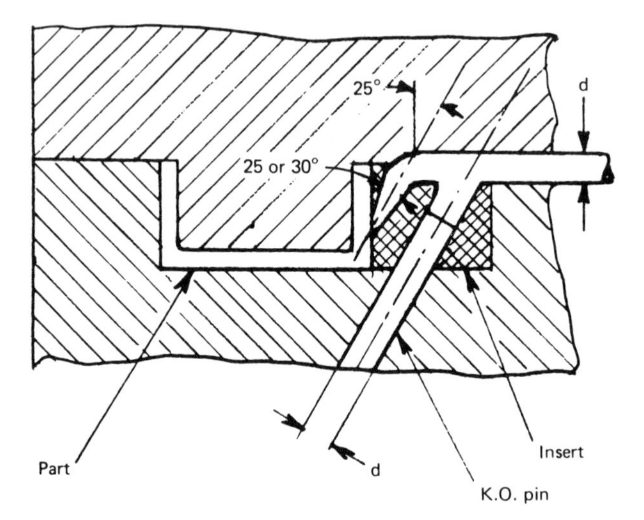

Fig. 21–5. Subgate for brittle materials at ejection.

shearing pin acting along the axis of the slug. This will insure ejecting the gate and slug, but may cause them to break off from the runner. This would cause no harm provided the sprue pin is available to eject the remainder of the runner.

Regardless of how the submarine gate is dimensioned, a block with the gate in it should be a replaceable insert because of the wear of the shearing edge as well as the gate diameter (Fig. 21–5).

Mold Vents

Inadequate venting to permit expelling air and volatile gases generated during plastication can cause mold staining, porosity in parts, unfilled parts, and poor appearance of parts. Vents are more important in thermosets than in thermoplastics. First of all, runners should be vented prior to approaching the gate. The vents should be the full width of the runner and 0.005 in. deep. The circumference of the cavity should be vented, and the vents should be spaced about 1-in. apart and should be 0.25 in. wide and 0.003 to 0.007 in. deep, depending on the flow characteristics of the material. A softer material would call for a lower value. K.O. pins should be as large as possible, and in most cases they should have 0.002-in.-deep flats, three or four of them ground on the circumference of the diameter and with the grinding lines parallel to the length of the pins. The grind should be with a fine-grit wheel. The end of the pin should have the corner broken by 0.005 in. so that if any flash is formed, it will adhere to the part.

Occasionally, it is necessary to place K.O. pins at the vent slots to insure that the flash from the vents is physically removed, thereby assuring open vents for the following shot.

Shrinkage

For parts with close-tolerance dimensions, the shrinkage information should be obtained from test bars made by the injection process. These shrinkages can be more than twice those of compression-molded test bars. For that matter, plunger transfer shrinkages are also higher than compression-molded test bars and in reality come very close to the injection-molded data. Injection-molding shrinkages for phenolics range from 0.007 to 0.012 in./in.

Warm Manifold Molding

Sprues and runners in thermoset materials are a waste, and their cost has to be included in the part cost. To eliminate this waste with each shot, a warm runner mold was designed and is being successfully used to conserve material. Figure 21–6 shows the diagrammatic arrangement of such a mold. The capacity of the runner system should be less than, or at most equal to, shot size in order to keep residence time to a low value. The economics of this runner system should be thoroughly evaluated before deciding on its adoption. The adjacent location of the warm manifold against the hot mold will cause a temperature gradient in the manifold-nozzle area. This gradient has been found to cause an increase in cure time ranging from 5 to 10 sec in comparison with a mold that has a runner of the same temperature as the mold. The cost of the machine seconds against the cost of the scrap phenolic, plus additional mold expense and possible additional power cost, have to be carefully analyzed and evaluated. For some jobs, it may be a worthwhile undertaking. Only analysis will indicate which method is to be pursued. Another design of runner saving systems is the Stokes Trenton patented cold mold, which is shown schematically in Fig. 21–7.

The smooth operation of the insulated runner depends on the care with which heat transfer to nozzle and runner plate is minimized. If airspace is to insulate the sealing area between nozzle and cavity, the contact of the seal should be only a 0.25-in.-wide ring, and the nozzle should be permitted to slide on a good contact surface for expansion reasons. The diametral clearance should be 0.045 in. per side. The clamping pressure should be distributed over the nozzle, and additional pressure pads should be strategically spaced to prevent plate deflection and at the same time minimize heat transfer between plates. The permissible stress level on the pads and plates should be 15,000 psi, which would have ample safety against embedding into the plates. The steel in the plates in such a mold base should be P-20 prehardened to 250 to 300 Bhn, and the nozzle should be H-13 steel prehardened to 44 RC. The airspace between the plates should be 0.090

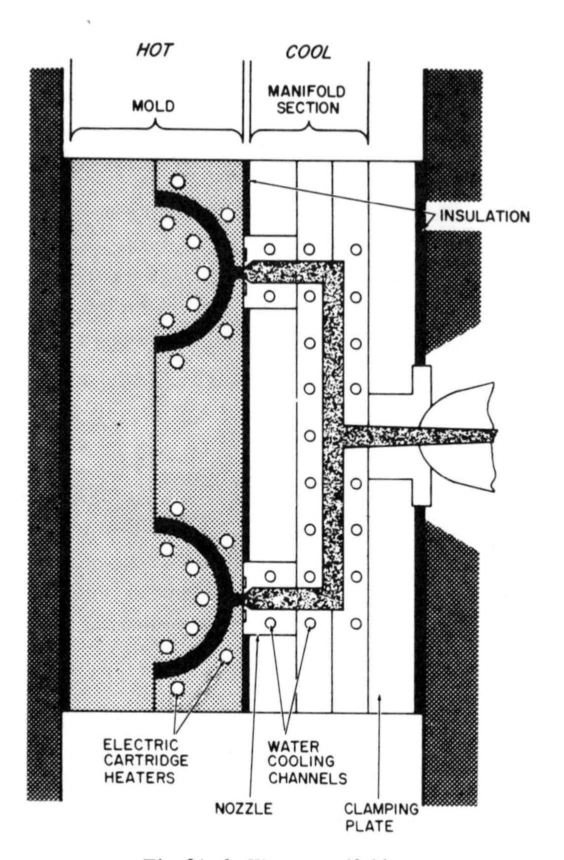

Fig. 21–6. Warm manifold.

to 0.100 in. The nozzle face should have about 0.005-in. taper toward the outer edge so that, when it is compressed about 0.002 in. in length, it will form a leak-proof seat against the smooth surface of the cavity plate.

Let us now determine the size of pressure pads and location with respect to cavity for a four-cavity mold run in a 150-ton press. The nozzles have an opening of 3/16 in. diam, and while they are 1 in. in diameter, the area of 0.25-in.-wide ring around the opening will absorb part of the clamp pressure, and the additional area will be provided by the pads. The total area for clamp pressure absorption will be $A \times S = F$

where A = area being compressed

S = allowable stress = 15,000 psi

F = clamping force or 150×2000 = 300,000 lb

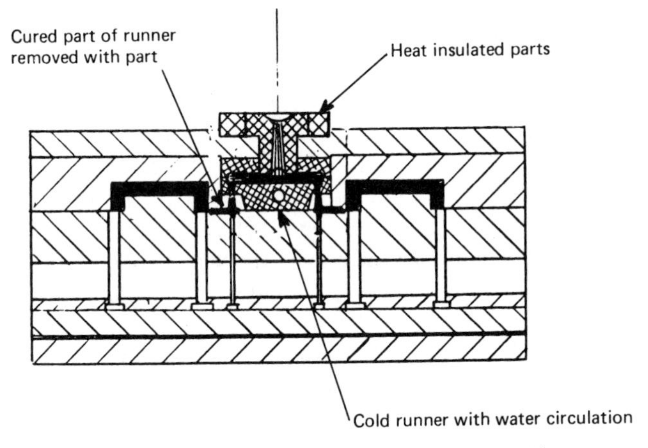

Fig. 21–7. Schematic principle of Stokes patented cold runner.

Thus,

$$A = \frac{F}{S} = \frac{300{,}000}{15{,}000} = 20 \text{ in.}^2$$

The contact area of the nozzles = $4 \times 0.7854\,(D^2 - d^2)$

$$D = 2 \times 1/4 + 3/16 = 11/16 = 0.6875 \text{ as outlined above}$$

$$d = 3/16 = 0.1875$$

$$A = 4 \times 0.7854(0.473 - 0.35)$$

$$= 4 \times 0.7854 \times 0.438 = 1.38 \text{ in.}^2$$

This leaves 20 - 1.38 = 18.62 in.^{2} for pads; per cavity = 4.65 in.^{2} If we select four buttons of 0.75 in. diam and place them under the edge of the cavity, they will provide support against deflection and also area for absorbing part of the clamp pressure. The area of the 0.75-in. pads = 4 $\times$ 0.4418 = 1.77 in.^{2} This leaves 4.65 - 1.77 = 2.88 in.^{2} for additional pads. A 2-in.-diam pad has an area of 3.14 in.^{2} and will provide more area than the needed 2.88 in.^{2}. The 2-in.-diam. pad will be used and placed in the vicinity of the return pins, so that heat transferred by the pad will be outside of the molding area and should therefore have no effect on temperature change in the plates that contain runners, cavities, and heaters.

This type of analysis can lead to a smooth-operating insulated runner mold. The use of insulating material of about 3/8-in. thickness between plates can lead to problems of compressibility, flatness, difference of expansion with steel, possible nozzle leakage, etc. When carefully engineered and properly selected, insulating material applied can work but would appear more expensive and of shorter life.

HEATING THE MOLD

Electrical heat by means of strip or cartridge heaters or by both in combination provide the greatest flexibility for injection molds. It is possible to have several heat capacities for heaters of the same dimensions and to apply the wattage to suit a particular area. The controls can be quite accurate and thus the maintenance of consistent temperature can be assured.

To arrive at the correct wattage requirements for a job, calculations can be made along the lines of the following example. The mold base used for the thermoplastic design shown on Fig. 5-6 was 11-7/8 X 15, and we will use the same size here for heater calculations. All other considerations such as mounting cavities, pillar supports, and K.O. pins will be essentially the same except that the overall thickness of cavities and cores will be reduced since there are no cooling passages.

What we need is to figure the Btu requirements for the job, and the number and size of heaters. The mold base will have A and B plates of 7/8-in. thickness, and all other dimensions will be standard.

The weight of the material to be molded will be 3 oz per cavity or 12 oz per shot plus 1 oz for runners and sprue. The molding temperature is to be 400°F, and the room temperature 70°F.

The amount of heat for the job is determined by the need for bringing the mold to operating temperature in the required time and by the amount necessary to maintain the temperature under operating conditions. Whichever of the two is larger will be selected for the job.

The steps for the calculation are as follow:

1. Required capacity for bringing the job up to operating temperature in the desired time which will be 0.5 hr.

$$\text{kW} = \frac{\text{Btu}}{3412 \times \text{hours allowed for heat-up}} = \frac{\text{Btu}}{3412 \times 0.5}$$

$$\text{Btu} = \text{Weight of cavity plates and backup plates} \times \text{Specific heat} \times \text{Temperature rise}$$

Temperature rise of mold is $400 - 70 = 330°\text{F}$

$$\text{Btu} = \text{Cubic inches of above plates} \times \text{lb/in.}^3 \times \text{Specific heat} \times \text{Temperature rise}$$

$$= 11\text{-}7/8 \times 15 \times 4.53 = 806.90 \text{ in.}^3$$

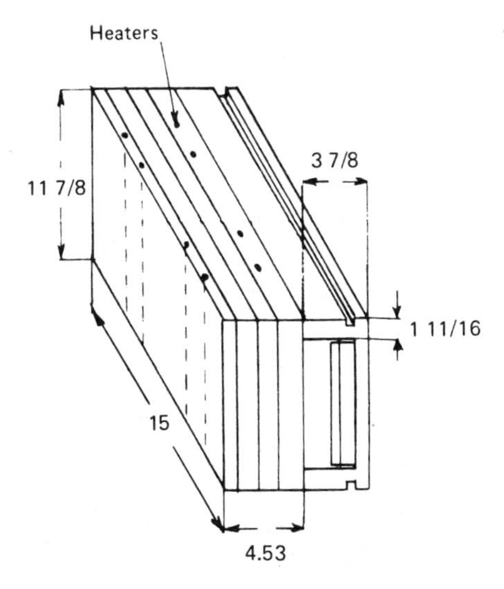

Fig. 21–8. Mold example for calculating heat requirements.

The thickness of plates is

$$7/8 + 29/32 + 7/8 + 1\text{-}7/8 = 4.53 \text{ (see Fig. 21–8)}$$

$$\text{Btu} = 806.90 \times 0.284 \times 0.12 \times 330 = 9074.79$$

For steel lb/in.^{3} = 0.284

Specific heat = 0.12

Temperature rise = 330

$$\text{kW} = \frac{9074.79}{3412 \times 0.5} = 5.32$$

2. Heat losses at 400°F
   a. Losses by convection and radiation

$$\text{Exposed surface area} = 2 \times 15 \times 4.53 + 2 \times 11\text{-}7/8 \times 4.53 + 2 \times 15 \times 3\text{-}7/8 + 4 \times 3\text{-}7/8 \times 1\text{-}11/16 =$$

$$= 135.9 + 107.58 + 116.25 + 26.16 = 385.89$$

$$= 2.68 \text{ ft}^2$$

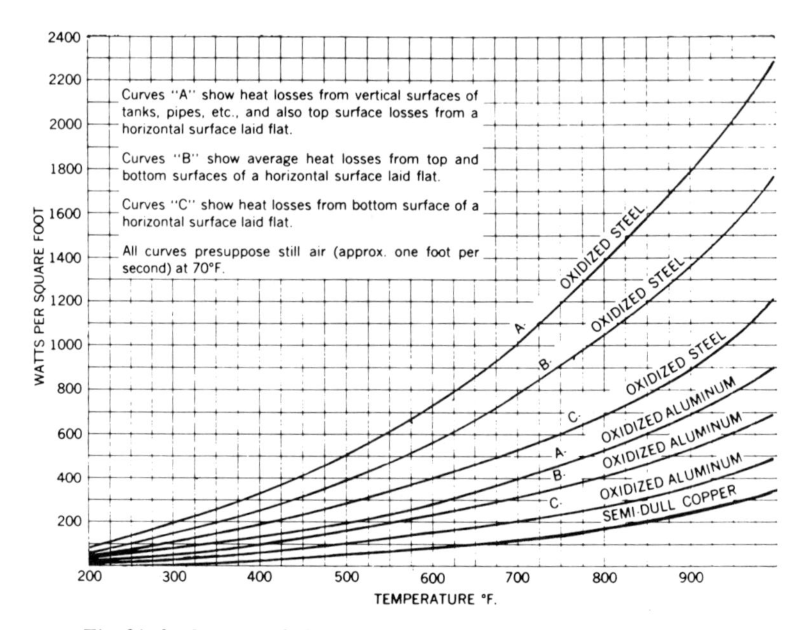

Fig. 21–9. Curve G-125S–Heat losses from uninsulated metal surfaces.

Losses by convection plus radiation at 400°F from Fig. 21–9 oxidized steel curve A is 350 W/ft^2. Therefore,

$$\text{kW lost through convection + radiation} = 2.68 \times 350 = 938 \text{ W}$$

b. Losses by conduction through insulation. Thermal conductivity of transite board is 0.087 (from tables) in Btu/hr/ft^2/ft.

$$\text{Area of conduction} = 2 \times 11\text{-}7/8 \times 15 = 356.25 \text{ in.}^2$$

$$= 2.47 \text{ ft}^2$$

$$\text{Thickness of insulation is } 0.375/12 = 0.031 \text{ ft}$$

$$\text{kW lost through conduction} = \frac{0.087 \times 2.47 \text{ ft}^2 \times 330^\circ\text{F}}{2 \times 0.031 \times 3412} = 0.47 \text{ kW or } 470 \text{ W}$$

c. Losses from (a) and (b) = 938 + 470 = 1408 W

$$\text{The average loss during heat-up is } \frac{1408}{2} = 704 \text{ W}$$

Total watts during heat-up is 5320 + 704 = 6024 W.

3. Heat needed for material during processing: Weight of material processed per hour with 60 shot/hr is

$$\frac{13 \text{ oz} \times 60}{16} = 48.75 \text{ lb/hr}$$

The temperature rise will be 400° - 225° (preheat temperature from cylinder) = 175°F

$$\text{kW} = \frac{\text{Weight of material} \times \text{Specific heat} \times \text{Temperature rise}}{3412} =$$

$$= \frac{48.75 \times 0.32 \times 175}{3412} = 0.8 \text{ kW or } 800 \text{ W}$$

The total heat requirement for operation will be

Heat for material plus heat losses = 800 W + 1408 (from above) = 2208 W

The total watts needed in (c) is greater than in (3); therefore, the value of 6024 W will be used. As a safety factor, the heater manufacturer recommends an addition of 15%, in which case the total will be 6928 W.

4. This wattage for the mold should be divided into eight heaters, 12-in long, with two heaters under a pair of cavities and two under a pair of cores. The heaters will be of high watt density with an alloy sheath that resists oxidation and corrosion. The wattage per heater is 865. We have to select a heater near this wattage and about 12 in. long. The most suitable heater has a 5/8-in. × 12 in. length and carries 1000 W and 240 V.

Watt density is the heater "wattage" divided by its surface in square inches. Heater manufacturers list the watt density as part of the specifications of a heater. The watt density of this heater is 45 W/in.^{2}, a favorable density for long life. Watt densities over 200 W/in.^{2} are usually a potential source of problems. These can be avoided by using more heaters, using larger diameter heaters, or having a closer fit between heater and hole, and, if possible, reducing heat requirements by shielding and insulation, thus making possible the use of lower-watt heaters.

HEATER INSTALLATION

Installing the heaters into the plates and connecting them into the circuit require a certain amount of attention to insure long and uninterrupted performance.

1. The holes for the heaters should be reamed to a smooth finish and size so that the clearance per side should be 0.005 in. or less.

2. The spacing of the heater should be such that the lead end is flush with the opening. The heater should be retained in place by a clip attached to the plate so it will not move out of position from either direction.

3. The leads should be protected by armored covering and held in position to keep from vibrating.

4. The wires extending to the leads should be the heat-resisting type, attached firmly with crimped tubing or similar method so there is no chance of poor contact. The extension wires should also be protected by armored covering to eliminate the dangers of wire breakage and damage to insulation. Stranded wire of ample size should be used. All cables should be clamped in place. Whenever insulating tape has to be used, it should be of the silicon type.

5. The wiring of heaters and their controls should have all the earmarks of good workmanship and permanency of arrangement.

6. Capacities of wires, terminal blocks, connector plugs, etc., should be carefully checked out for current-carrying ability and correct voltage.

7. In conclusion, every phase related to heater life should be carefully planned and carried out by qualified and experienced workers.

When heaters of proper wattage are selected and their mounting is carried out along suggested lines, the heating system will perform in a trouble-free manner.

There are many other elements pertinent to an injection or transfer mold such as strength of cavities and the retaining plates, consideration of mold and press, details inherent to molds, types of molds, materials for molds, cavity fabrication, etc. All of these have been discussed in connection with thermoplastic molds in previous chapters. It is suggested that they be carefully reviewed for application to thermoset molds. Thermoset molds have to be heated to a relatively high temperature, and whenever high heats are present there is always the problem of thermal expansion. Anyone working with heated molds should always be on the lookout for the expansion factor that may have an influence on any device that works with the mold or that may have a temperature considerably lower than that of the mold.

Index

Index